ASTRONOMICAL PHOTOMETRY
A GUIDE

T0332626

ASTROPHYSICS AND SPACE SCIENCE LIBRARY

A SERIES OF BOOKS ON THE RECENT DEVELOPMENTS
OF SPACE SCIENCE AND OF GENERAL GEOPHYSICS AND ASTROPHYSICS
PUBLISHED IN CONNECTION WITH THE JOURNAL
SPACE SCIENCE REVIEWS

Editorial Board

R. L. F. BOYD, *University College, London, England*

W. B. BURTON, *Sterrewacht, Leiden, The Netherlands*

C. DE JAGER, *University of Utrecht, The Netherlands*

J. KLECZEK, *Czechoslovak Academy of Sciences, Ondřejov, Czechoslovakia*

Z. KOPAL, *University of Manchester, England*

R. LÜST, *Max-Planck-Institut für Meteorologie, Hamburg, Germany*

L. I. SEDOV, *Academy of Sciences of the U.S.S.R., Moscow, U.S.S.R.*

Z. ŠVESTKA, *Laboratory for Space Research, Utrecht, The Netherlands*

VOLUME 175

ASTRONOMICAL PHOTOMETRY
A GUIDE

by

CHR. STERKEN

Astronomisch Instituut,
Vrije Universiteit Brussel, Belgium
and Belgian Fund for Scientific Research

and

J. MANFROID

Institut d'Astrophysique,
Université de Liège, Belgium
and Belgian Fund for Scientific Research

KLUWER ACADEMIC PUBLISHERS
DORDRECHT / BOSTON / LONDON

ISBN 0-7923-1653-3

Published by Kluwer Academic Publishers,
P.O. Box 17, 3300 AA Dordrecht, The Netherlands.

Kluwer Academic Publishers incorporates
the publishing programmes of
D. Reidel, Martinus Nijhoff, Dr W. Junk and MTP Press.

Sold and distributed in the U.S.A. and Canada
by Kluwer Academic Publishers,
101 Philip Drive, Norwell, MA 02061, U.S.A.

In all other countries, sold and distributed
by Kluwer Academic Publishers Group,
P.O. Box 322, 3300 AH Dordrecht, The Netherlands.

Printed on acid-free paper

All Rights Reserved
© 1992 Kluwer Academic Publishers
No part of the material protected by this copyright notice may be reproduced or
utilized in any form or by any means, electronic or mechanical,
including photocopying, recording or by any information storage and
retrieval system, without written permission from the copyright owner.

Printed in the Netherlands

Table of Contents

Preface

This book was written in order to provide an introduction to practical astronomical photometry. It is intended to help the astronomer and the astronomy student to make measurements of quality, and to extract the maximum of information out of these observations.

Photometry is now one of the principal branches of observational astronomy, with applications to almost all classes of celestial objects. Yet, photometry is most intensively used in variable star research. The photometrist now faces a unique situation: contemporary detector systems offer the observer measurements of higher precision—in terms of linearity, repeatability and sensitivity—than was the case one or two decades ago; still, the precision of the resulting photometric data lags behind that which is technologically possible because of the systematic application of ill-designed reduction techniques, especially with respect to extinction calculations and color transformations.

Intensive astronomical photometry is mostly done by young scientists at telescopes with relatively small apertures (50 cm to 100 cm). Many of those young observers have not received proper training, and it often takes them many years to acquire a level of skill that combines the production of high-precision results with efficient use of telescope time. Small and large telescopes equipped with photometers are being put into operation all around the world. Besides several excellent specialist papers, there are some undergraduate text books available on the subject of astronomical photometry. However, few books are at hand that give an introduction to basic principles and, at the same time, cover fundamental issues.

This guide is based on our own observing experience accumulated during numerous observing runs with various instruments at different observatories, and on our experience in writing and using a variety of photometric software. The level of presentation and the approach of the book is based on a course of lectures entitled "Observational Astrophysics" given in 1986 by one of the authors (C.S.) at the University of Canterbury, Christchurch, New Zealand, for the Honorary M. Sc. degree in Stellar Astrophysics. The book is intended for the astronomy student and the professional astronomer, but serious amateurs may find it a useful reference work.

Since the photomultiplier is now a solid piece of technology, the main focus is on this type of detector. Some general principles and techniques, however, are applicable as well in the field of two-dimensional photometry. Two-dimensional electronic detectors—essentially Charge Coupled Devices (CCDs)—rapidly evolve, and their principles of functioning and the techniques of application are also discussed. There is also a short chapter on the application of photographic photometry.

This book would not have eventuated had it not been for several people who have played an important role at various stages in the career of the authors. Though these people have not had any direct involvement in the preparation of this book, their contribution to the formulation of ideas have been invaluable. The authors are very indebted to N. Cramer, M. de Groot, H.W. Duerbeck, R. Garnier, M. Jerzykiewicz, E.H. Olsen, B. Wolf and A.T. Young.

Special acknowledgements are due to Jan Van Mieghem (Vrije Universiteit Brussel) for the careful reproduction of all the figures contained in this book, and to Claus Madsen (ESO) for providing the cover photograph.

The authors express their gratitude to the Belgian Fund for Scientific Research for the financing of several research projects which, through the years, provided the opportunity to acquire the know-how necessary to prepare this book.

Cover photograph

The photograph of the center part of the open cluster NGC 3293 is copyright of the European Southern Observatory. The authors are grateful to ESO and in particular to Dr. R.M. West for their consent to reproduce it as cover illustration.

Chapter 1 Introduction

The word *photometry* derives from the Greek *photos* for light and *metron* for measure. According to *Webster's Third New International Dictionary* (Merriam & Merriam 1976), photometry is "a branch of science that deals with measuring the intensity of light". It is also "the practice of using a photometer". When measuring monochromatic light quantities, photometric methods are straightforward, but when comparing light sources with different spectral energy distributions, measurements become more problematic.

In the 16th and 17th centuries, the study of light was mainly in the field of geometrical optics, which deals with the interaction of light rays with mirrors, lenses and prisms, without any relevance to quantitative measurements.

The common—and original—object of photometry is light in everyday life. Photometry in general is a technique that helps us to design windows for a schoolroom or for a kitchen, and assists us when choosing lamps for outdoor lighting. Most useful photometric quantities have been defined in that context. However, the *measurement* of the intensity of light is not restricted to such applications. Many engineering and scientific disciplines are very dependent on photometry, especially when the range of wavelengths is not limited to the visual spectrum. The development of devices such as video displays, projection apparatuses, photographic equipment, ovens, furnaces and solar plants requires photometric studies, as do satellite surveying and meteorological observing apparatuses. It was unavoidable that such unrelated fields would give birth to different approaches to photometry, resulting in a wide variety of definitions, often characterizing the same concepts.

Astronomical photometry refers to:

- the measurement of the spatial distribution of the light emitted by a celestial object in different spectral regions;
- the monitoring, in a specific spectral region, of the variations of brightness of such objects;
- the understanding of the astrophysical significance of these measurements.

The first photometrist to make quantitative measurements in astronomy was Bouguer (1698–1758), who compared the apparent brightness of celestial objects to that of a standard candle flame (Bouguer 1729). He made tables of atmospheric refraction, investigated the absorption of light in the atmosphere, and formulated his law regarding the attenuation of a light beam upon passage through a translucent medium. His results were published in a posthumous work (Bouguer 1760) by M. l'Abbé De La Caille. Bouguer was followed by Lambert (1728–1777), who developed the basic system of photometric concepts and nomenclature (Lambert 1760).

Light is the aspect of radiant energy which man perceives through visual sensation: the human eye is the detector which transforms radiant power into luminous sensation.* We do, however, use the term "light" in its broadest sense, i.e., for visible as well as for non-visible radiation. In the past, only the visible part of the spectrum was of practical use in everyday life, and in its original meaning, photometry is thus the measurement of visible light, or, more precisely, the measurement of visual perception. In the strict sense, the modern definition of photometry still is the measurement of radiation evaluated in accordance with some visual impression. *Colorimetry* is more specifically aimed at measuring the quality (color) of the light. It is more appropriate to denote by *radiometry* (from the Latin *radius*, "ray") the study of the energetic aspect of radiation (without regard to its visibility) but, in this book, we shall often speak of *photometry* for both aspects. For more extensive discussions of the photometric definitions, in optics, physics and engineering, the reader is referred to Walsh (1958), Smith (1960), Roig (1967), Ditchburn (1963) or Biberman (1971).

A *radiometer* is a general term that denotes any device used to measure radiation over a specific wavelength interval. A radiometer can be called a *spectrometer* in the limiting case of measurements at high spectral resolution. Strictly speaking, *photometers* are devices that measure the amount of visible light, but in astronomy, radiometers are called photometers.

1.1 The quantum nature of light

Isaac Newton described light as a stream of particles associated with certain periodic properties. Light, in its interaction with matter, behaves as though it is composed of particles (photons) which, just as material bodies, carry energy and momentum. The frequency (ν) is proportional to the energy of the photons through a fundamental constant of nature called Planck's constant $h = 6.6256 \times 10^{-34}$ J sec,† and light is emitted and absorbed as discrete quanta of specific energy content given by

$$E = h\nu = hc/\lambda \tag{1.1}$$

where c is the velocity of light, and λ the wavelength.

---

* Hence, one may say that a burning candle emits radiation, but that radiation does not become light, unless a human perceives it.

† Most often we use units in the international system (SI), but we shall make a few exceptions to this rule: nm (or Angstroms) for wavelength, cm^{-1} for wavenumber, etc.

E can be expressed either in wavelength units or in frequency units, according to the equation

$$E_\lambda = (\nu^2/c)E_\nu \tag{1.2}$$

and

$$E_\nu = (\lambda^2/c)E_\lambda \tag{1.3}$$

(E_λ and E_ν are absolute values).

Photons are absorbed by a receptor (photosensitive elements in the eye, the photographic plate, or the electronic imaging device) in quanta of such fixed amounts of energy. A quantum of blue light of wavelength 400 nm has an energy of 4.9×10^{-19} J, or 3.1 ev. One can write E (ev) = $1240/\lambda$, with λ expressed in nm. 1 mW (milliwatt) of light power corresponds to an energy rate of 6×10^{15} ev per second, and is carried by 2×10^{15} photons of blue light or 3×10^{15} photons of red light per second. When a photon strikes a detector, it produces a *photoevent* (photoevents are not necessarily all of the same size). In the visible region (around 550 nm), each second about 2×10^{12} photons per nanometer arrive from the Sun at each square centimeter of the Earth. For the faintest stars visible, the dark-adapted human eye intercepts about 20 photons per second per nanometer. This photon arrival rate is still reduced by transmission losses in the eye. Figure 1.1 is a schematic representation of the terminology corresponding to the different wavelength bands. Associated wave number* (cm^{-1}), frequency (Hz), and energy level (in ev) are given. Note that the ratio between the highest and lowest visible frequencies is less than 2, against a range of more than 10^{16} for the full range of frequencies that are detected nowadays by modern instruments. The outer band gives the temperature of a black body (see Chapter 8) of which the spectral energy distribution peaks at the associated wavelength.

Despite the eventual constancy of a light-emitting source, photons arrive at a random rate, like rainfall. A theoretical limit to the accuracy of any detection is set by the random statistical variation in the time of arrival of photons and in their location of impact. This is called *temporal* and *spatial noise* or, in general, *photon noise*.† Two close surfaces illuminated by the same steady light source will not necessarily receive the same amount of photons. Because of the tendency of photons to group themselves in packets, the statistical distribution (in space and time) of their arrival is rather complex. Neglecting this effect, which tends to increase the fluctuations, one may assume a Poisson distribution for the probability that in a time interval t the number of photons arriving will be exactly i (probability of i occurrences):

$$P(i) = e^{-n} n^i / i! \tag{1.4}$$

where n is the mean photon arrival rate in the time interval t. It is a well-known property of the Poisson distribution that the variance σ^2 is equal to the mean value n; in other words: the rms deviation from the mean value is

$$\sigma = n^{1/2} \tag{1.5}$$

---

* The wave number, which is the reciprocal of the wavelength, is mostly used in the far infrared region of the spectrum.

† Noise is a general term used for describing degradation of information in a signal.

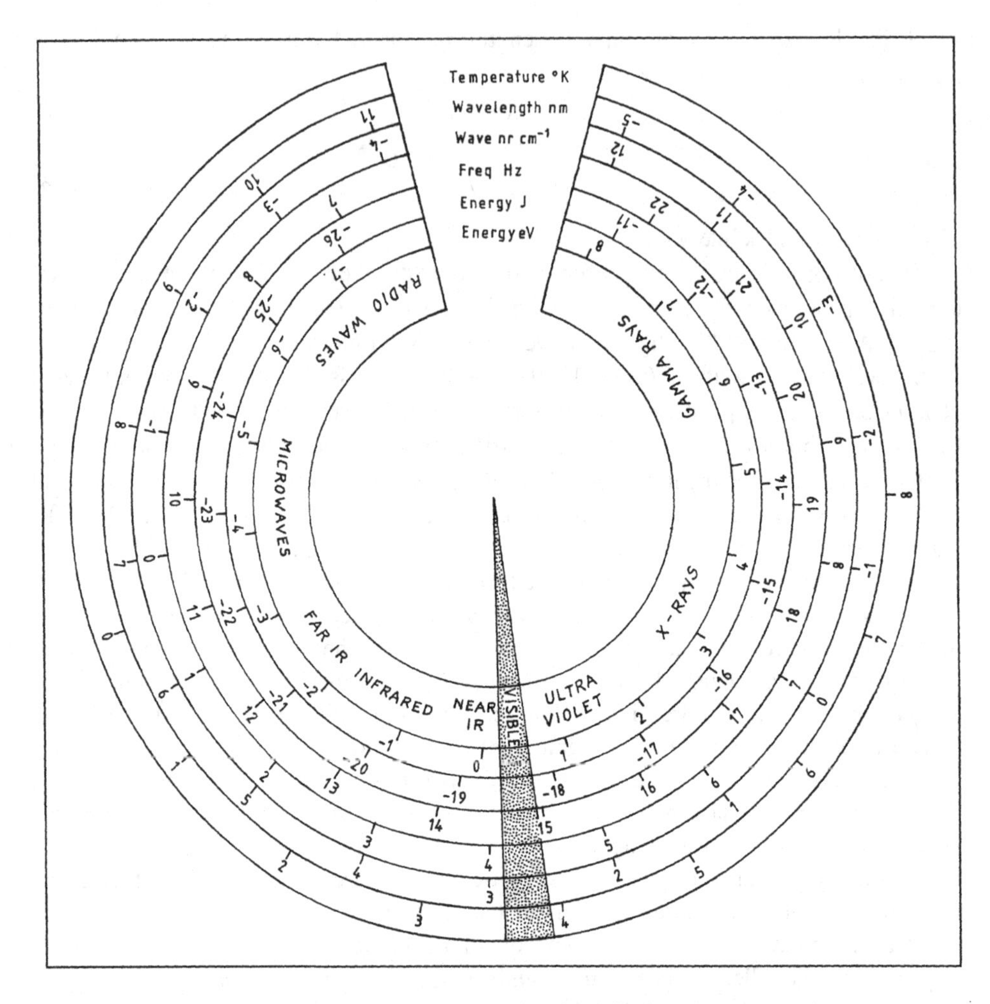

Fig. 1.1 *Electromagnetic wave spectrum. The full electromagnetic spectrum is represented with the terminology corresponding to the different wavelength bands. Logarithms of the associated wave number (in* cm^{-1}*), frequency (Hz), and energy level (J and ev) are given (see text).*

This fluctuation in the photon arrival rate is called photon noise or *shot noise*. Thus an average flow of n photons has associated deviations from the average with a rms value of $n^{1/2}$. These deviations are a measure of the precision with which the average n can be determined from observations performed during the time interval t. Photon noise is the ultimate source of noise that cannot be overcome. A visible effect of photon noise is a certain granularity introduced into any image.

1.2 Photometric concepts and radiometric quantities

1.2.1 General definitions

We now discuss four fundamental quantities in radiometry: *radiant flux, radiant intensity, irradiance* and *radiance*. The corresponding quantities in photometry are distinguished by the root "lumi-" in place of "radi-": *luminous flux, luminous intensity, illuminance* and *luminance*. Monochromatic—or "spectral"—versions of those quantities can be defined (in terms of "densities", where density does not refer to a volume, but to a value per unit spectral interval, which is taken to be the same as the unit of frequency or the unit of wavelength). In general the same symbols are used for radiometric as well as for the corresponding photometric quantities. When there is a risk of ambiguity, the symbols for photometric quantities may be followed by the subscript v, and the symbols for radiometric quantities by the subscript e.

- The *radiant flux*, F, is the rate of energy flow (also called *total* radiant power emitted by a source), and is measured in watts (W). It is often useful to analyze the spectral content of radiation; the spectral radiant flux F_ν (or F_λ) is defined as the radiant flux emitted per unit of frequency (or wavelength).

  To the radiometric definition of radiant flux corresponds the photometric concept of *luminous flux*, the unit of which is the *lumen*; the lumen (lm) is defined in terms of another unit, the *candela* (see below for a definition). The lumen is the luminous flux emitted in a solid angle of 1 steradian by a *point source* having an intensity of one candela. Just as watt-second and watt are units, respectively, for quantity of energy and rate of flow of energy, lumen-second and lumen are the units for quantity of light and rate of flow of light, respectively.

  The luminous flux couples the spectral distribution of radiant power with the spectral response of the eye: luminous flux is related to the integral of the product of the spectral radiant power, and the spectral response function of a normal human eye. Radiometric quantities, when properly weighted with this visual response curve, become photometric quantities.

- The *radiant intensity*, I, of a source S (see Fig. 1.2) in a direction $\mathbf{k}$ is the radiant flux emitted by S in an infinitesimal cone of solid angle $d\Omega$ divided by $d\Omega$, $I = \lim_{d\Omega \to 0} dF/d\Omega$. Radiant intensity gives the angular distribution of radiant flux. The unit of radiant intensity is the watt per steradian (W sr^{-1}). The spectral radiant intensity is either I_ν (W Hz^{-1} sr^{-1}), or I_λ (W nm^{-1} sr^{-1}). As an example, the radiant flux of the Sun is 3.86×10^{26} W; its radiant intensity is 3.05×10^{25} W sr^{-1}.

  The corresponding visual unit is the lm/sr or candela (cd) which measures the *luminous intensity*; it is the basic photometric unit, and was adopted as late as 1948. As all visual quantities, it has a built-in dependence on the spectral sensitivity of the eye and on the spectral content of the light. Originally the unit

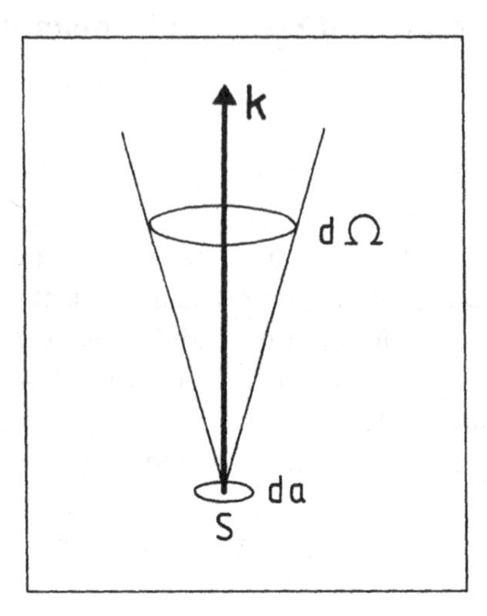

Fig. 1.2 Radiant intensity.

of luminous intensity was the *British candle,** derived from observing a burning candle of standard weight and combustion. This unit has been replaced by the candela which does not depend on the burning of any lamp or candle, but is defined as $1/600,000$ of the luminous intensity in the perpendicular direction, of a blackbody radiator having an area of one square meter at a temperature of 2046 K (solidification point of platinum under standard atmospheric pressure). From this follows the definition of the lumen: *the luminous flux emitted within one unit solid angle (steradian) by a uniformly radiating source of luminous intensity of one candela.* The luminous flux of the Sun (though the Sun is not a point source) is 2.6×10^{28} lm; its luminous intensity is about 2×10^{27} cd.

- The *irradiance*† (*illuminance* for the corresponding photometric quantity), E (E_ν, or E_λ for the spectral quantity), is an aspect of incident flux at a surface: it is the value of the radiant flux (or luminous flux) incident on an infinitesimal surface da, divided by the area da, $E = \lim_{da \to 0} dF/da$. The unit is the W m^{-2}, or the *lux* (1 lx=1 lm m^{-2}). One lux—also called meter-candle—is the illumination resulting when one lumen of luminous flux is uniformly distributed over an area of one square meter.

  The illuminance on the retina and the state of adaptation of the eye determine the subjective brightness of objects viewed under identical conditions. In the corresponding wavelength domains, the irradiance is also the factor which determines the response of detectors (photographic plates, CCDs). Photographers

---

* The British candle was formed from about 60 gram of spermaceti wax shaped in such a way that it would burn during an 8-hour period.

† Irradiance and illuminance refer to a quantity; irradiation and illumination refer to the process itself.

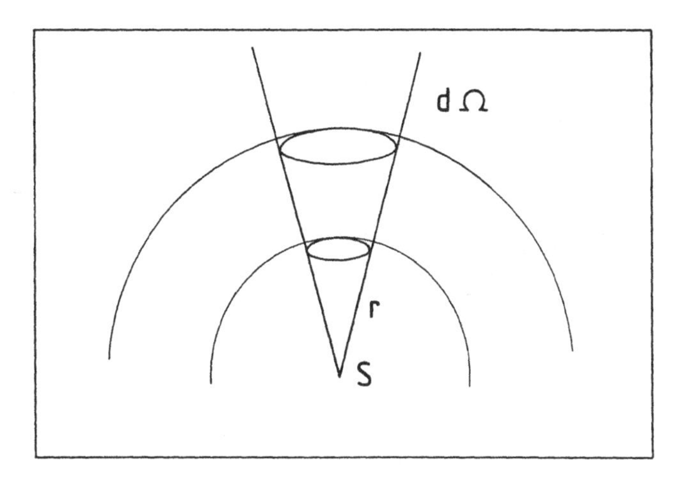

Fig. 1.3 Irradiance. The radiant flux within a given solid angle is conserved. The irradiance varies as $1/r^2$.

often use a light meter with its scale in lumens per square meter. One should remember that such a meter incorporates a spectral filter which corrects the spectral sensitivity of the cell in such a way that the resulting meter deflection is proportional to the response of the eye. Visual quantities and their units are matched to the human eye (lumens, for example, can only be converted to watts when the response of the eye and the spectral distribution of the light are taken into account), and they are of no significance for other detectors (photographic films, photomultipliers, CCD) if these are not equipped with appropriate filters to match the response function of the eye. Some manufacturers of photo-optical devices, nevertheless, quote the performance of their products in visual quantities, such as micro-amperes per lumen. So, an unrelated, and needless, response function is introduced. It is obvious that the lumen is of much better use in the field of window design and electrical lighting than in a physical context.

The irradiance (or the illuminance) on a surface, of course, depends on the distance to the source. Consider an isotropic point source (Fig. 1.3) and concentric spheres of radii r. The total radiant (or luminous) flux crossing the spheres is conserved if there is no absorbing medium. Because of the symmetry we have $F_\nu = 2\pi r^2 E_\nu(r)$. Hence the irradiance (illuminance) varies inversely with the second power of the distance.

Note that E_ν does not refer to the solid angle through which the flux leaves or reaches the surface: in the case of irradiance, the solid angle may vary from a minute amount (collimated light beam) up to a hemisphere.

- The *radiance,* L, (also B for *brightness* or *photometric brightness* according to Born & Wolf 1964), is the fourth important definition of radiometry and refers to an extended surface. The corresponding photometric quantity is called *luminance.*

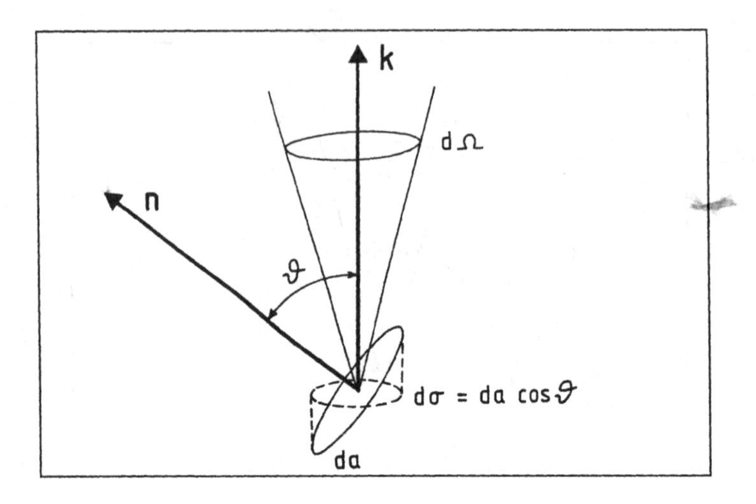

Fig. 1.4 *Radiance. The radiance in direction* **k** *is the quotient of the radiant intensity in that direction, divided by the projected area* $d\sigma = da\cos\theta$ *of* da.

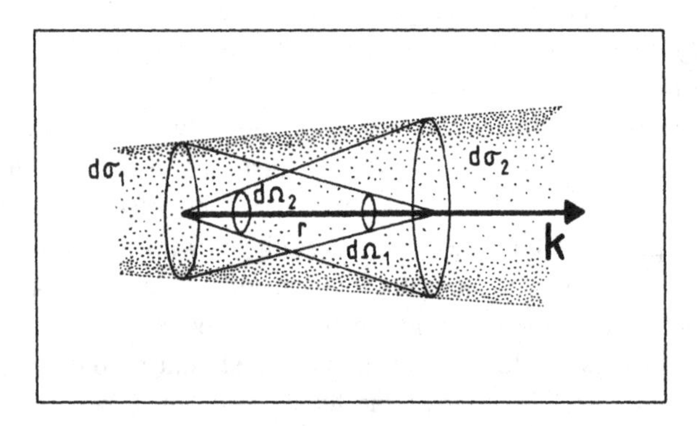

Fig. 1.5 *Invariance of radiance. In the absence of an absorbing medium, the radiance along a pencil of light—defined here by* $d\sigma_1$ *and* $d\sigma_2$*—is a conserved quantity.*

The symbol for the spectral quantity is L_ν or L_λ. The radiance (luminance) should not be confused with the above-defined irradiance (illuminance) of a surface element. Radiance describes the geometrical distribution of radiant flux with respect to position and direction: it equals the flux per unit projected area and solid angle (the area is always the apparent area of the emitting surface as seen by the observer). Consider the surface element da (see Fig. 1.4). The radiance in a given direction **k** is the radiant (or luminous) intensity in that direction, divided by the projected area $d\sigma = da\cos\theta$ of da on a plane perpendicular to **k**. $L =$

$\lim_{da\to 0} dI/da\cos\theta = \lim_{da\to 0\, d\Omega\to 0} d^2F/d\Omega da\cos\theta$. Units are W sr^{-1} m^{-2} and cd m^{-2}. Sometimes the *lambert* ($10^4/\pi$ cd m^{-2}) is used. The unit cd m^{-2} is also called *nit* (from the Latin *nitere*, "to shine").

The radiance (luminance) has the interesting property that it *is invariant along a light beam.* Consider in Fig. 1.5 a direction of propagation **k** and two surface elements $d\sigma_1$ and $d\sigma_2$ perpendicular to it, defining a light pencil. The fluxes through both ends of the pencil are equal, $L_1 d\sigma_1 d\Omega_1 = L_2 d\sigma_2 d\Omega_2$. But $d\sigma_1 d\Omega_1 = d\sigma_2 d\Omega_2 = d\sigma_1 d\sigma_2/r^2$, so that $L_1 = L_2$. *Along a light pencil the decrease in divergence ($d\Omega$) is exactly compensated by the increase of cross-sectional area $d\sigma$.*

This leads to the definition of *throughput*, or *étendue* (French for "extent"). Consider (Fig. 1.6) the flux (radiant or luminous) carried from a surface of area da_1 at S to a receiving surface of area da_2 at R. We have

$$\begin{aligned} dF &= L d\sigma_1 d\Omega_2 \\ &= L da_1 \cos\theta_1 da_2 \cos\theta_2/r^2 \\ &= L d\sigma_2 d\Omega_1 \end{aligned} \tag{1.6}$$

The quantity

$$d\sigma_1 d\Omega_2 = d\sigma_2 d\Omega_1 = da_1\cos\theta_1 da_2\cos\theta_2/r^2 \tag{1.7}$$

is called throughput and is a purely geometrical notion. One can show that in a medium of refractive index n, the throughput multiplied by n^2 is constant. The irradiance (illuminance) on da_2 due to da_1 is

$$\begin{aligned} dE &= dF/da_2 \\ &= L\cos\theta_1 da_1\cos\theta_2/r^2 \\ &= L\cos\theta_2 d\Omega_1 \end{aligned} \tag{1.8}$$

and is directly proportional to the solid angle under which the source is seen from R. The irradiance of an image of the source formed on a receptor at R will also be proportional to $d\Omega_1$.

Besides these four quantities there are also the *radiant energy* (in joules), with the corresponding photometric *luminous "energy"* (quantity of visible light, expressed in lumen-second, lm s, or talbot), and quantities related to the absorption, emission and transmission of light by materials.

1.2.2 Spectral or monochromatic quantities

As stated above, spectral quantities make sense only in radiometry. The frequency densities of radiant flux, radiant intensity, irradiance and radiance are F_ν (W Hz^{-1}), I_ν (W Hz^{-1} sr^{-1}), E_ν (W Hz^{-1} m^{-2}) and L_ν (W Hz^{-1} m^{-2} sr^{-1}). Wavelength densities $F_\lambda \ldots$ are defined in a similar way. Integrated over the whole electromagnetic spectrum,

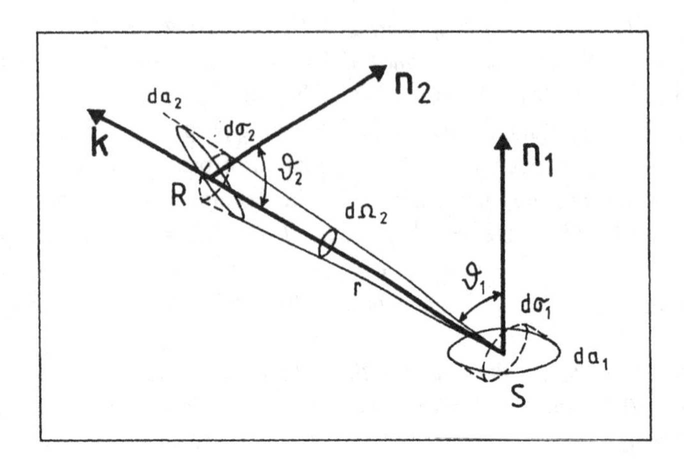

Fig. 1.6 *The throughput, or étendue, of a light beam,* $d\sigma_1 d\Omega_2 = d\sigma_2 d\Omega_1$*, is conserved.*

they yield F, I, E and L. For instance, the radiant flux is $F = \int_0^\infty F_\nu d\nu = \int_0^\infty F_\lambda d\lambda$. When integration is carried over a range of frequencies, one deals with *heterochromatic* quantities. Taking into account the sensitivity curve s_ν of the eye or of other receptors, it is then possible to set up various *photometric systems* (visual, photographic..., see Chapter 16)). In particular, the luminous flux is proportional to $F = \int_0^\infty s_{\mathrm{eye},\nu} F_\nu d\nu$.

1.2.3 Usages in engineering and astrophysics

Photometry is beset with numerous problems of nomenclature. We have introduced the notations used by photometrists and by physicists. However, when it comes to engineering (in applications such as radiative transfer in industrial furnaces and solar energy) and astrophysics (radiative transfer in stellar or planetary atmospheres and in nebulae) the usage is often different (see for instance Pecker & Schatzman 1959, Léna 1988). The main disagreement relates to what engineers and astrophysicists call the intensity, which is equivalent to the concept of radiance of radiometrists. The spectral radiance is called *specific intensity* by astronomers. To add to the confusion, they often write the monochromatic radiant flux (F_ν) as L_ν and call it *monochromatic luminosity*. As if this were not enough they use *monochromatic flux* for the monochromatic irradiance, and denote it by F_ν, and they have also a specific unit for this, the Jansky (1 Jy = 10^{-26} W Hz^{-1} m^{-2}). Table 1.1 presents a brief summary of the various definitions together with the principal units. Throughout this book we shall try to consistently use the radiometric nomenclature.

1.2.4 Point sources and extended sources

The brightness of an object is generally measured by the irradiance it produces at the sensitive surface of a detector (eye, photographic plate, CCD...). When an image is

Table 1.1 Photometry and radiometry. Symbols differing from the standard ones are placed between brackets. Spectral quantities are written in the frequency domain; they could be equally specified in wavelength units.

Photometry	Radiometry		Astrophysics
Luminous flux F (lm)	F	radiant flux F (W)	Luminosity [$\mathcal{L}$]
		spectral radiant flux F_ν (W Hz^{-1})	monochromatic luminosity [$\mathcal{L}_\nu$]
Luminous intensity I (cd)	$\frac{dF}{d\Omega}$	radiant intensity I (W sr^{-1})	
		spectral radiant intensity I_ν (W Hz^{-1} sr^{-1})	
Illuminance E (lx=lm m^{-2})	$\frac{dF}{da}$	irradiance E (W m^{-2})	flux (density) [πH, q]
		spectral irradiance E_ν (W Hz^{-1} m^{-2})	monochr. flux [F_ν, πH_ν, ε_ν] (Jy = 10^{-26} W Hz^{-1} m^{-2})
Luminance L (cd m^{-2})	$\frac{d^2F}{d\Omega da \cos\theta}$	radiance L (W sr^{-1} m^{-2})	intensity [I]
		spectral radiance L_ν (W Hz^{-1} sr^{-1} m^{-2})	specific intensity [I_ν]

formed on a 2-D detector, a point source (star) will, in theory, illuminate one or more picture element (called *pixel*) whatever the distance r to the source may be. The amount of light focused on that pixel(s) is proportional to the radiant flux from the point source, collected by the instrument, i.e., it is proportional to the area of the *entrance pupil** of the instrument, and to the irradiance on the entrance pupil; for sources of equal luminous flux this quantity will follow the inverse-square law. Extended sources, on the other hand, have an image area which also depends on r^{-2}. These effects exactly balance each other. The net result is that *the irradiance on the detector does not change when an extended source is seen from different distances.* This explains why galaxies or nebulae of similar types but at vastly different distances give images with the same photographic density in the same exposure time.

1.3 Natural sources of diffuse illumination

The dark sky, even on a moonless night, is never perfectly black, but is faintly luminous. This diffuse light (if of natural origin) comes from four sources: integrated

* The amount of light reaching the focus is determined by an *aperture stop*. In a camera it is a diaphragm inside the lens; for a telescope it normally is the primary mirror. The image of the aperture stop formed by the elements preceding it is called the *entrance pupil*, while the image formed by the elements following it is the *exit pupil*. Rays directed towards the boundary of the entrance pupil will be redirected to the edge of the aperture stop by the optical system. In the case of most telescopes the aperture stop coincides with the entrance pupil since no optics precede it.

starlight (the combined effect of the many stars and nebulae, including those which are too faint to be seen individually), the zodiacal light (sunlight scattered by fine interplanetary dust particles), the aurorae (flickering light caused by energetic particles entering the atmosphere along the lines of force of the magnetic field of the Earth), and the airglow (photochemical process of luminescence in the upper layers of the atmosphere where atoms and molecules emit light due to excitation, ionization, dissociation and recombination under the influence of the Sun's ultraviolet radiation). Part of this light is scattered by the atmosphere and gives an overall blueish contribution (the equivalent of the daytime blue sky, although the hue is imperceptible to the eye at night, see Section 1.5). The integrated night-sky radiation sheds more light on the Earth than does the combined effect of all the resolved stars and nebulae. On a moonless night it is of the 4th magnitude per square degree, or the equivalent light of a star of magnitude 22 in a circular area with diameter $15''$.* Figure 1.7 illustrates the extent of the different natural levels of illuminance.

The four components of the night sky light differ in relative brightness. The integrated starlight is brighter along the Milky Way. The zodiacal light is brightest along the ecliptic, with a maximum close to the position of the Sun, and a secondary maximum at the antisolar position (the "gegenschein"). The aurorae are negligible at geomagnetic latitudes below 40°, but at extreme northern and southern latitudes their intensity, and variability, can cause severe problems to the photometrist. The airglow increases in brightness towards the horizon.

The spectral energy distributions of these components differ widely. Whereas integrated starlight has a spectrum of relatively early type (i.e., corresponding to young hot stars), zodiacal light has an absorption spectrum which is almost identical to sunlight. Aurorae and airglow have spectra with intense emission lines. The effect of the night sky on photometric measurements will depend on the filters used (bandwidth and central wavelength), and on the detector (spectral response). In addition, the Moon may bring in a variable component with a solar spectrum.

1.4 Detectors

Detectors are used to study radiation. They transform the radiation (light) energy into another form of energy, such as thermal energy, photochemical energy, or electrical energy. Such detectors are selective: their sensibility strongly depends on the wavelength of the infalling light (the eye, for example, is a very selective detector).

1.4.1 The ideal detector

The light detector is a sensing device that generates (and possibly amplifies and/or records) a signal when exposed to a pattern of incoming light. The assistance provided

* For a definition of the term magnitude, see Section 1.7.

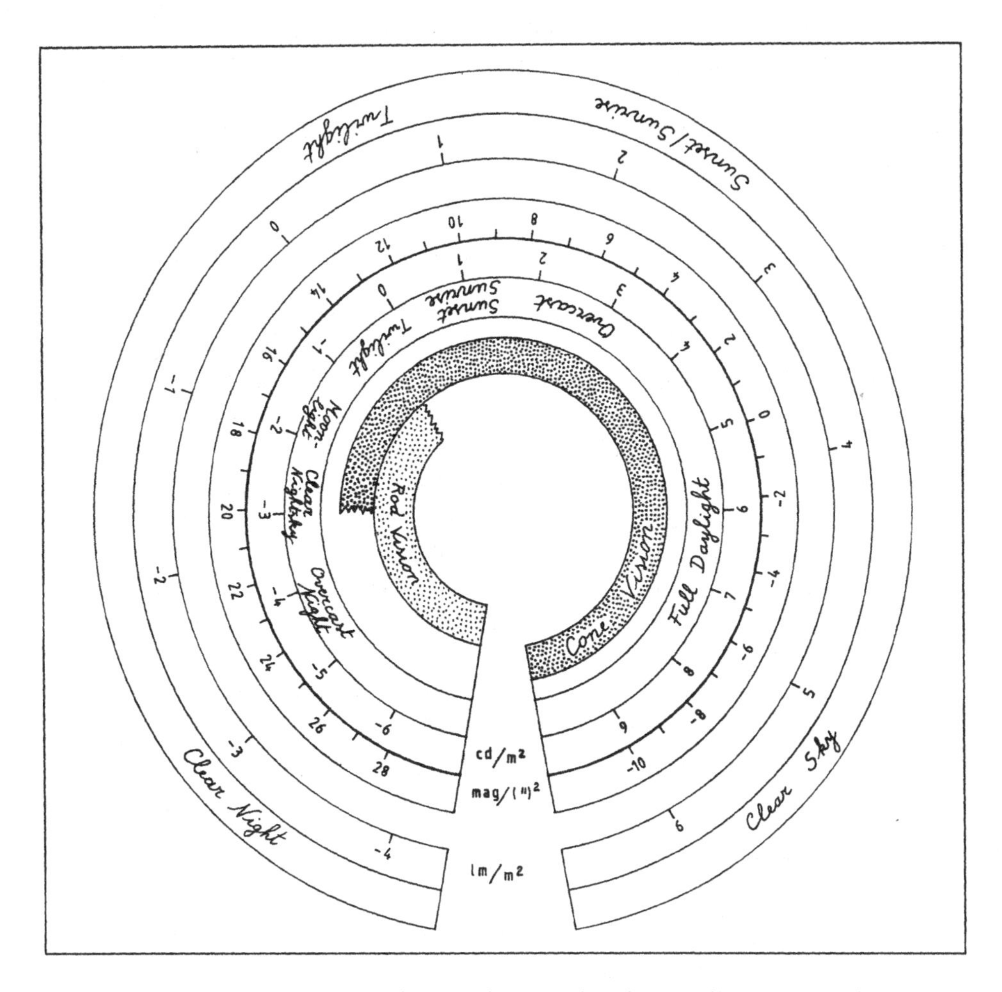

Fig. 1.7 *Aspects of natural lights. The outer band gives illuminances, the inner band luminances of the indicated sources of natural light. Radiances are also expressed in magnitudes per square arc second.*

by a telescope consists in collecting the light over a large area and in concentrating it at a much higher level of irradiance on the detector.

An ideal light or radiation detector is one in which every photon produces an observable event: it counts every one of the incident quanta of the input signal at every point of the image, and it also gives no response when that signal is absent. Such a detector should be perfectly stable over long time intervals, should never be saturated (i.e., overloaded by too many events), should have an infinite spatial resolution (infinitely small picture elements), a perfect time resolution (instantaneous response to the signal and ability to count all photons, even if many arrive in a very small time interval), and a perfect spectral resolution (perfect assessment of wavelength of incoming photons). In addition, the detector surface should be as large as possible.

If such a detector would exist, it would not mean that a performed measurement would have infinite precision: detector performance will always be limited by the random fluctuations in the rate of arrival of photons. The quality of the image and its associated noise level are determined by the telescope optics, by the atmosphere, and by the unavoidable Poisson photon noise.

1.4.2 The common detector and its characteristic parameters

Photometric detectors are either thermal or photon detectors. In *thermal detectors* the energy of the photons is absorbed and converted into thermal agitation of the atoms and electrons, thereby causing a temperature change in the absorbing layer of the detector element. This temperature change can produce measurable effects, such as pressure changes of a gas (Golay cell), a voltage change (thermocouple), or a change in resistance of a semiconductor (bolometer). Due to the involvement of warming and cooling of a substantial amount of material, such detectors respond slowly to changes in radiation. In *photo-emissive* detectors, the incident photon flux interacts with the detector matter and causes emission of electrons. All photo-emissive detectors receive light on a photocathode. *Photo-conductive* detectors are based on the principle that when the sensitive surface absorbs light, its electrical conductivity changes.

Existing detectors always add noise to the already present photon noise. There is noise on the level of detection, on the level of amplification, and also on the level of registration, counting or reading-out of the signal.

1.4.3 Geometrical properties

- The *sensitive area* is the area over which the detector responds to radiation. The pixel is the smallest element of a detector to which a definite signal level can be attributed (for example individual cells in a charge coupled device). The sensitive area may have any number of pixels: from one (photomultiplier tube) to several million (CCD).

- The *spatial resolution* is determined by the *point-spread function* (PSF), a bidimensional (often approximately Gaussian) profile that represents the distribution of light in the image of a point source. The distributions in real images can be calculated by convolving idealized (perfectly sharp) images with a PSF. Spatial resolution is the minimal distance between two equal point sources that can be resolved. *Resolving power* is the reciprocal of spatial resolution, and is often quoted as the number of lines per mm in the image that can be resolved. Note that spatial resolution is not necessarily identical with the geometric separation of the receptor cells. For example, in photopic vision (see Section 1.5) the average resolution of the eye is about one arc minute, and is comparable to the spacing of the cones in the fovea. The resolution of the dark-adapted eye strongly depends on the illumination level and on the contrast with the background (see Clark 1991). It ranges from about one arc minute for relatively bright and contrasted objects (which is also comparable to the rod spacing where they are most densely packed, at 20° from the fovea), to several tens of arc minutes for the faintest objects. In this case a summation effect over many receptor cells allows detection, but yields lower resolution.

1.4.4 Spectral and dynamical properties

- The *quantum efficiency* (*QE*), or *responsive quantum efficiency* (*RQE*), is the number of photons recorded by the detector divided by the number of photons that would have been recorded by a perfect detector under the same conditions (see Eccles et al. 1988). In the case of countable events, *QE* can easily be determined. For other detectors the concept of *QE* is less clear (in the case of the photographic plate, for example, the output event may be a silver atom in the latent image, or a grain made developable; in human vision the output event may be the firing of a nerve cell connected to a rod, or the perception of a light flash).
- The *responsivity* or *sensitivity* is the ratio of output signal to input signal.
- Quantities such as quantum efficiency or sensitivity are defined at a given wavelength or frequency. Thus they display a *wavelength (frequency) dependence* or a *spectral response*. Those functions are needed to estimate the domain of the electromagnetic spectrum where the detector can work: this is the useful *spectral range*.
- The *noise-equivalent power* (*NEP*) is defined either at a given wavelength, or for a blackbody at a given temperature. The *spectral NEP* at wavelength λ is the rms radiant power at λ of a sinusoidally modulated signal falling on a detector which gives rise to an rms output voltage equal to the rms noise voltage of the detector in a bandwidth of 1 Hz. This technical specification depends on the chopping frequency, the detector area, and the electrical bandwidth. The *blackbody NEP* is the same as the spectral *NEP*, but for a blackbody at a given temperature. The spectral and blackbody *detectivities* are the reciprocal of the corresponding noise-equivalent power ($D = 1/NEP$).
- Lower and upper *thresholds* exist for the useful detection of a signal. The lower threshold corresponds to the minimum signal (ultimately a single photon) which can trigger a response. In practice, various effects (such as the *readout noise*, see Chapter 13) will move this threshold to higher values. The upper threshold corresponds to the saturation of the detector (a black photographic negative, or a CCD frame with saturated pixels). Both upper and lower limits may be fixed by the physical detector itself, or by the auxiliary equipment, or by the processing

(e.g., the number of bits of the counter and the digitization procedure may set those limits; high- or low-contrast photographic developers give quite different results). The useable domain between the upper and lower thresholds defines the *dynamic range* of the detector, which is often expressed as a ratio (for example, glossy-print photographs have a dynamic range which can exceed one hundred; high-contrast line-art photographs have a much lower range; the range of CCDs may reach millions).

- The *resolving power* of a spectrometer (eventually used as a photometer) is given by $R = \lambda/\delta\lambda = \nu/\delta\nu$, where $\delta\lambda$ or $\delta\nu$ is the width of the resolution element expressed, respectively, in wavelength or in frequency units. This concept is often called *spectral resolution.*

1.4.5 Temporal properties

- The *speed of response* characterizes how fast the detector reacts when a luminous excitation starts or stops. If the output signal decays exponentially when lighting ceases, one sometimes defines the *time constant* as the time needed to reach e^{-1} of the equilibrium illuminated level, in other words, to about 37% of this value. For non-exponential decays, the time constant could be defined as the time for the signal to drop from 90% to 10% of its value at illumination. A more conventional definition of the time constant is $1/2\pi\nu$, where ν is the chopping frequency at which the responsivity falls to $1/\sqrt{2}$ of its maximum value.

  In thermal detectors, where the measurable effect is produced by a temperature change in the absorbing layer of the detector, the response time is very long. Photon detectors have a much faster response.

- The *bandwidth* is related to the time constant, and is used in the frequency domain (note that frequency means here the rate of modulation of the input illumination, not the frequency of the light waves). The concept may be of interest when rapid variations (due to a celestial object or to the Earth's atmosphere) have to be studied, when the input signal is chopped, or when individual photons are detected. Fluctuations more rapid than the time constant are not recorded and this sets the high limit of the bandwidth.

The acquisition system is a set of (often electronic) devices which register and eventually analyze the signal. The eye has an acquisition system which analyzes in real-time. A CCD and a photomultiplier have acquisition systems which only partially analyze in real-time.

The recording device has also a time constant: if it is too long, a large pulse is passed over before the recorder has time to respond fully. For instance, in the case of a strip-chart recorder this response is limited by the speed of the pen, hence the time constant of the electronics must be large enough to hold a signal until the pen can respond. Strip-chart recorders have now virtually disappeared as recording devices in photometry, since recording of data is now universally done on magnetic storage. Chart recorders, or video displays, however, are very useful when used in a parallel setup. They have unsurpassed value during the process of optical centering of photometers, and for monitoring the quality of atmospheric transparency, and they are still kept in operation at infrared telescopes.

1.4.6 Noise

Noise is any modification to the signal, limiting the accuracy and reliability with which the input can be determined from the measured output. Noise may be due to the radiation itself, or it may arise in the detector, in the recording electronics, or in the telescope (for instance noise may be introduced by focal-plane chopping if the modulating system presents erratic fluctuations, see Chapter 12).

All detectors have a *background noise* of some kind. The mean background response can be eliminated by subtraction, but its random fluctuations do remain and make up noise. To make possible the detection of a small signal, it is necessary to integrate during sufficiently long intervals of time: the deviations of the mean values of both signal and background may then become sufficiently small and both means will then be distinguishable. This operation is more delicate for signals which vary themselves, especially if these variations have a quasi-random behavior.

Background signals of different types exist:

- Internal background, or dark noise, which is introduced by the detector itself when it is not illuminated (for example thermionic emission in photomultipliers). This major cause of noise is present in photometry and in spectroscopy of faint objects.

- Sky background, scintillation noise and seeing: noise arising from fluctuations in the transmissive medium between the source and the detector.

- Thermal radiation background from the detector and surrounding apparatus. This radiation is of no importance in visual detection, but is a source of problems in infrared experiments.

- Cosmic radiation background, which is caused by impacts of cascading cosmic rays. This problem becomes more serious as more sensitive detectors are being developed (visual detection of such radiation at high-altitude sites has occasionally been reported).

The *detective quantum efficiency* (*DQE*) is defined not by comparing input and output counts, but by comparing input and output *signal-to-noise ratios*. *DQE* is given by

$$DQE = (S/N_{out})^2/(S/N_{in})^2 \qquad (1.9)$$

DQE takes into account any loss of detected photons and the *QE* at each stage of the system and allows unambiguous evaluation of the detector efficiency. For two detectors it may be that $QE_1 > QE_2$ and $DQE_1 < DQE_2$. *QE* and *DQE* are only equal if the detector does not introduce any noise of its own, i.e., when the detector conserves the exact distribution of the incident photons.

Some detectors (e.g., CCD, see Chapter 13) measure accumulated charges. Each charge transfer has an associated fluctuation, with standard deviation measured in electrons per reading.

1.5 The most versatile detector of light images

The versatility of the human eye as a light detector is beyond any imagination. The eyeballs in fact are two small cameras, which work in parallel and have several degrees of freedom. They have a focal length of about 15 mm, and a maximum aperture of about 8 mm (decreasing with increasing age). They are about 25 mm in diameter and weigh approximately 7 grams each.

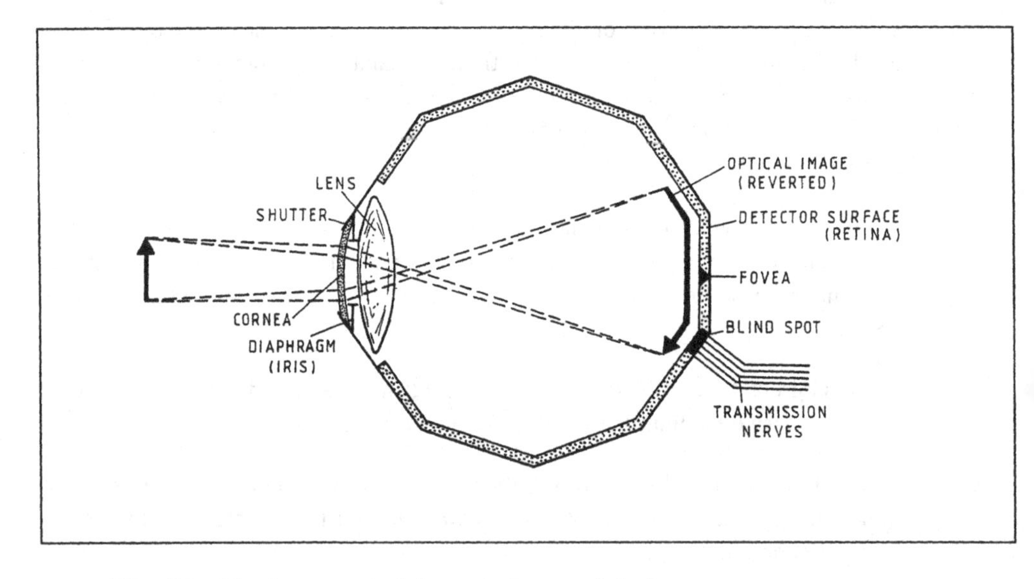

Fig. 1.8 *The human eye. Schematic layout of the human eye, seen as an optical detector.*

Light passes through the cornea (a transparent shield in front of the eye, which constitutes the main refracting surface, contributing most of the focusing) and through the iris (a self-adjusting diaphragm) into the biconvex lens of slightly adjustable power. Lens and cornea are separated by a waterlike fluid, called the *aqueous humor*. This optical system, made of three refracting surfaces, focuses the inverted image on the retina, the thin photosensitive surface at the back of the eye, which contains photoreceptors and a net of interconnecting nerve tissues. There the light is converted into nerve currents which are transmitted for interpretation to the brain. The receptors detect the infalling light at each position of the image, and this image is transferred as a large array of data which represent the relative illuminance at the various pixels. From the differences in information collected at the retinae of each eye, the mind adds the third dimension, and furthermore "ignores" the chromatic aberration (the phenomenon in which a lens, like a prism, disperses white light into a spectrum of colors), and fills in the sensation of color. Every image is thus processed first by the retina, where the image is transformed into a chemical signal, and then by the brain where it is analyzed ("reduced") and interpreted. We use the same word ("image") for both the optical pattern on the retina and for the mental representation. Note that both images are fundamentally different: where mental perception usually is accurate and stable, the actual image on the retina continuously changes, and a new mental image is being formed continuously. For more details on how the color signals are transmitted to the brain we refer the reader to Lennie (1991) and Brou et al. (1986).

The lens of the eye changes its curvature continuously and autofocuses on nearby objects as well as on distant objects (a quality known as "accommodation"). The transparency of the lens decreases with increasing age, and this is often accompanied by a noticeable yellowing at approximately sixty years of age.

The cells composing the core of the lens are denser than those at the edge, which partially corrects the spherical aberration effect which is so common in ordinary glass lenses (the spherical aberration is totally corrected when the eye accommodates at a distance of about 0.5 m. It is positive for larger distances and negative for smaller ones).

The retina consists of a mosaic with millions of individual biological photodetectors of two different types: rod cells (about 100 million) and cone cells (about five million). Cones and rods contain different pigments, which absorb radiation. The cones are mostly situated in the fovea (a yellow spot in the central region of the retina, with an angular diameter of about two degrees that is used when looking straight ahead), while the rods are situated in the peripheral regions. In the fovea the cones are densely packed together, and this is the region with the highest spatial resolution, about one arc minute. Outside the fovea, cones and rods are mixed, and farther out, rods dominate. The cones in the fovea each have a fiber of the optic nerve, and furthermore signals from individual cones are influenced by those from nearby cones.

Rods have higher sensitivity than cones, and are involved in the detection of light at low levels of illuminance (night, or *scotopic vision*); they are capable of signalling the absorption of a single photon. Their high sensitivity causes saturation in ordinary daylight. The high sensitivity of rod vision is accompanied by a lack of color vision: rod vision is achromatic. Individual rods are interconnected in groups, so that the effective surface of each elementary detector is larger.

Cones function at higher light levels (daylight or *photopic vision*) and provide very sharp visual definition and color sensitivity (the eye switches between cone and rod vision at a luminance level of about 10^{-3} cd m^{-2}). The retina has three kinds of cones, only differing by the specific spectral response of the pigment they contain. Modern nomenclature refers to the *S, M* and *L* classes* with peak sensitivities, respectively, at short (430 nm), medium (530 nm) and long (560 nm) wavelengths. The cone action-spectra overlap, but *M* and *L* do most strongly. *S* contributes little, if at all, to the luminance sensation.

The response of a cone to a photon is is about 10 femtoampères, which is about 100 times smaller than a rod's quantal response, but a cone's response is about four times as fast as that of a rod (Schnapf & Baylor 1987). The time constant of the eye is a function of light level: whereas it is about 0.1 sec (ten frames per second) at low luminances, it shifts to about 0.02 sec (50 frames per second) at high light levels. Because of this greater speed of response, cones can encode rapidly changing images and fast changes in intensity.

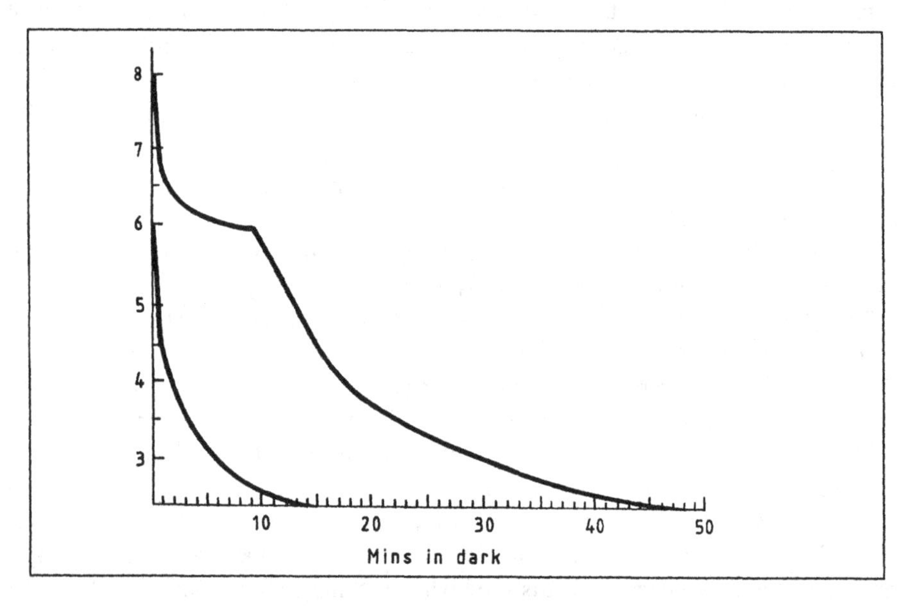

Fig. 1.9 *Dark adaptation strongly depends on the previous level of exposure of the eyes to light.*
Top: *dark adaptation thresholds measured after leaving a brightly lit room. The upper left part of the curve represents cone vision, the lower part rod vision. The vertical axis shows the logarithm of the test intensity required for visibility.*
Bottom: *dark adaptation threshold measured following a weak retinal illumination (about 2% of the value in the upper diagram). Based on Fig. 5.2b of Lamb 1990.*

* By convention, older nomenclature referred to the blue *(B)*, green *(G)* and red *(R)* cones, despite the fact that 560 nm light appears yellow, not red.

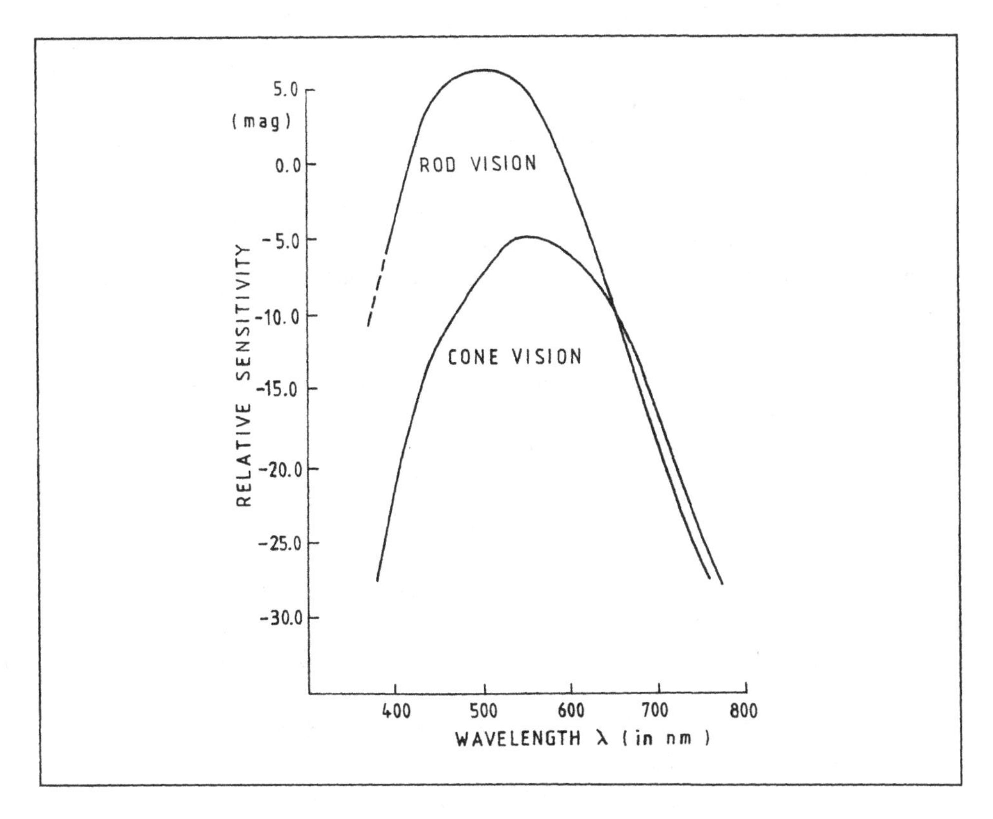

Fig. 1.10 *Spectral response of the eye. Variation of the relative sensitivity of the eye as a function of wavelength for cone and rod vision.*

The human eye is designed to see light levels ranging from the midday sunlight to starlight (which is about equivalent to a candle at a distance of about 20 km): the eye's working range is more than a billion to one in light intensity. Also the *DQE* of the human visual system is substantially constant over the lower part of this range of light intensity. Neither TV nor photographic film matches the ability of the eye to record pictures in such drastically different conditions. The nature of our photosensitive detector offers a tradeoff between sensitivity and resolution in time. The sensitivity of rod vision saturates when the background light intensity exceeds the level of the midday blue sky; at higher light levels only cone vision is active. The eye is also provided with some kind of automatic gain control. A first, very fast (a few seconds) but minor effect, is the opening up (or closing down) of the iris (the solid angle of infalling light is variable over a factor 1 to 16). Most of the adaptation is due to the production of rhodopsin ("visual purple") which causes switching between cone vision and rod vision. Adjustment to total darkness takes about 30 minutes, which approximately coincides with the duration of the gradual darkening during sunset and twilight (see Fig. 1.9). Adaptation to bright illumination happens much more quickly. These mechanisms are negatively influenced by increasing age and by physical fatigue or poor health condition.

Dark adaptation strongly depends on the previous level of exposure of the eyes to light. After exposure to very intense illumination, dark adaptation is composed of two components, a fast cone-dominated phase followed by a slow rod-dominated phase. Dark adaptation is adversely affected after exposure to short-wavelength light sources (fluorescent tubes), which bleach the rod pigments. On the other hand, deep-red illumination of moderate level improves the speed of adaptation to darkness. One therefore finds red-filtered illumination in domes and adjacent laboratories in most observatories.

Because rod vision is colorless, the dark adapted eye sees the sky (with the exception of the brightest objects) and the landscape in gray tones during the night (a photograph reveals a blue sky and a colorful landscape under moonlight, whereas the eye only sees a black-and-white scene).

The peak quantum efficiency of the eye is a few per cent (depending on the observer) and occurs in yellow-green light, very close to the peak spectral irradiance of the Sun. Figure 1.10 gives the spectral response curve of the human eye. It is a function of light level. Cone vision has highest sensitivity at a wavelength of about 550 nm; the peak of the sensitivity curve shifts to 505 nm for rod vision. This shift in sensitivity is known as the *Purkinje effect.* When looking through a spectroscope to a hot star producing a weak spectrum, one will see it colorless, with the brightest range around 505 nm. A bright star of similar spectral type on the other hand shows a colored spectrum; the maximum brightness is around 555 nm, and its total length is much greater than the weak spectrum perceived in the former case.

Light with wavelengths exceeding 750 nm is poorly absorbed by the cone pigments, and light with wavelengths shorter than 400 nm is absorbed by the lens. Note that Griffin et al. (1947) give the reddest detection by human eye at 1050 nm; sensitivity is then about 3×10^{-13} times its value at 505 nm. De Groot (1934) confirms seeing down to 310 nm.

The spectral resolution of the eye—the threshold value at which two hues of different color can be differentiated—is about 5 nanometers for a normal eye; the error in discriminating red and blue tones is substantially larger than for the yellow-green spectral domain.

Under the conditions of dark adaptation, where most astronomical observations are done, one notices that the visibility of objects is larger in the peripheral region than in the central region of the retina. Due to the absence of rods, the fovea is blind in scotopic vision. The technique of using the rod-rich periphery to see faint objects is known as *averted vision,* and requires some practice to be mastered. However, averted vision cannot yield the same details as direct or central vision (lower spatial resolution at low light levels).

For the visual observation of celestial objects, the eye has major drawbacks: its inability to quantify accurately and objectively the received information, its poor performance in measuring colors at low light levels, and its inability to accumulate and store the received information. After all, one may not forget that the eye has an optimal design for operation in our direct environment which is predominantly populated by volumes and where point sources are much less prominent.

1.6 Ancient catalogues

In 120 B.C. Hipparchus classified the naked-eye stars according to their brightness in six groups, called magnitudes. The first class contained the brightest stars, and the sixth class included the faintest stars. They formed a uniform scale where each magnitude difference represented an equal step in visual perception: as such, the relation between magnitude and illuminance was a power law. This classification was recorded by Claudius Ptolemy in his *Almagest* three centuries later, and this elementary qualitative numerical scale was used during more than 2000 years. The Almagest became known in Europe through Latin translations in the 12th century, and was first printed in 1515 (*Almagestum Cl. Ptolemei*, Venice). The origin of the catalogue of stars found in Books 7 and 8 of the Almagest has long been a matter of dispute, and it has frequently been mentioned that the observations were not made by Ptolemy at all, but by Hipparchus (for extensive discussions of this issue, see Evans 1987a,b, and Grasshoff 1990). Ptolemic magnitudes were integer values, to which qualifiers "greater" or "smaller" (respectively, *e.m.* or *e.l.* in the printed version of 1515) were added. These qualifiers are commonly associated with 1/3 or 1/2 magnitudes (thus, a star of magnitude 3 with label "greater" is a star with magnitude around 2.7). Note that in the above-mentioned translation the qualifier "greater" occurs about in 8% of the cases, whereas the indication "smaller" is found in 4% of the cases (in Toomer's translation (1984), the respective occurrences are 10% and 5%).

Empirically, the old magnitude scale was based on the progressive visibility of stars in the advancing evening twilight. After sunset, stars of the first magnitude were seen first, stars of the sixth magnitude were seen last. Apparently, the time interval between beginning and end of the astronomical twilight had been divided in six parts of equal length, and the stars which became visible to the naked eye during each part were assigned the same magnitude (Zinner 1939, Widorn 1955). This procedure clarifies why stars of increasing brillance have decreasing magnitudes. The choice of six for the number of classes, as well as our sexagesimal time and angle measuring systems, may have been influenced by the mysticism of Babylon, where six was regarded as a sacred number.

1.7 Definition of the magnitude scale

Originally, magnitude referred to the intensity of the visual sensation produced by a star, and magnitudes were not more than estimates of visual sensation, without quantitative definition. The response of the eye to the illuminance is not linear: we do not sense the absolute illuminance levels of stars, but we approximately perceive the relative differences. This seems to have been first noticed by Steinhel and then by Pogson who assumed a logarithmic scale.*

* Fechner (1859, 1860) asserted that this also holds for sensations of pressure, noise, and taste. In other words, if γ is the sensation, and β the stimulus, then $\delta\gamma \propto \delta\beta/\beta$. Thus, the relation between brightness and observed magnitude would follow the law that the sensation (observed magnitude) is proportional to the logarithm of the stimulus (illumination): if stimuli constitute a geometrical progression, the sensations will follow an arithmetic sequence.

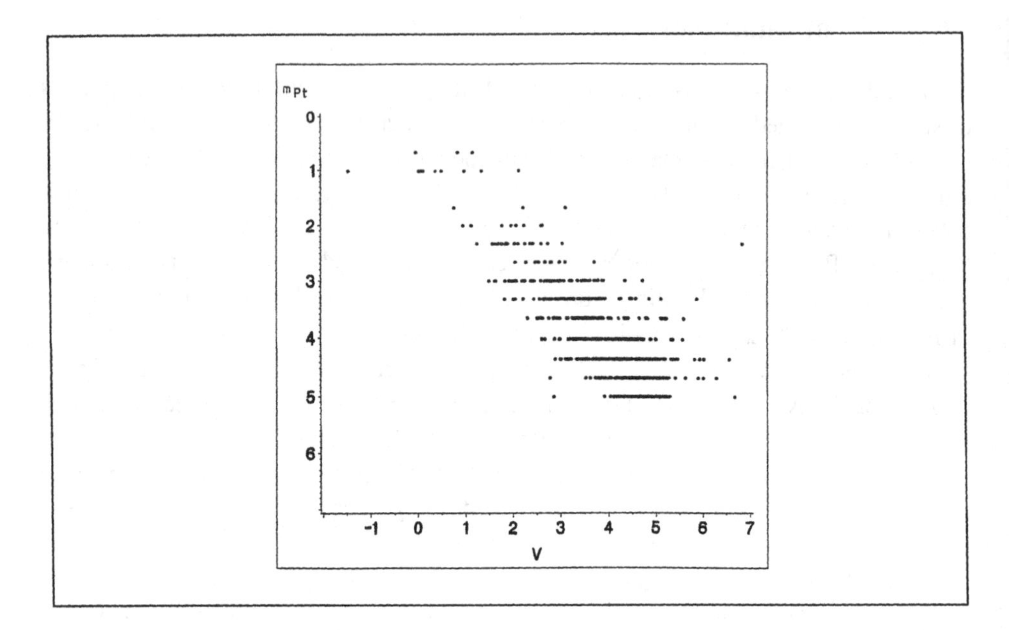

Fig. 1.11 *Magnitudes. Correlation between Ptolemy's magnitudes* m_{Pt} *and the corresponding V-magnitudes taken from the Catalogue of Bright Stars (Hoffleit 1982).* m_{Pt} *are taken from the catalogue of Zinner (1926), who corrected Ptolemy's original magnitude estimates for the sky background (depending on the distance to the Milky Way), for atmospheric extinction, and also for stellar color (see Section 1.10). We have represented* m_{Pt} *in subclasses of 1/3 mag, according to a common interpretation of the "subclasses" given in Ptolemy's Almagest (see text).*

The logarithmic character of such a magnitude scale had computational advantages over a power law, and one had only to experimentally determine the light ratio which corresponds to one magnitude step (the "scale" of the magnitude system). Halfway the last century there was a consensus for an experimentally determined value of about 2.5, but Pogson (1856) *selected* the value 2.512, the fifth root of 100 (the logarithm of which equals 0.4) for convenience of calculation, and so *defined* the modern magnitude scale where the ratio of illuminance of two stars corresponds to their magnitude difference according to the formula

$$E_1/E_2 = 2.512^{m_2 - m_1} \tag{1.10}$$

where E_1, E_2, respectively, represent the recorded illuminances, and m_1 and m_2 the corresponding magnitudes. This leads to the formula

$$m = -2.5 \log E + C \tag{1.11}$$

This expression is often referred to as *Pogson's formula*, and the coefficient (−2.5) is called the *Pogson scale*. C is an arbitrarily chosen constant, and is generally called the zeropoint. The scale and the zeropoint were the first and second problems to be solved in astronomical photometry. Note that magnitudes can be negative: the apparent visual magnitude of the Sun is equal to −26.78.

For small illuminance differences δE one can expand $\log((E+\delta E)/E)$ in series, so that

$$\delta m = -2.5\,(\log e)\,\delta E/E = -1.086\,\delta E/E \simeq -\delta E/E \tag{1.12}$$

In other words *a small magnitude difference is numerically equal to the (opposite of) the variation of the illuminance ratio.* This rule-of-thumb is extremely useful to photometrists during observations and allows estimation of the precision directly in magnitudes.

Let us notice that the magnitude scale is remarkably close to a natural logarithmic scale since Eq. (1.11) can be written

$$m = -2.5(\log e)\ln E + C = -1.086 \ln E + C \tag{1.13}$$

Despite claims to the contrary, the modern and the old magnitude scales appear to be globally in rather close agreement. Figure 1.11 shows the correlation between Ptolemy's magnitudes (corrected for sky background, atmospheric extinction and color, by Zinner 1926) and the corresponding V-magnitudes taken from the Catalogue of Bright Stars (Hoffleit 1982). The large scatter, due to the inaccuracy of ancient magnitude estimates and to actual variations of the stars, makes comparison between the catalogues valid only on a statistical basis. This is also true for catalogues of later date, for instance Al-Sufi and Brahe (Weaver 1946).

The magnitude scale is known to have three major faults (see Hearnshaw 1991):

- it is an inverse scale, with larger numbers pertaining to fainter stars;
- it is a logarithmic scale;
- the base of the logarithm is not 10, but 2.512.

The principal disadvantage of the eye as a detector, its inability to quantify the received information, must lead to inconsistent estimates. Even if one expects that each observer's habit of estimation contains a reasonable amount of internal consistency, the individual differences between results from different experienced observers had led to the establishment in the 19th century of widely different scales of magnitude. This not only created confusion when using estimates from separate sources, but also prevented any transformation between magnitudes made by one observer to measures made in another scientist's scale.

The application of the refracting telescope at high magnifying power to extend the magnitude scale to fainter stars furthermore revealed a remarkable fact, namely that there was a tendency, common to almost all observers, to underrate the magnitudes of stars when seen through a telescope.

1.8 The perception of colors

The second problem of the eye, its poor performance in measuring colors, is noticeable in old catalogues: yellow and red stars are recorded fainter than blue ones, and a color correction is needed (Zinner 1926).

Faint stars excite rods, not cones, and all stars fainter than cone threshold look colorless. Once cone threshold is reached, the light gets chromatic appearance. Laboratory experiments confirm that for observers with normal color vision, the color and brightness sensation are very reproducible. The relation between wavelength and hue is complex, however; it not only varies from person to person, but also varies in time. It also seems that, when luminance is increased, the response of the yellow and blue sensitive cones is stronger relative to the red, so that the violet part of the spectrum becomes less reddish. Prolonged vision also leads to adaptive effects that produce drastic changes in appearance. It also seems that the region of the retina stimulated has a powerful effect upon chromatic appearance. Moreover, experiments during manned space missions have shown that a cosmonaut's eyesight becomes less sharp during the first days of flight, with full adaptation occurring only after 2 to 3 weeks in orbit. The spectral resolution of a cosmonaut's vision seems to degrade at the red end of the spectrum, whereas it produces smaller errors in discrimination at the blue end of the spectrum than do ground-based observers. This effect largely depends on the duration of the mission (Vasyutin & Tishchenko 1989). Experiments carried out at high altitude (Jungfraujoch Scientific Station, 3450m), on the other hand, seem to indicate that visual sensitivity increases during a stay of several days at such an elevated site (Posternak 1948).

The presence of the three different S, M and L sensitivity curves can be interpreted as if indicating that the eye functions as a three-color photometric system (see Chapter 16). The resulting spectral response is a composite of the individual spectral responses of all (types of) cones. This provides the basis for trichromatic color vision. The problem here is that we do not register the three responses separately: the output is a single sensation of luminance. Ideally one should be able to match every color sensation with a mixture of three primary colors and by summation of the individual luminances. However, because of the overlap in the pigment spectra, there are *no* sets of primaries that can additively reproduce all color sensations.

It is well known that about 8% of the male population suffer from chromatic deficiency (only 0.4% of the females suffer from this defect). Three forms of dichromacy are recognized, corresponding to the absence of one of the three dimensions possessed by trichromats. A quarter of these cases are called dichromatic color mixers: they use only two primary colors to make a color match. The remaining 75% of chromatic deficient people are called anomalous trichromats, who have deviating sensitivity curves mostly in the red spectral region. Visual observations performed by people with particular chromatic vision must be regarded as the equivalent of photoelectric observations made with a non-standard filter system (a two-color, or a non-standard three-color system). It is very well-known that such measurements can never be transformed to a standard system, and such data should therefore not be compared to, or combined with other measurements.

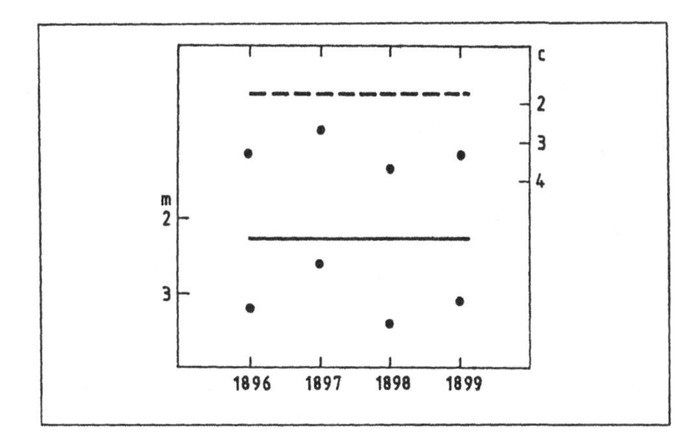

Fig. 1.12 *Color estimates. Variations of color estimate (color class c) associated with light variations (in magnitudes) of Algol. The full line represents Algol's light level outside eclipse; the dashed line represents Algol's color outside eclipse (based on data from Osthoff, 1900).*

1.9 Colors and visual observations

Osthoff (1900, 1908) did very careful visual color estimates of thousands of stars during the time span 1885 to 1907. He performed estimates using a 34 mm and a 108 mm refractor, with magnifications of 18× and 40×, respectively. His observing procedure was very careful and meticulous: during the visual estimates he would cover his head and also the eyepiece with a black cloth, and he would write down the colors in complete darkness. Only rarely would he switch on a lamp to inspect a finding chart. The mean duration of a color estimate was 1.5 to 2 minutes, during which he looked at the star until the impression of color ceased to alter. He would never observe during bright moonlight, nor on occasions where the atmosphere was hazy. He used a scale of 10 color "classes", going from 0 for white, 4 for yellow, to 10 for pure red. With this scale he could classify the brightest stars visible in the telescope. With the naked eye he could see yellow stars up to the second magnitude, and orange stars up to mag 2.5.

The color classification was clearly systematically different when observing with one of his telescopes or with the other: color differences of up to 1.5 color class for the reddest stars were apparent. He also noted that, when using higher magnifying powers with a single refractor, the whitest stars would be more affected than the yellow ones. The presence of moonlight would also lead to a systematic color difference in the same sense. Osthoff observed variable stars such as Algol, Mira and χ Cygni in order to estimate the influence of the brightness of the star on its estimated color. A redder color (for the same star) was associated with a lower brightness,* as is clearly shown by Fig. 1.12 which has been drawn from Osthoff's original data.

* Note that some of these stars do show intrinsic color variations as well.

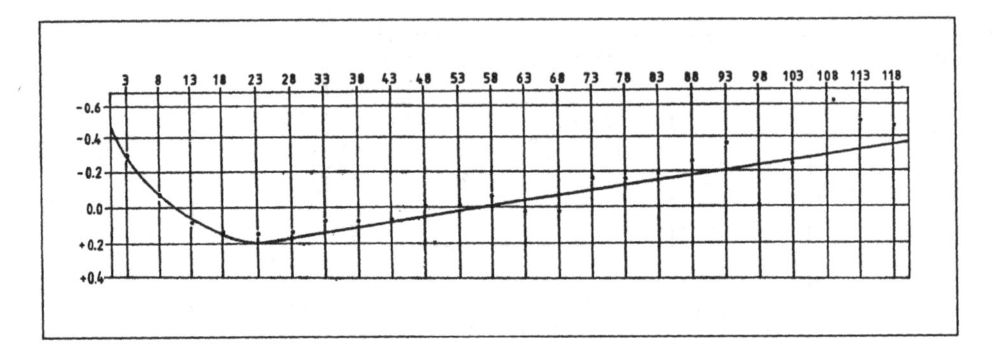

Fig. 1.13 *Variation of the color sensitivity of the eye. Evolution of color sensitivity of the eye during one night. The X-axis indicates the sequential number of the color estimate; the Y-axis gives the color shift (expressed in units of Osthoff's color class). Source: Osthoff (1900).*

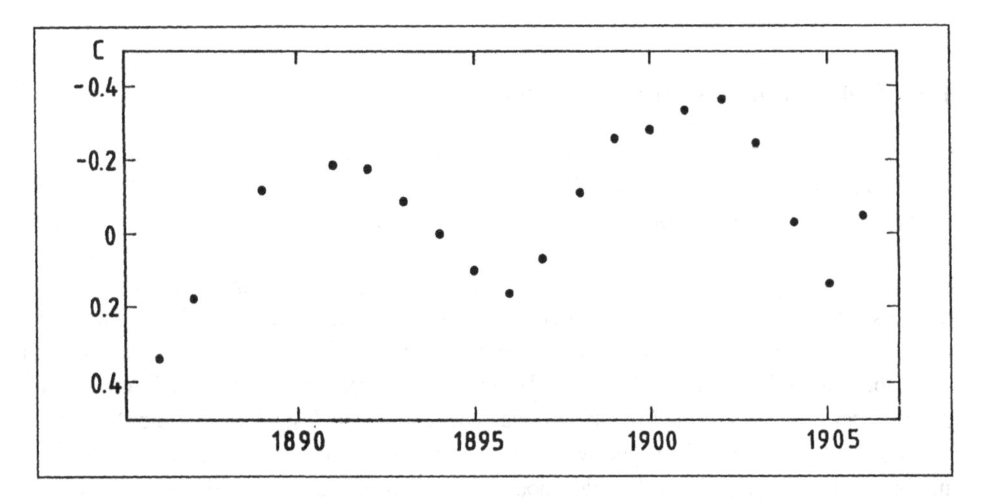

Fig. 1.14 *Annual variation of the color sensitivity of the eye. Variation of the annual mean drift of the red sensitivity of the eye. The Y-axis gives the color shift (expressed in units of Osthoff's color class). Based on data from Osthoff (1908).*

We have seen that the color response of the eye is not constant: Osthoff's data not only reveal a variable response over the course of a single night, but also night-to-night variations, and even variations on a time scale of several years. The evolution of color sensitivity during the course of a night is very interesting: in Fig. 1.13 we see how the color of the first stars observed is underestimated by about half a class. The color is correctly estimated after about ten measurements, but a trend to overestimate continues to reach a maximum around the 20th estimate. The individual scatter seems to increase, and in the long run, star colors are registered "paler" than they really are. This effect is most pronounced for orange and red stars, and the amazing point is that the defect is not time-dependent, but only depends on the number of previous estimates. It is clear that we are dealing with minute differences in sensitivity, and they only stand out because the eye becomes extremely sensitive after hours of dark adaptation.

Osthoff furthermore noticed that on some nights all color estimates were systematically redder or bluer than they were on other nights, as if the zeropoint of the eye's color response would change from night to night. Such shifts really manifested themselves as sudden jumps between consecutive nights. A similar effect was also present on a year-to-year basis. Figure 1.14 illustrates this variation of the annual mean drift of the red sensitivity of the eye. The curve has an amplitude of about half a color class, and a characteristic time of about ten years. Surprisingly, Osthoff considers that seasonal temperature-dependent effects did not appear. But he did point out that the observer's mental state, physical condition, heart rhythm, blood pressure and stress are important factors in the behaviour of the color sensitivity of the eye.

His extensive data allowed him to compare his results with other catalogues. He gives the following systematic differences between his own catalogue and other catalogues: Dunér 0.83; Krüger 1.3; Schmidt −0.23. Systematic differences of up to one full color class thus exist between color catalogues.

One should not forget that the magnitudes of these systematic differences, and also the individual variations, as discussed above, come from data collected by very experienced observers, and that in the case of poorly trained observers, the effects may be much larger.

To our knowledge, catalogues of color estimates made by female observers do not exist. Levander (1889) remarked: "It is also a well-established fact that the education of the color sense is much neglected, especially in the case of our own sex". Taking into account the minute occurrence of deviant chromatic vision within female's eyes, it is a matter of considerable regret that so few female visual observers exist: among the top-25 AAVSO (American Association of Variable Stars Observers) observers—with more than 33,000 recorded estimates each— there is not a single female observer.

1.10 Major sources of errors in ancient visual observations

The earliest star catalogues suffered from some common effects which contributed to the deterioration of accuracy, such as:

1. *The error in magnitude estimate caused by color.* Since the color sensitivity of the eye varies from one observer to the other, two observers may agree in their magnitude estimates for stars of the same color, but their estimates may systematically deviate for stars of different colors. The correction for this effect, the color equation, is a function of the apparent magnitude of the stars.

2. *The background effect.* Magnitude estimates obtained during a moonlit night will deviate from estimates obtained during a dark night. Also estimates made in a populous field like the Milky Way may systematically differ from measurements in a scanty region. This effect is also observer-dependent.

3. *The position error.* The sensitivity of the retina varies over its surface (this is a common property of 2-D detectors), and comparisons of illuminances of two stars will depend on the position of the star images on the retina.

4. *The decimal preference.* This is a very well known psychological error, which is also present when one makes estimates of decimals using a ruler with only unit divisions. Some observers tend to avoid decimals like 1, 4 or 6, and their estimates are crowded into preferred classes.

5. *The extinction correction.* Correction for atmospheric extinction is not always properly done, and stars observed low in the sky are estimated too faint.

The evolution of the degree of precision of one single photometric measurement or estimate during the last decades is given in Fig. 1.15. Young's statement (1984):

> "...I cannot think of any other physical measurement that has become only two orders of magnitude more precise in two thousand years, and has improved by little more than a factor of ten in the past century..."

clearly illustrates the point that photometric accuracy routinely achieved today is of the same level as the expert work which was done more than fifty years ago. This means that hardware technology has reached a level of solid reliability at which only small improvements may be expected, and that real progress must come from systematic removal of errors on the different levels which add up to final deterioration of accuracy.

Systematic errors may be caused by the construction of the telescope, the design of the photometer and the electronics, and by the conception of the data reduction routines and algorithms. Each of these components is discussed in the next chapters.

1.11 Magnitude

For a long time, magnitude meant nothing but a label attached to right ascension and declination, and which was useful only for some statistical considerations. In modern photometry, magnitudes are physical quantities used to derive fundamental stellar parameters.

All information concerning the physical characteristics of celestial objects is embedded in the light which is emitted by these objects. Between the moment of emission of the light and the moment of detection, the starlight undergoes several modifications—for example, the absorption and scattering by interstellar material, by the Earth's atmosphere, or by the individual optical components of the light collector on which the receiver is mounted.

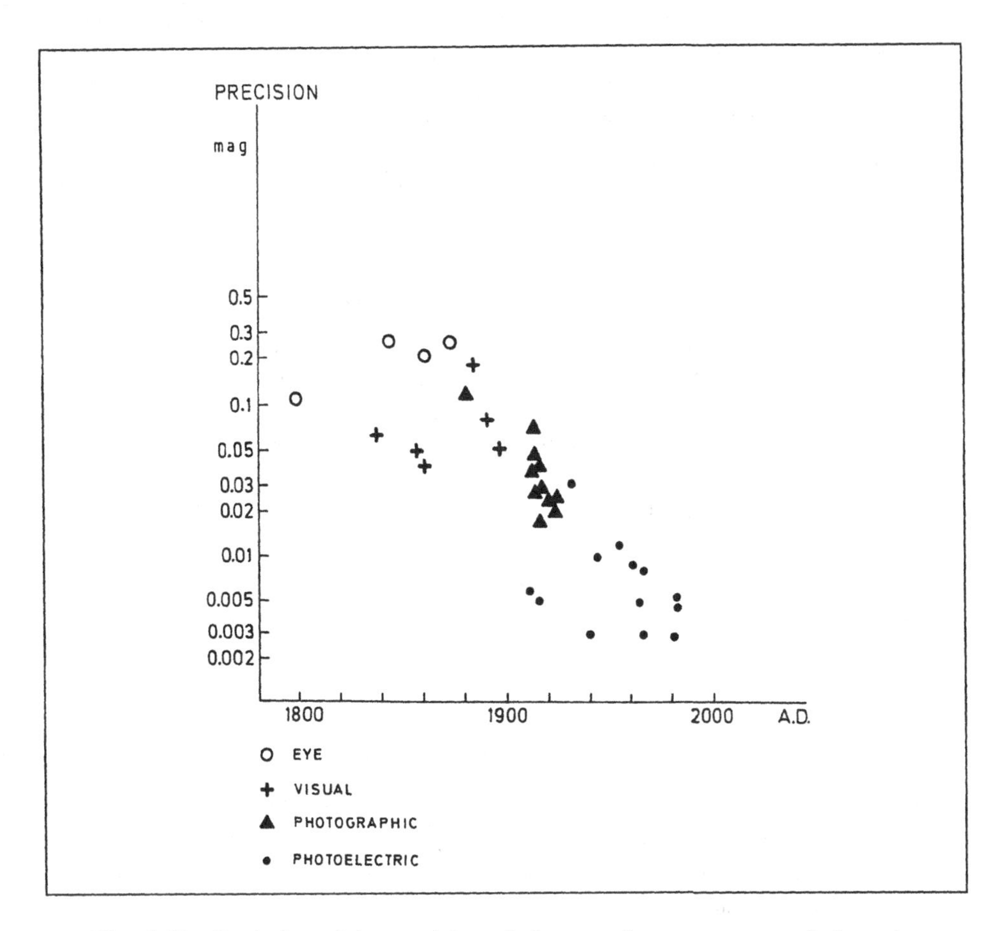

Fig. 1.15 *Evolution of the precision of photometric measurements (adapted from Young 1984). "Eye" means visual estimates aided by telescope only. "Visual" refers to all methods where the eye, as a detector, is assisted by other means (attenuating wedges, comparison lamps...).*

The measured heterochromatic irradiance due to a celestial object is given by

$$E_{\rm m}(\lambda_1, \lambda_2) = \int_{\lambda_1}^{\lambda_2} E(\lambda) s_{\rm i}(\lambda) s_{\rm e}(\lambda) s_{\rm t}(\lambda) s_{\rm s}(\lambda) s_{\rm r}(\lambda) d\lambda \tag{1.14}$$

where $s_{\rm i}$, $s_{\rm e}$, $s_{\rm t}$ and $s_{\rm s}$, respectively, represent the spectral transmissions due to the interstellar medium, the Earth's atmosphere, the telescope and the photometric system. $s_{\rm r}$ is the spectral sensitivity of the receiver. E is the irradiance outside the Earth's atmosphere (it is assumed that the various $s(\lambda)$ are independent of $E(\lambda)$, a condition which, for example, is not met by $s_{\rm r}$ in visual and photographic photometry).

So one can write

$$E_{\rm m}(\lambda_1, \lambda_2) = \int_{\lambda_1}^{\lambda_2} E(\lambda) s_{\rm i}(\lambda) s_{\rm e}(\lambda) s(\lambda) d\lambda \tag{1.15}$$

where $s(\lambda) = s_t(\lambda)s_s(\lambda)s_r(\lambda)$ is the response curve of the observing system and λ_1 and λ_2 define the outer edges of the passband. The interstellar contribution is an astrophysical problem on which the observer has no control (see Chapter 9). We will assume that s_i is merged with E, which is equivalent to treating the stars as intrinsically reddened, so that

$$E_m(\lambda_1, \lambda_2) = \int_{\lambda_1}^{\lambda_2} E(\lambda)s_e(\lambda)s(\lambda)d\lambda \tag{1.16}$$

In principle one measures $E_m(\lambda_1, \lambda_2)$, or rather

$$m(\lambda_1, \lambda_2) = m_0 - 2.5 \log E_m(\lambda_1, \lambda_2) \tag{1.17}$$

where m_0 is an arbitrary constant which defines the zeropoint of the magnitude scale.

In the case of unaided visual estimates, $s(\lambda)$ is reduced to the spectral response curve of the eye. Whenever one uses another detector, such as a photographic plate or an electronic device, $s(\lambda)$ will be different, and the recorded magnitude will deviate significantly. This gives rise to a variety of magnitude systems.

The monochromatic magnitude is the limiting case where λ_1 and λ_2 differ by a couple of nm only, with

$$s(\lambda) = s(\lambda_1)\delta(\lambda - \lambda_1) \tag{1.18}$$

δ being the Dirac distribution. One has

$$m(\lambda_1) = m_0 - 2.5 \log E_m(\lambda_1) \tag{1.19}$$

with

$$E_m(\lambda_1) = E(\lambda_1)s_e(\lambda_1)s(\lambda_1) \tag{1.20}$$

As shown by Eq. (1.16), the contribution of the Earth's atmosphere s_e affects the measurements. One must correct for this effect in order to get clean, out-of-the-atmosphere, magnitudes, and this is a difficult problem since the function $s_e(\lambda)$ appears under the integral. Various assumptions, which are made to solve this *extinction* problem, are discussed in Chapters 6 and 10.

For extended objects (nebulae, galaxies, comets, sky backgrounds) the apparent size is important, and one defines the concept of magnitude per square arc second as the magnitude of an area of that size. Of course, magnitudes are not additive and one should not divide total magnitudes by areas.

1.12 Magnitude systems

The light-adapted eye (cone vision) is most sensitive to radiation with a wavelength of about 550 nm. This maximum shifts towards the blue with decreasing intensity of illumination (see Fig. 1.10 for the spectral sensitivity curves of the eye). The magnitude corresponding to the sensitivity curve of the eye* is called the visual magnitude m_v.

$$m_v = -2.5 \log \int_{\lambda_1}^{\lambda_2} E(\lambda)s_{eye}(\lambda)d\lambda + C \tag{1.21}$$

* The visual magnitudes correspond to *mesopic vision*, i.e., between photopic and scotopic vision.

where λ_1 and λ_2 bracket the blue and red spectral thresholds of the eye (for reasons of simplicity we consider the extra-atmospheric magnitude, i.e., $s_e = 1$). The illuminance due to an astronomical object of magnitude m_v is approximately $E(\text{lx}) = 10^{-.4\,(m_v+14)}$. Photographic plates are often more sensitive towards the blue side of the spectrum. The combination of an emulsion and a yellow filter may yield a sensitivity curve which is similar to the eye's response curve. The corresponding magnitudes are called photovisual magnitudes m_{pv}. They differ slightly from visual magnitudes. Blue-sensitive plates yield photographic magnitudes m_{pg} (see Section 16.3). Photometric detectors, combined with specific filters, define other magnitude systems, which are generally called photometric systems (*UBV*, *uvby*, etc.; see Section 16.4).

Magnitude, in general, is a measure of irradiance, expressed logarithmically. The associated unit for radiance (also in a logarithmic scale) usually is magnitude per square second of arc.

1.13 Bolometric magnitude

In theoretical astrophysics one introduces the concept of *bolometric magnitude*, which is, according to Eddington (1926),

> "a measure of the heat-intensity of a star in the same way that the visual magnitude is a measure of its luminous intensity or the photographic magnitude is a measure of its photographic intensity".

It should be stressed that, in this context, the concept of heat-intensity is not restricted to infrared radiation, but means radiation over all wavelengths.

The (extra-atmospheric) bolometric magnitude of a star is given by

$$m_{\text{bol}} = -2.5 \log \int_0^\infty E(\lambda) d\lambda + C \tag{1.22}$$

This magnitude cannot be measured from the ground because of the strong atmospheric absorption.

The *bolometric correction* is the difference between the measured visual magnitude and the apparent bolometric magnitude:

$$m_{\text{bol}} = m_v - BC \tag{1.23}$$

(note that some authors give the opposite sign to *BC*). Hence we can write

$$BC = 2.5 \log \alpha + C \tag{1.24}$$

with

$$\alpha = \frac{\displaystyle\int_0^\infty E(\lambda) d\lambda}{\displaystyle\int_{\lambda_1}^{\lambda_2} E(\lambda) s_{\text{eye}}(\lambda) d\lambda} \tag{1.25}$$

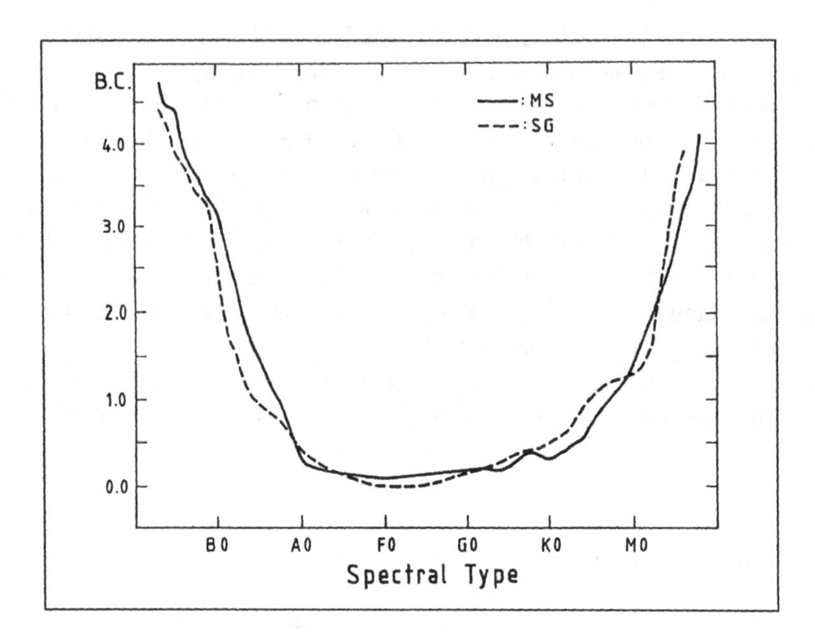

Fig. 1.16 *Bolometric correction. The bolometric correction is a function of the spectral type. Main-sequence (MS) stars and supergiants (SG) have slightly different values (data from Schmidt-Kaler 1982).*

The bolometric correction is strongly dependent on the spectral energy distribution of the star (see Fig. 1.16). Since the sensitivity curve of the eye is optimum for sources having the color temperature* of our Sun, it is not surprising that the minimum value of α for main-sequence stars is reached for stars with an effective temperature of about 6500 K, i.e., close to that of solar-type stars. Frequently, the constant C is chosen so that $BC = 0$ for unreddened (i.e., unaffected by interstellar extinction, see Chapter 9) main-sequence stars of spectral type G2, which yields $C = -2.5 \log \alpha_{\rm min}$. We have

$$BC = 2.5 \log \alpha / \alpha_{\rm min} \tag{1.26}$$

Thus, BC is positive for almost all stars of other spectral type. For the Sun, $BC_\odot = 0.07$. F supergiants have lower BC than main sequence stars. Hence some of them have negative values. Tables of bolometric corrections can be found in Schmidt-Kaler (1982).

1.14 Apparent and absolute magnitudes

The *apparent magnitude* of a star depends on the distance between the Earth and the star. The *absolute magnitude* M is defined as the apparent magnitude a star would

* Color temperature is the temperature of a black body radiator that has approximately the same spectral radiance as the mentioned source of radiation.

have at a distance of 10 pc (1 pc, parsec = 3.086×10^{16}m), in the absence of interstellar absorption. The irradiance due to a star at distance r is inversely proportional to r^2 so that

$$\begin{aligned} m - M &= -2.5 \log(10/r)^2 \\ &= 5 \log r - 5 \end{aligned} \tag{1.27}$$

where r is measured in pc. $m - M$ is the *distance modulus*.

Absolute magnitudes are defined in the various magnitude systems: visual (M_v), photographic (M_{pg}), bolometric (M_{bol}), etc.

The absolute bolometric magnitude allows direct comparison of stellar "luminosities", $\mathcal{L}$ (luminosity is radiant power in astronomer jargon, see definitions in Section 1.2.3 and Table 1.1). One has

$$M_1 - M_2 = -2.5 \log \mathcal{L}_1 / \mathcal{L}_2$$

and hence

$$M_{bol} - M_{bol\odot} = -2.5 \log \mathcal{L} / \mathcal{L}_\odot$$

Since $M_{bol\odot} = 4.72$ and $\mathcal{L}_\odot = 3.83 \times 10^{26}$ W, one obtains

$$M_{bol} = 4.72 - 2.5 \log \mathcal{L} / \mathcal{L}_\odot \tag{1.28}$$

$M_{bol} = 0$ corresponds to $\mathcal{L} = 3 \times 10^{28}$ W.

1.15 Color index

The study of stellar colors was somewhat neglected during the early times of astronomical photometry, and only became a subject of importance after the introduction of the photographic plate, which was sensitive to the blue part of the spectrum only. The color of a star as seen by the eye is a qualitative measure of the star's spectral type: as a rule blue stars have early spectral type, while red stars have late spectral type.

Stars having the same visual magnitudes, but different colors, have greatly different photographic magnitudes. Pickering (1888) pointed out that as long as the spectral distributions of two objects are the same, the relative intensity will appear to be the same whether it is measured by the eye or by the photographic plate. But when the spectra differ, and the colors are unlike each other, no single number will properly express the ratio of the two lights. Blue stars will appear comparatively much brighter in a photograph, and red stars brighter to the eye. Pickering was also the first to suggest that the difference between the visual magnitude of a star and that obtained from ordinary photographic plates would give an estimate of the star's color. This concept led to a more general definition of the color index, *CI*: the difference between the apparent magnitudes of a star in two different spectral regions

$$\begin{aligned} CI &= m_1 - m_2 \\ &= -2.5 \log \frac{\int E(\lambda) s_1(\lambda) d\lambda}{\int E(\lambda) s_2(\lambda) d\lambda} \end{aligned} \tag{1.29}$$

s_1 and s_2 are the response curves in the two different spectral intervals in which the radiation is detected.

In general, color indices may be defined as more complex linear combinations of magnitudes:

$$\mathbf{c} = \{c_i\} = \{\sum_j a_{ij} m(\lambda_j)\} \qquad (i = 1 \text{ to } n) \tag{1.30}$$

with

$$\sum_j a_{ij} = 0 \tag{1.31}$$

so that effects of purely geometrical origin are eliminated.

Color indices, together with magnitude, are a substitution for the energy distribution over the spectrum; they are very important and valuable because of their physical significance.

1.16 Characteristics of the photometric response curve

The measured spectral irradiance $E_m(\lambda)$ incident on a photometric system with response curve $s(\lambda)$ is obtained from the integral equation (1.16)

$$E_m(\lambda_1, \lambda_2) = \int_{\lambda_1}^{\lambda_2} E(\lambda) s_e(\lambda) s(\lambda) d\lambda$$

Along with Strömgren (1937), King (1952) and Young (1988) we group the non-instrumental part of this expression in a single term $S(\lambda) = E(\lambda)s_e(\lambda)$, which changes from one observation to the next. If $S(\lambda)$ is continuous and has continuous derivatives in the interval $[\lambda_1, \lambda_2]$, it can be expanded in a Taylor series about some wavelength $\lambda_0 \in [\lambda_1, \lambda_2]$

$$S(\lambda) = S(\lambda_0) + (\lambda - \lambda_0) S'(\lambda_0) + \frac{1}{2}(\lambda - \lambda_0)^2 S''(\lambda_0) + \ldots \tag{1.32}$$

where primes denote derivatives. Equation (1.16) becomes

$$E_m(\lambda_1, \lambda_2) = S(\lambda_0) \int_{\lambda_1}^{\lambda_2} s(\lambda) d\lambda + S'(\lambda_0) \int_{\lambda_1}^{\lambda_2} (\lambda - \lambda_0) s(\lambda) d\lambda + \frac{1}{2} S''(\lambda_0) \int_{\lambda_1}^{\lambda_2} (\lambda - \lambda_0)^2 s(\lambda) d\lambda + \ldots \tag{1.33}$$

The integrals represent the moments of increasing order of the instrumental function $s(\lambda)$ about λ_0. λ_0 is chosen so as to cancel the first-order moment, $\int_{\lambda_1}^{\lambda_2} (\lambda - \lambda_0) s(\lambda) d\lambda = 0$, hence

$$\lambda_0 = \frac{\int_{\lambda_1}^{\lambda_2} \lambda s(\lambda) d\lambda}{\int_{\lambda_1}^{\lambda_2} s(\lambda) d\lambda} \tag{1.34}$$

This centroid wavelength is also called *mean wavelength* by Golay (1974), or *effective wavelength* (King, 1952),* or *equivalent wavelength* (Strömgren 1937), and depends only on the charcteristics of the instrumental system.

The normalized second moment is denoted by

$$\mu_2^2 = \frac{\int_{\lambda_1}^{\lambda_2} (\lambda - \lambda_0)^2 s(\lambda) d\lambda}{\int_{\lambda_1}^{\lambda_2} s(\lambda) d\lambda} \tag{1.35}$$

(μ_2 is a wavelength interval that is approximately half the passband width of the instrumental function; μ_2 also depends only on properties of the instrumentation). Equation (1.33) becomes

$$E_{\rm m}(\lambda_1, \lambda_2) = [S(\lambda_0) + \frac{1}{2}\mu_2^2 S''(\lambda_0)] \int_{\lambda_1}^{\lambda_2} s(\lambda) d\lambda + \ldots \tag{1.36}$$

If the development is limited to the second order, the measured irradiance $E_{\rm m}$, is expressed in terms of

- an—in principle—invariant instrumental factor $\int_{\lambda_1}^{\lambda_2} s(\lambda) d\lambda$ and
- a non-constant part involving the stellar spectral distribution (continuum and lines)—modified by the atmospheric extinction—, its second derivative over λ, and the second moment of the response function $s(\lambda)$ around λ_0.

This means that *if second-order and higher terms can be neglected* (i.e., when $S(\lambda)$ is linear), the relation between $E_{\rm m}$ and $E(\lambda_0)$ is fully determined by a passband-defined factor and by the atmospheric transmission.

The measured heterochromatic irradiance is approximately

$$\begin{aligned} E_{\rm m}(\lambda_1, \lambda_2) &= S(\lambda_0) \int_{\lambda_1}^{\lambda_2} s(\lambda) d\lambda \\ &= E(\lambda_0) s_{\rm e}(\lambda_0) \int_{\lambda_1}^{\lambda_2} s(\lambda) d\lambda \end{aligned} \tag{1.37}$$

while the measured monochromatic irradiance is

$$m(\lambda_1, \lambda_2) = m_0 - 2.5 \log E_{\rm m}(\lambda_1, \lambda_2) \tag{1.20}$$

or

$$E_{\rm m}(\lambda_1) = E(\lambda_1) s_{\rm e}(\lambda_1) s(\lambda_1) \tag{1.38}$$

Thus, the heterochromatic magnitude of a given star is equivalent to the monochromatic magnitude at some wavelength which depends on the instrumental equipment. However, λ_0, as given by (1.34), is not a true "equivalent" wavelength, i.e., a wavelength λ_0 such that,

$$E_{\rm m}(\lambda_1, \lambda_2) = S(\lambda_0) \times \text{const.} \tag{1.39}$$

---

* Some authors denote by effective wavelength the wavelength given by Eq. (1.34).

with a factor independent of the atmosphere and of the stellar spectrum. Only in the first-order approximation is Eq. (1.39) verified, with const. = $\int_{\lambda_1}^{\lambda_2} s(\lambda)d\lambda$, and neglecting the second-order and higher terms is clearly wrong, because it implies small wavelength intervals, over which the spectral energy distributions do not show strong emission lines, nor deep absorption lines,* nor discontinuities (such as the Balmer discontinuity), and over which the atmospheric transmission function is smooth (no molecular bands).

King (1952) realized that the second-order term must be conserved. However, keeping this term in Eq. (1.36) does not even yield a satisfactory approximation, unless drastic conditions are imposed on the photometric system (see Young 1974b, 1988). Those problems will be discussed in Sections 6.5, 6.5.2, and 8.3. Consequently, Eq. (1.39) cannot be valid.

It is, of course, possible to find, for every observation, a particular wavelength where monochromatic measurements would reproduce the heterochromatic data. The *isophotal wavelength* λ_i is precisely that. It is defined by,

$$E_m(\lambda_1, \lambda_2) = S(\lambda_i) \int_{\lambda_1}^{\lambda_2} s(\lambda)d\lambda \tag{1.40}$$

λ_i depends on the stellar energy distribution as well as on the instrumental response and on the atmospheric transmission. In other words *it is different for every observation*, and thus of little use. This wavelength is not necessarily unique, since every λ_k will be acceptable as long as $S(\lambda_k) = S(\lambda_i)$ (remember that $S(\lambda)$ usually is not monotonic). If the second- and higher-order terms in (Eq. 1.33) are small enough to be neglected, the mean wavelength equals the isophotal wavelength.

Nicolet (1991; see also Rufener & Nicolet 1988) introduced the *quasi-isophotal* frequency ν_1, which is the mean between the mean frequency ν_0 and their *effective frequency* ν_{eff} defined by

$$c/\nu_{\text{eff}} = \lambda_{\text{eff}} = \int_0^\infty \lambda S(\lambda)s(\lambda)d\lambda / \int_0^\infty S(\lambda)s(\lambda)d\lambda \tag{1.41}$$

This concept of quasi-isophotal frequency is, according to Nicolet, very close to the characteristic of the isophotal frequency. Rufener & Nicolet (1988) derived linear expressions, in terms of Geneva photometric indices, that give a good approximation for the effective frequency and the quasi-isophotal frequency.

Figure 1.17 illustrates the relative importance of the different transmission and response curves. The mean wavelengths for the instrumental systems illustrated in Fig. 1.17 left and right, respectively, are 439.7 and 433.8 nm; it is obvious that the Taylor expansion is carried out at significantly different wavelengths. The "effective" wavelengths (Eq. (1.41)) are 431.1 nm and 451.5 nm, respectively, for the O and M stars measured in the leftmost system, and 424.1 nm and 449.0 nm for the same stars measured in the system at right. The large differences between the characteristic wavelengths are apparent.

---

* such as narrow absorption lines of H_2O in the infrared, see Chapter 12

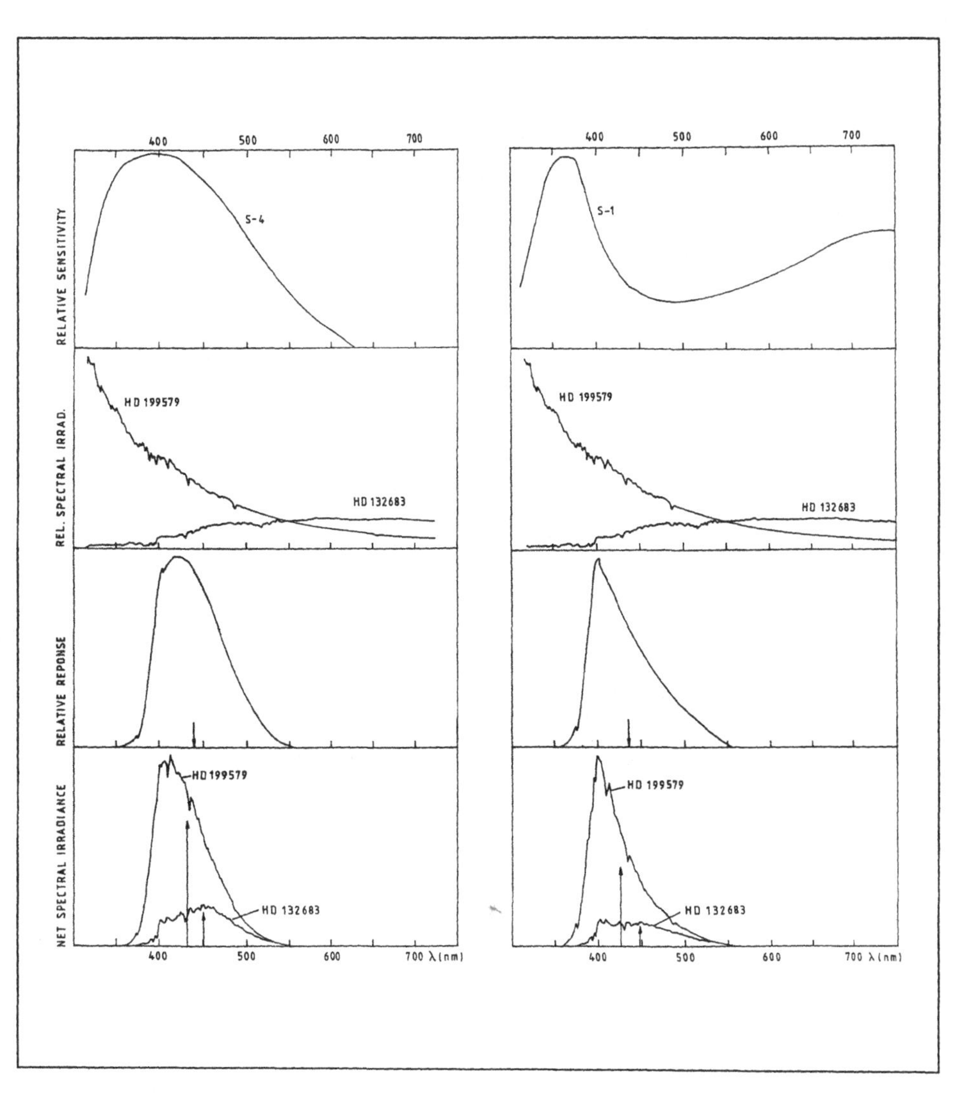

Fig. 1.17 *Response curves. Net response of the Johnson-Morgan* B *passband combined with the response of two photocathodes of importance in astronomy. From top to bottom: response of detector, spectral energy distribution of an O star (HD199579) and an M star (HD132683), combined response of filter and detector, and the resulting response. Left panel is for the well-known RCA 1P21 (S-4) tube; right panel is for a S-1 cathode (see Chapter 4.3.2, and Fig. 6.7). The downward arrows indicate the position of the mean wavelength (Eq. 1.34) of the instrumental system. The upward arrows in the bottom panels indicate the position of the wavelength resulting from Eq. (1.41). for the two observed spectral energy distributions. The spectral data used are from Gunn & Stryker (1983).*

Chapter 2 The telescope

A telescope is an optical system for collecting and focusing radiation. The optical system as a rule modifies the apparent area of the light source (i.e., changes the magnification), but it should be remembered that radiance of a light beam cannot be increased by optical means (see Section 1.2.1). As a matter of fact, radiance is effectively reduced due to transmission losses in and at the surfaces of the optical components, which modify the function $s_t(\lambda)$ in Eq. (1.14). This is especially true for spectral radiance at wavelengths where transmission or reflection losses in the optics are important.

On the other hand, irradiance is vastly increased, and this is a fundamental property of the telescope (see Section 2.2.1).

2.1 Types

Photoelectric photometry is possible with a broad range of telescopes. In the case of a reflector the most common optical configuration is the Cassegrain telescope, because it offers easy access for the observer. Photometers mounted at the Nasmyth or Newton focus exist, but they are less common. Especially photometers designed to work in Newton focus should be light, and have a fair degree of automation. Whatever the type of configuration is, one should be sure that the principal focus extends outside the telescope tube. Specific imaging telescopes, such as Schmidt or Kutter telescopes, are unsuitable for photometric measurements.

Most telescopes used for photometry are reflectors, where the principal optical interfaces are glass mirrors with highly reflective coatings. A very common mirror coating is a deposit of evaporated aluminum, eventually overcoated with silicon monoxide for protecting the soft aluminum layer from corrosion and from abrasion during cleaning. Aluminum has good reflection properties from the ultraviolet to the infrared, and shows good chemical stability. Silver coatings give an extremely high reflectance across the spectrum for wavelengths longer than 400 nm (see Fig. 2.1). They are easily made by chemical means, and were much in use in the second half of the 19th century and the beginning of 20th century. Unfortunately they rapidly deteriorate, especially in the blue part of the spectrum. When performing photometry of early-type stars without using

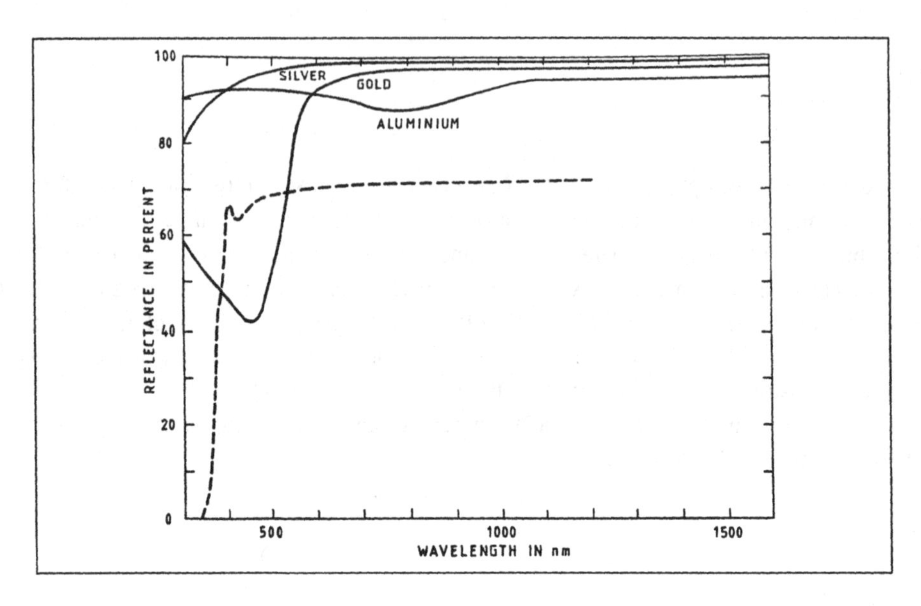

Fig. 2.1 *Spectral reflectances of various metallic coatings. The dashed line represents the transmission of a 15-inch achromat (after Wood 1963).*

a filter, the aluminum reflector will yield a gain in limiting magnitude relative to the silvered reflector. Gold coatings are only used to cover the infrared spectrum from 1 μm to beyond 25 μm. Note that different coatings may drastically modify the passbands of photometric systems.

Lens telescopes (refractors) are suitable, but they have two drawbacks:

- The transmission of glass is generally restricted to a spectral range smaller than the reflection of the common mirror coatings. Unless it is made of special ultraviolet-transparent glass, the objective causes a short-wavelength cutoff, and this influences the ultraviolet magnitudes in those photometric systems which have such a band (see Fig. 2.1). This disadvantage is also present in telescopes with refracting elements such as the nowadays very popular commercial Schmidt-Cassegrain telescopes, or more classical Newtonians and Cassegrains with field correctors.

- The refractive index of glasses is a function of wavelength, so that contrarily to what occurs in catoptric telescopes (reflectors), the optical combination using lenses is only suitable in a narrow spectral domain. The *chromatic aberration* causes a variation of the focal length—and consequently of the image scale—with wavelength, e.g., the blue and the red focus do not coincide. Since photometric studies generally are carried out in a variety of colors, the chromatic aberration may be a problem. Multi-lens objectives exist, made of glasses with different refractive indices, which bring into coincidence the foci at two different wavelentgths ("achromats"). "Apochromats" are still better corrected and force three different wavelengths at a same focus. With these lenses, the chromatic aberration is very well controlled over a whole range of colors, and usually the bandpass of the filter will be narrow enough to avoid any problems.

2.2 Requirements and specifications

2.2.1 F-ratio

The diameter D of the objective determines the light-gathering power of a telescope. The focal length f fixes the image-scale and the space needed around the focal plane. The F-ratio, (f/D) determines the *optical speed* of the telescope. The designation "optical speed" is borrowed from photography. It describes the geometrical gain in irradiance in the focal plane. More precisely, this gain is inversely proportional to the F-number squared. A small ("fast") F-ratio produces a strongly convergent light beam, so that the photometer must have a compact design. Another disadvantage of extremely small F-ratios is that the transmission of the filters may change because the transmission properties of some filters are strongly dependent upon the angle of incidence of the infalling light beam (see Chapter 5). A too large ("slow") ratio defines a slowly converging beam, and produces too small a spot on the filter surface (unless the filters are placed prohibitively far from the focus). Dust on the filters, or small scratches and manufacturing defects, will then cause problems. Most professional photometric telescopes have F-ratios of 10 to 15.

2.2.2 Scale in the focal plane

The image scale, in arc sec per mm, is approximately equal to $200/f$, where the focal length f is expressed in meters. Large focal lengths give high spatial resolution, which allows the study of crowded fields. Such designs also facilitate the manufacturing of the small diaphragm holes which correspond to small angles on the sky and which are used in "aperture photometers" to isolate the object of study from its surroundings.

2.2.3 Size of the telescope

In principle photometry can be done with telescopes of any size. Large telescopes tend to be used with CCD cameras (see Chapter 13) to reach the faintest objects, but they occasionally serve as light collectors for a photometer. Telescopes of the one-meter class and over are often equipped part-time with a photometer; on smaller-class telescopes, a photometer is often permanently mounted. For larger telescopes the atmospheric effects determine the lower limit of the diaphragm. With very small telescopes the image shows diffraction rings, and the Airy disk (the central round disk of light) has a radius (in rad)

$$\theta = 1.22\lambda/D \tag{2.1}$$

D and λ being expressed in a same unit. A substantial amount of light is distributed farther out in the diffraction rings. The transition in the appearance of the image as a function of the telescope size occurs, in optimal atmospheric conditions, at diameters of about 25 to 30 cm.

There is no simple formula that gives the limiting magnitude obtainable with a photoelectric photometer for a given size telescope. The limiting magnitude depends on the desired accuracy, the quality of seeing, the light level of the sky background, the optical qualities of the instrument, the type of detector, and the wavelength at which the observations are done.

2.2.4 Stability of mount

A sturdy mount is absolutely necessary for a telescope in general, and particularly for a photometric instrument. The commercially available small telescopes are in this respect inadequate, because their mounts are too light and cannot prevent flexure and vibrations, and they are not designed to carry heavy accessory equipment. Their mounts need to be replaced before doing any serious photometry (this is, of course, worthwhile only if the optical parts are of good quality, which is not always the case). The telescope should be especially stable in windy circumstances, since wind gusts may induce large-amplitude vibrations which cause the star to move outside the diaphragm (see Fig. 2.2) or give blurred images. Large-amplitude wandering of the image inside the diaphragm may even cause variations in the recorded intensity or color (see Fig. 2.3). Image wandering can be fatal for the quality of observations of a star embedded in a crowded field or in an extended nebula (for example η Carinae), since one is sampling a variable sky-background.

2.2.5 Accuracy of tracking

Manual or automatic guiding of the exposure is, of course, essential for 2-D photometry, for which perfect images are needed, particularly in crowded fields, and also if one wants to apply reduction techniques based on the PSF profile (see Chapter 13). Note, however, that photometry is also possible on blurred or out-of-focus images as long as the images of the stars do not overlap. It is sometimes necessary to degrade the focus for bright stars which would otherwise saturate the detector.

Guiding is not generally possible with conventional photometers equipped with focal-plane diaphragms. Not only the star should stay within the diaphragm, but, as stressed above, it should stay close to the center. Tracking problems due to misalignment of the telescope can be easily eliminated for any telescope. The telescope drive, however, should be precise enough to keep an object well-centered for at least ten minutes. Photometry of faint stars benefits from uninterrupted integrations for such long time intervals. An

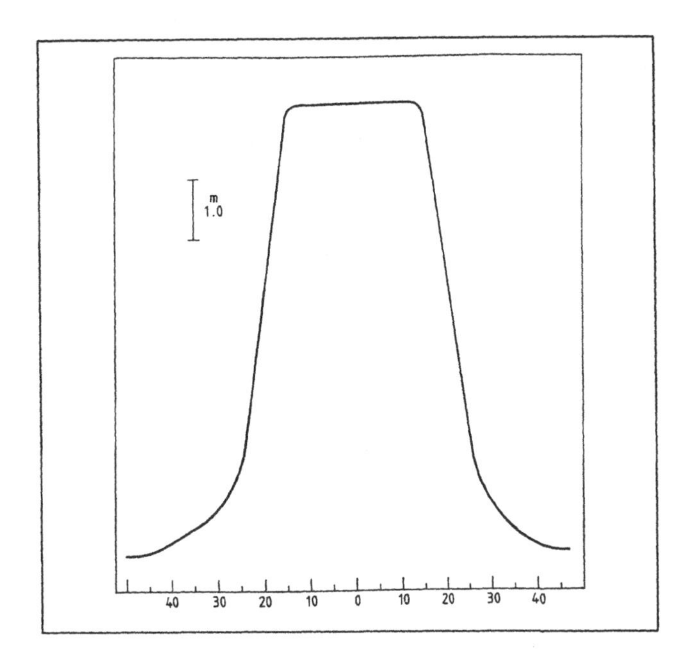

Fig. 2.2 *Recorded signal when a star moves across a 30″ diaphragm. Based on data obtained at the Danish 50cm telescope at ESO. The horizontal scale is given in arc sec.*

image, after having been centered visually, will remain centered at other wavelengths (as long as the atmospheric dispersion is not too large, i.e., at not too large zenith distances). Together with the wavelength-dependent sensitivity variation of the detector, this effect may increase the detection noise level, and also cause spurious variations (see, for example, Figs. 2.2 and 2.3).

The telescope drive often displays a regular oscillation superimposed on the continuous rotation, i.e., the telescope periodically goes slower and then faster than the diurnal motion of the sky. This is generally due to the eccentricity or to other defects of a worm wheel (this is a very common design deficiency of commercially available small telescopes that use small worm gears in the clock drive). If the error is large, a star is seen oscillating around its average position. This will blur the images or will introduce fluctuations of the signal recorded in aperture photometry. In the latter case, the effect is amplified when the star comes close to the edge of the aperture. Hence a telescope's drive error is easily detected by recording the brightness of a star deliberately placed near the edge of the diaphragm. Analysis of the signal will yield the period. This period will be present in all other data sets obtained at that telescope with long integration times, and it is important not to confuse it with an intrinsic period of a star. Warner (1988) points out that this introduces blindness to periods at or near the period of the drive error. He also stresses that, although the drive error itself may be strictly periodic, the Fourier frequency spectrum may not show this so clearly.

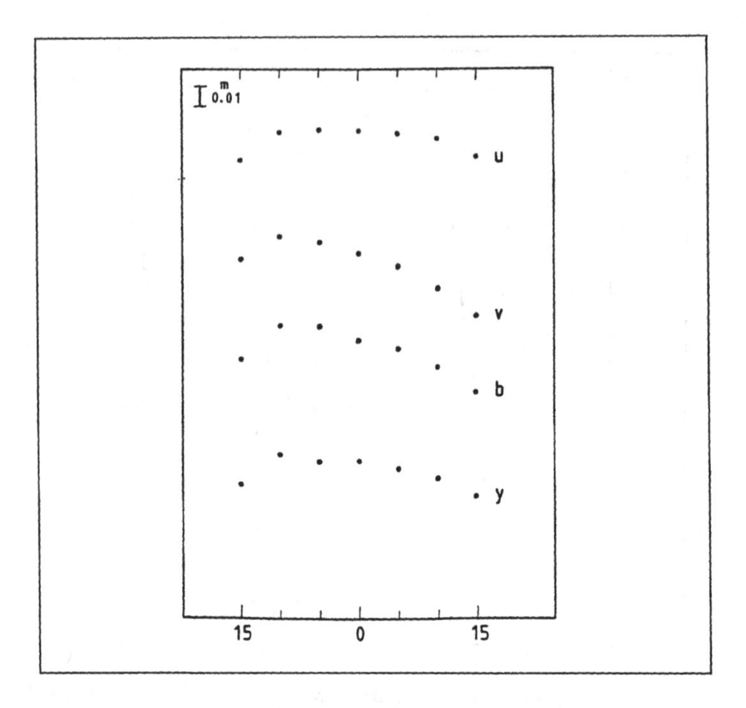

Fig. 2.3 *Variation of sensitivity across the diaphragm. Signal from a bright star in the* u, v, b *and* y *bands, in function of the distance (in arc sec) from the center of the diaphragm of the four-channel photometer of the Danish 50cm telescope at ESO. For distances exceeding 3 to 5 arc sec, the deviation is larger than the nominal precision of measurements.*

2.2.6 Tracking of moving objects

For specific applications, such as photometry of moving objects (i.e., asteroids, comets) the automatic tracking option is required. This is especially important in the case of comets, where individual features in the tail or in the coma have to be analyzed separately. For most clock drives, the tracking in right ascension can be easily adjusted, but tracking in declination is not so commonly available.

2.2.7 Setting

For some kinds of work, like aperture photometry of stars in a cluster, or photometry of faint field stars, the time needed for setting is always short compared to the integration time. Slow setting, or even manual setting, will result in only a slightly smaller nightly production of photometric results. On the contrary, work on bright stars does not need long integration times. In that case higher efficiency and better accuracy can be reached with fast switching from one star to another. This minimizes all changes of atmospheric or instrumental nature that can occur. For instance, in variable star photometry, where one continuously switches from a program star to one or several comparison stars, fast setting speed is crucial for obtaining high-precision data (see Young et al. 1992). Variable star observations profit enormously from computer-controlled setting where the telescope autonomously moves to a preselected position. Photometry of field stars which are widely separated in the sky profits from automated setting, provided it is fast (at least 90° per minute of time). More standard and constant stars can be observed, and this in turn leads to a better determination of instrumental and atmospheric parameters (mainly the extinction and its variations), and ensures higher accuracy overall. Thus the ratio of the time spent outside integrations to the time spent during integrations is a good assessment of the efficiency of the equipment. Very often, automated setting is coupled with autocentering and with automated dome rotation. Sometimes the dome is slower than the telescope and is the main limitation on speed.

A particular nuisance is caused by backlash, the dead space occurring when reversing one of the directions of motion of a telescope. Backlash can be compensated by always approaching the preset position from the same direction, or, better, by pre-loading the gearing.

When working with detectors such as CCDs, integrations are usually long compared to the setting time. Moreover, a certain amount of time is lost between each integration (reading out the frame and storing it on disk or on some other support). The speed of setting is then less critical, at least if moving the telescope and dome do not introduce noise in the readout!

2.2.8 Offset guiding

This means the possibility of setting on an object that is not visible to the eye looking through the photometer viewfinder, and guiding on a nearby object which is visible. Offset guiding is an absolute necessity for measurements of infrared sources which are not visible in the viewing eyepiece of the photometer.

2.2.9 Optics

In aperture photometry, on-axis images are used, and they should be as small as possible, and relatively free of coma and astigmatism. The off-axis image quality is of importance for 2-D photometry, as well as for multiple-beam photometry, or when offset guiding is used.

The quality of astronomical images, obtained with the best optical instruments, has for a long time been severely limited by the impact of turbulence of the Earth's atmosphere (see Chapter 7). A recent break-through in optical technology (see Merkle et al. 1989) permits neutralization of the atmospherically induced smearing of a stellar image and yields images where sharpness is almost only limited by the diffraction produced by the telescope aperture. This technology is called *adaptive optics* and is based on an optical/electronic feed-back loop that controls a small deformable mirror in the optical path in such a way that its surface profile exactly compensates for the distortion of an incoming wavefront (see also Fig. 7.1). It is obvious that such sophisticated systems, which need very fast and very powerful computers, are available only at the largest and most modern telescopes. These systems are of no use in classical photometry, but prove very useful in imaging photometry at visible and infrared wavelengths.

2.2.10 Automated versus manual telescopes

Automatic telescopes represent a novel concept leading to a radically new way of planning and conducting observations. This is best illustrated in photoelectric photometry where the human factor is one cause of errors. Man, with his slow reaction time and high proneness to fatigue, cannot compete with a computer and with ultrafast equipment. In manually conducted photometric observations, most of the time is spent with the photometer in idle status, when the observer moves the telescope to the next star, when the field is being identified, or when one is forced to change the planning of the rest of the night. Above all there is the problem of manpower: for each telescope in operation a skilled observer is needed all year round, and this is a major limitation for the total number of measurements that can be made.

Specifically designed telescopes are developed which operate much faster than conventional ones. The Automatic Photometric Telescopes (APTs, see, for example, Genet & Hayes 1989) take only a few seconds for pointing and centering a star. They do not need conveniently placed eyepieces, auxiliary scopes, etc., so that for a similar light-collecting power, they have a lower inertial momentum and can work at greater angular speed.

Automatic telescopes, specifically built for observing without human assistance, will always have an edge over conventional telescopes, even over those which are computer controlled, because automatized telescopes cannot get rid of their large inertial momentum and their low setting speed. This does not mean that renovating existing telescopes with automatic control is not interesting, but there are many arguments against doing that, and the costs of refurbishing should be compared to those of buying cheap, reliable APTs which are commercially available. The advantage of the APTs in term of efficiency but also of floor space (several 1m APTs use as little space as one 50cm conventional telescope) and maintenance should be considered. On the other hand, these telescopes should not be so close to each other that they can only work near the zenith, and cannot determine extinction well!

Installation and operation costs of automatic photometric telescopes of the one-meter class are negligible compared to the scientific return. Besides more classical programmes, they can perform tasks that are too tedious to be undertaken by astronomers, such as the continuous monitoring of objects during several months.

Programming a robotic telescope in an efficient way, however, requires a thorough knowledge of the programming language, and an evaluation of all possible situations that can be expected during the night. Since observing runs usually are of relatively short

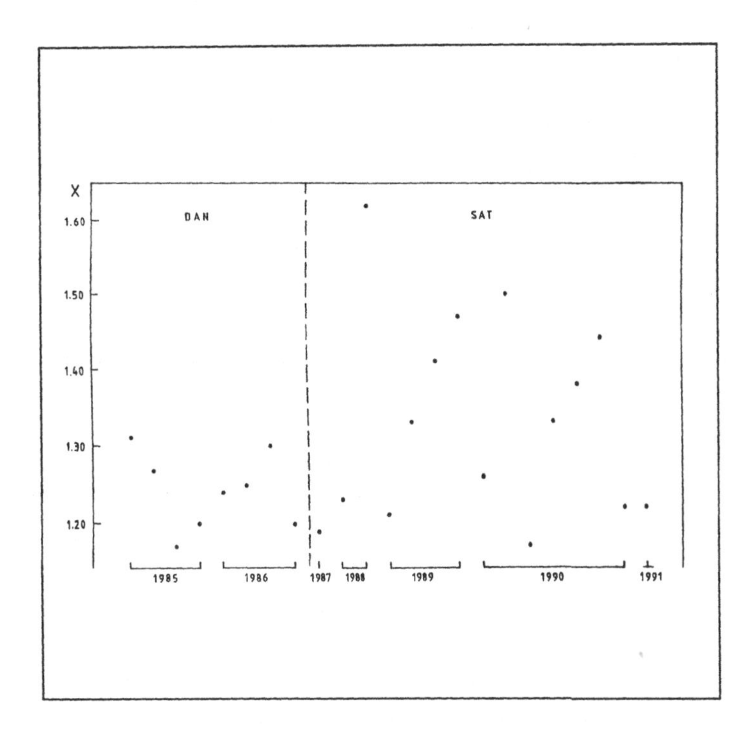

Fig. 2.4 *Average air mass during various observing runs. The air mass X (averaged over complete observing runs) for DAN50 and SAT (Strömgren Automatic Telescope; see Florentin Nielsen et al. 1987) telescopes (from Sterken & Manfroid 1992).*

duration, few astronomers make the effort to thoroughly study the programming language, and this may lead to inferior results when stars are not observed at the right moment, for example when standard stars are observed at too high air masses,* or because a critical phase in the lightcurve of an eclipsing binary is missed. Since automatic photometric telescopes as a rule are small instruments, so that many stars have associated photon noise in excess of 5 mmag, the approach is to increase the number of integrations for the fainter stars. And this is a rule which is easily forgotten when programming observing sequences in automatic mode, especially if the telescope is used by people with limited observing experience in manually-conducted photometry. An example of such a situation is given in Fig. 2.4, which gives the average air mass (during an observing run of several weeks) at which observations were conducted at the 50cm Danish telescope at ESO before and after automation of the telescope. It is obvious that data collected in the SAT configuration have systematically higher air masses than is the case when observing with the manually operated DAN50 telescope—even during a completely identical observing programme. Humans apparently are more careful to observe close to the meridian than a computer is. Of course the computer can be instructed to do equally well or even better,

* For a definition of air mass, see Section 6.3.

but a much larger programming effort is required. Because of the accumulation of the difference between the estimated and real duration of each measurement, observations are done ahead of schedule or, more often, they tend to lag more and more behind schedule.

Another damage to quality comes from a poor choice of the sky area where the sky background is being measured. The practice of visual inspection of the field is not feasible in automatic mode, and the result is that rather often background measurements are contaminated by the light from field stars. This can occur at every measurement of the same star and the observations can be affected by systematic errors. The numerical specification of the offsets in right ascension and/or declination is an absolute must. It is clear that this unfavourable circumstance can, with little effort, be turned into a benefit: once sky locations have been carefully selected, an automatic telescope will invariably point to the specified sky-background spot, a situation which certainly is not always the case in manual mode.

Periods of relatively poor transparency, or bad seeing (see Chapter 7), may also go unnoticed by the robotic control system, whereas a qualified photometrist would write down a comment or, better, would stop observing. In a few cases data obtained in less than perfect (but stable) conditions may go through the reduction procedure unnoticed, and find their way to the final table of results.

All those effects can be avoided by (i) an adequate software and a correct programme, (ii) by a careful planning of the observations and cautious selection of accurate positions for the sky measurements, and (iii) by an intelligent data acquisition system that controls the stream of output data.

So far, automatic photometric telescopes have taught us some lessons:

- Automatic telescopes are only as good as the software that runs them: the programming language must be highly sophisticated to allow for very flexible operation during the observations. However, such automatic telescopes are an improvement only when they are being programmed by observers who have much experience in manually conducted observations. Photometric robots in fact are an outstanding illustration of a situation where expert-systems or "artificial intelligence" is needed (the similarity with satellite operation is striking), but it is clear that such systems must be designed with the help of "experts" in the field in the sense mentioned above, i.e., by people having become skillful in manually conducted observations.

- A good programming language is not enough. Not only the instructions given by the astronomer must make sense, but also the command files written by the user should be complete and well-tested.

- Refurbishing an old telescope for automatic operation is not the only solution. The cost of retrofitting may even be comparable to the cost of building or buying a very compact specifically-designed photometric telescope.

Chapter 3 Photoelectric photometers

Astronomical photometric observations are carried out with two different techniques: *imaging photometry* (also called 2-D photometry), and *aperture photometry*.

Imaging photometry consists of measuring images of the sky obtained with photometrically accurate 2-D detectors. After a long (and still going on) period with photographic plates (see Chapter 14), the technique is now in full expansion thanks to the more accurate electronic devices, particularly CCD cameras (see Chapter 13). Imaging photometry generally gives a spatial resolution of the order of one arc sec, or better, over fields of a few arc min (CCDs) up to degrees (photography).

Photoelectric photometers represent the classical variety of aperture photometers in use since the beginning of the twentieth century. Because of the non-imaging nature of their detector—normally a photomultipier—a diaphragm is needed to isolate the field or the object of interest. Global measurements are then obtained of that particular area. Imaging or charting of a larger field is possible by raster-scanning, a very slow process. Diaphragms typically have diameters ranging from 10 to 100 arc sec, providing a much coarser spatial resolution than is produced by any 2-D photometer.

3.1 Types of photoelectric photometers

A photoelectric photometer is designed according to the type of telescope available and according to the astrophysical problems under investigation. There is a great deal of confusion in the nomenclature of photometers. One can differentiate between single-star photometers and multi-star photometers, depending on the design which allows one either to observe one star at a time, or several stars simultaneously. Some people call multi-star photometers multichannel photometers; others label them simultaneous photometers. And then there are photometers which measure one color at a time, and others which make it possible to measure several colors simultaneously.

In a photometer of the sequential type, light sequentially passes through each filter. In an instrument of the simultaneous type, the measurements through different filters are all performed at the same time. The simultaneous design is more difficult to apply in the case of overlapping passbands such as those of the *UBV* system (see Fig. 5.1), but it can be achieved by splitting the light beam (with dichroic filters) before separating the colors, or even by suitable spectrographic techniques (both methods involve a corresponding loss of efficiency). There is a third type which one could name "pseudo-simultaneous", which is in fact a modified version of the sequential model, and which has a filterwheel that rotates at high frequency: light sequentially passes through the different filters, so that numerous measurements in each band are made in a short interval of time, i.e. practically simultaneously.

The next section describes the basic designs of the major types.

3.2 The sequential photometer

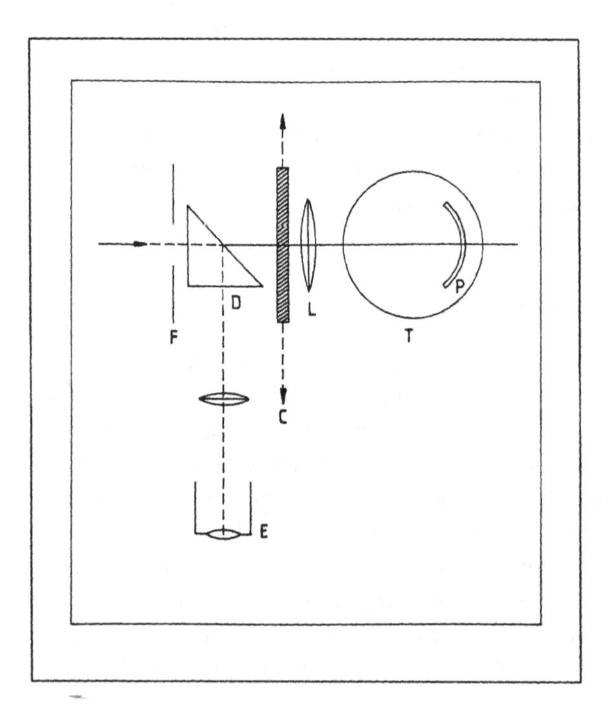

Fig. 3.1 *The sequential photometer. Basic layout (adapted from Evans 1968): Light passes through a diaphragm in the focal plane F and goes through a movable prism D into the viewing eyepiece E or through one of several filters mounted on a slide or wheel C. The Fabry lens L images the entrance pupil onto the photocathode P. The photomultiplier envelope is indicated by T.*

Fig. 3.1 is the schematic layout of a sequential photoelectric photometer attached at the Cassegrain focus of a $f/10$ reflector. The essential components are indicated.

Light falling on a flat field-viewing mirror, or on a prism, is deflected to an eyepiece with a field of, typically, half a degree. The viewing of this field is necessary to allow identification of the stars to be observed. Combined with the usual Cassegrain (or Newton) optics of the telescope, the field mirror brings the number of reflecting surfaces to an odd number (3) and there is an inversion of the field (this minor inconvenience can be avoided with additional optics). Flipping the mirror enables the light to pass through a focal-plane circular diaphragm. Diaphragms of different sizes are formed by apertures of different diameter in a slide or in a circular plate.

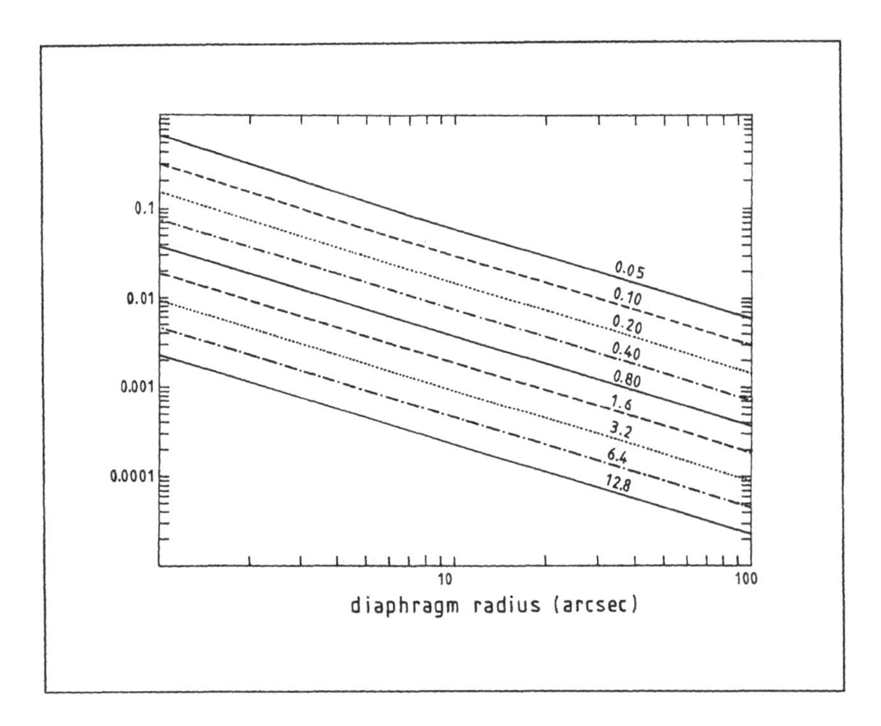

Fig. 3.2 *Fraction of light excluded from the diaphragm for a perfect telescope. The proportion ξ of excluded energy is computed with Eq. (3.1) for diameters D in geometric progression from 0.05 m up to 13 m. The wavelength is 560 nm.*

The role of the diaphragm is to isolate the light of a star from that of its neighbors. Background light unavoidably falls into the aperture, in an amount that is proportional to the area. Both reasons call for small diaphragms. But too small a diaphragm cannot admit the "entire" image of the star, which has quite extended wings due to various phenomena.

A first contribution to the image size is of course the diffraction pattern. The study of the diffraction pattern of a perfect telescope (Young 1970, 1974b) shows that the fraction

ξ of the radiant flux excluded from a diaphragm of radius r decreases as $1/r$. Taking into account an obstruction by a secondary mirror of diameter ϱ times the primary diameter D, one has

$$\xi(r) = \lambda[5rD(1-\varrho)]^{-1} \tag{3.1}$$

with λ and D expressed in the same units, and r in radians. In order to reduce the excluded light to 1% in the green region, it is necessary to use a 35 arc sec diaphragm with a 20 cm telescope having a secondary mirror of 8 cm. Relation (3.1) shows that very large telescopes are needed if one wants small diffraction losses and diaphragms of less than 15 or 10 arc sec (see Fig. 3.2). As stated by Young (1974b), "the importance of large telescopes for accurate photometry is not generally appreciated".

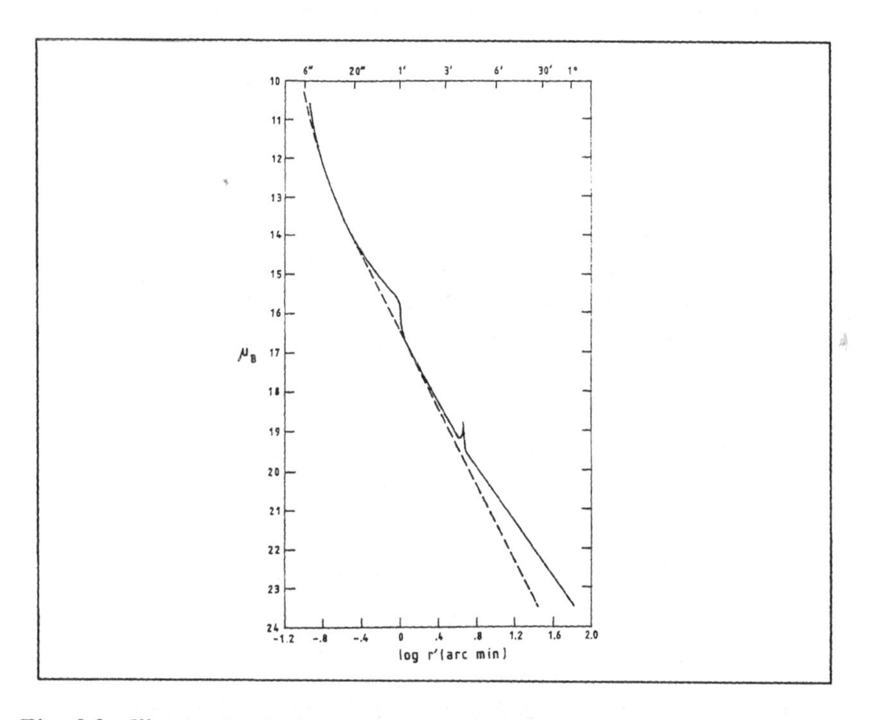

Fig. 3.3 *Illumination in the focal plane. The illumination (in magnitudes per square arc sec) is given as a function of the logarithm of the angular distance to the star. The data can be fitted with two power laws. At large distances the decrease is almost an inverse square (source: Kormendy 1973); the dashed curve is due to King (1971).*

The actual fraction of stellar light that is lost can be evaluated empirically by measuring the brighter stars with a variety of large and small diaphragms. When this is done (see Fig. 3.3 from Kormendy 1973; see also KenKnight 1984), one arrives at the conclusion that relation (3.1) largely underestimates the light losses.

Effects other than the diffraction pattern affect the radial distribution of the illumination on the focal surface. Any residual aberration in the optics modifies the diffraction pattern and increases the image size. The turbulence, or *seeing*, affects the image core (see Chapter 6). The main additional contribution, however, is surface scattering due to imperfect polishing. Small hills and valleys of 1–2 nm height and lateral dimensions of the order of mm to cm are responsible. The resulting effect is that at a few tens of arc sec from the image, relation (3.1) can be off by a factor of 100. Bad tracking and drive errors also contribute to enlarge the image size.

The optimal size of the diaphragm is then a compromise which has to take into account the light level of the object and of the background, the atmospheric conditions, the crowdedness of the field, and the error which can be tolerated (for the fainter stars, the light of the sky becomes comparable with that from the star itself, and too large a diaphragm unduly increases the photon noise). On some nights, or for particular objects, a smaller diaphragm may then be used than on others, but this is a rather dangerous approach which makes comparison between the observations very delicate (see Chapter 10). An aperture of 5 arc sec seems to be the minimal working size in perfect atmospheric conditions, with high-quality optical surfaces and perfect tracking (possibly aided by an offset guider).

The systematic component of the errors due to the use of a variety of diaphragms may be accurately estimated by careful calibrations. This means additional observations of reference stars and, consequently, implies consumption of valuable observing time. Adequate reduction procedures may keep that waste to a minimum (see Section 10.3.5). Those systematic light losses may be a severe problem in photometry of crowded fields, because light from neighboring stars does enter too. The same problem appears in the case of relatively close visual binaries, but it has an easy solution: the background corresponding to component A has to be evaluated at a point exactly symmetrical on the other side of component B; the contribution of B is then exactly reproduced if the unwanted contribution of A is sufficiently reduced by the increased distance.

After the diaphragm wheel there is another pivotable mirror that feeds the light beam that passes through the diaphragm into the diaphragm-viewing periscope, so that one gets a magnified view of the diaphragm. This enables the observer to center the stellar image into the diaphragm, and eventually to select the appropriate diaphragm size. Although this viewer is not absolutely indispensable, it is a very useful device. A diaphragm illumination, with adjustable intensity, helps the observer to see the diaphragm. This illumination should automatically shut off when the diaphragm viewing mirror is pivoted, so that unwanted light contribution and even damage of the detector by inadvertent exposure of the receiver can be avoided.

The diaphragm-viewing periscope is an optical system which is independent from the telescope optics, and it must be focused separately. This can be done on the illuminated edge of the diaphragm (even when the principal mirror or the secondary mirror is obscured). When observing, the stellar image must then be brought into focus at the diaphragm plane, by adjusting the telescope focus, i.e., the position of the secondary mirror.

When the diaphragm-viewing mirror is removed, the light passes through a filter, which is usually mounted in a filterwheel. The filter can be a glass filter, an interference filter, or even a gelatine filter. The properties of the different kinds of filters are described in detail in Chapter 5.

The light then passes through a simple but very important lens, commonly known as the *Fabry lens*, or *field lens*. Fabry (1910) placed a small auxiliary lens with short focal length near the focal plane of the telescope. In doing so, he obtained extrafocal images of uniform light distribution, even from irregular sources such as star clusters. The images were then photographed and measured photometrically. The Fabry lens in a photoelectric photometer has exactly the same function: it forms a uniform image of the entrance pupil on the sensitive surface of the detector. The need for such a lens lies in the fact that the active surfaces of detectors are not uniformly sensitive, and small excursions of the stellar image in the diaphragm would otherwise yield large spurious variations. These motions are instead converted by the lens into variations of the angle of incidence. These too cause variations of the sensitivity but they are usually much smaller, except at inclined photocathodes.

The Fabry lens in a photometer is sometimes ill designed or mounted in a wrong place. There should be no optical parts (like filters) which could disturb the beam size between the Fabry lens and the detector's sensitive surface. This is not always the case, as can be seen in Johnson (1962). Filters located between the Fabry lens and the detector are so close to the detector that any irregularity of the filter, or any dust, will disturb the homogeneity of the Fabry image.

The focal length F_{Fab} of the Fabry lens is given by

$$F_{\mathrm{Fab}} = bf/D \tag{3.2}$$

where D is the diameter of the objective, f the focal length of the telescope, and b the diameter of the light spot on the detector. The Fabry lens is mostly a fixed component of the photometer, and is therefore not compatible with a large range of F-ratios. Needless to say, the Fabry lens should transmit ultraviolet or infrared radiation if one plans to measure in these bands.*

Another point is that the diameter of the Fabry lens should be large enough to intercept all the light coming through the diaphragm. It sometimes happens that the diaphragm wheel has one extra large aperture of several arcminutes. Such an aperture is not meant as a real diaphragm, but rather as an extra help to find the star in the field, and hence the Fabry lens is not designed to that dimension. If such a large diaphragm is used, for instance, for photometry of comets, or of extended nebulae, a large fraction of the transmitted light falls beside the Fabry lens.

Much of the time, the light beam then enters a *coldbox*, which is a thermally insulated container with a cooling system that holds the detector at low temperature in order to keep the noise at a low level (see Chapter 4). The coldbox is sealed by a transparent window which is usually made of quartz to allow the transmission of ultraviolet light.

* Accordingly, Fabry lenses are sometimes mirrors.

Sometimes the window is slightly heated to prevent condensation. The light finally falls on the photoelectric detector.

The coldbox is often designed in such a way that it can easily be interchanged without dismounting the delicate receiver itself. The coldbox can be closed with a shutter slide. This slide is a mechanical safety which protects the receiver from inadvertent illumination during times when the photometer is not in use. It is also a very useful measure to close the shutter whenever the coldbox is not mounted on the photometer: photocathodes, even with voltage off, are sensible to bright light which would increase the dark noise (apparently because of photo-desorption of residual gas atoms).

The receiver used in a photometer as it is described here generally is a photomultiplier. A complete description of the photomultiplier is given in Chapter 4.

Some photometers use a solid-state detector known as photodiode (or PIN photodiode for "positive-intrinsic-negative photodiode"), which is a quantum sensor without internal gain. It has linear response, very high quantum efficiency, wide spectral range; it is very stable, is light, has small size and is cheap. Excellent discussions of photodiode applications are available in Henden & Kaitchuk (1982), Schaefer (1984), Eppeldauer & Schaefer (1988), Genet & Hall (1989). In spite of these nice specifications, photodiodes are restricted to the measurement of bright stars, unless one can build a complex electronic chain and provide cryogenic cooling for amplifiers and detectors.

A photometer of the quasi-simultaneous type often has a design very similar to the sequential photometer, with the difference that complete measurements are not made in one full sequence, but a filter wheel rotating at high frequency is driven by a stepping motor, ensuring quasi-simultaneous color measurements. Some of these photometers allow variable dwelling times in each spectral band, so that stars of very different spectral types can be accurately measured with the same efficiency.

3.3 The simultaneous photometer

In the simultaneous photometer, the colors are separated by a dispersive element (prism or grating, but a dichroic mirror can also be used). The principal components of such a photometer are virtually the same as for the sequential photometer. The light beam, however, is deflected towards the dispersive element, which produces a spectrum of the star. Instead of using a filterwheel, the passbands are defined by slots located on a panel plate which is parallel to the dispersion of the first-order spectrum in such a way that they intercept the spectrum at the appropriate locations, as is shown in Fig. 3.4. Filters can be added to correct the profiles. Several such photometers based on designs by B. Strömgren, E.H. Olsen and R. Florentin Nielsen have been built at Copenhagen Observatory, and are now in use at several observatories in the northern and southern hemispheres.

The photometer for the National Danish 1.5m telescope at La Silla uses solely the exit slots of the spectrograph section to define the passbands. As such, an enhanced throughput is obtained.

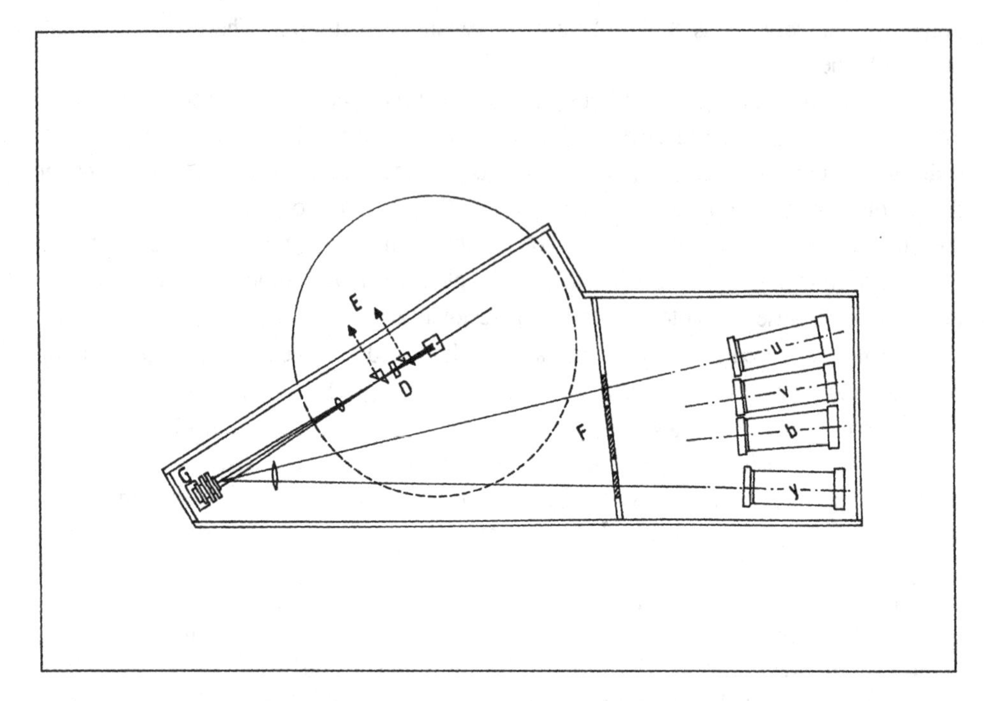

Fig. 3.4 *The simultaneous photometer. Basic layout (from Grønbech et al. 1976). E designates the eyepieces, G is the grating. F indicates the position of the filters and slots. u, v, b and y give the locations of the four photomultipliers.*

The layout of a *uvby*/Hβ photometer installed at La Silla (Florentin Nielsen 1983) is given in Fig. 3.5. Behind the focal plane a tilted mirror can be inserted to deflect the beam of light towards a set of Hβ filters and a separate photomultiplier. This enables the observer to switch from the *uvby* configuration to Hβ mode in virtually no time, which represents a real telescope-time saving concept, especially during nights of poor photometric quality (Hβ photometry in simultaneous mode is practically unaffected by variations in the atmospheric transparency). In four-color mode, the spectrum is formed on a curved surface at which exit slots select the spectral bands. The exit slots can either *define* the spectral bands, or can be used in conjunction with optical filters. The light is then directed to the four photomultipliers, which are uncooled EMI 9789QA tubes.

Fig. 3.6 gives the transmission curves for the *u*, *v*, *b* and *y* filters; the wavelengths under the shaded areas are excluded from the passbands by the exit slots. Unfortunately, the steep sides of the transmission curves, together with the rippled tops, cause some problems in the standardization of the data (see Chapter 11 and also Manfroid & Sterken 1987).

The Hβ-wide (with a FWHM* of 15 nm) and the Hβ-narrow (FWHM= 3 nm) spectral

* FWHM: Full Width at Half Maximum (see Section 5.1).

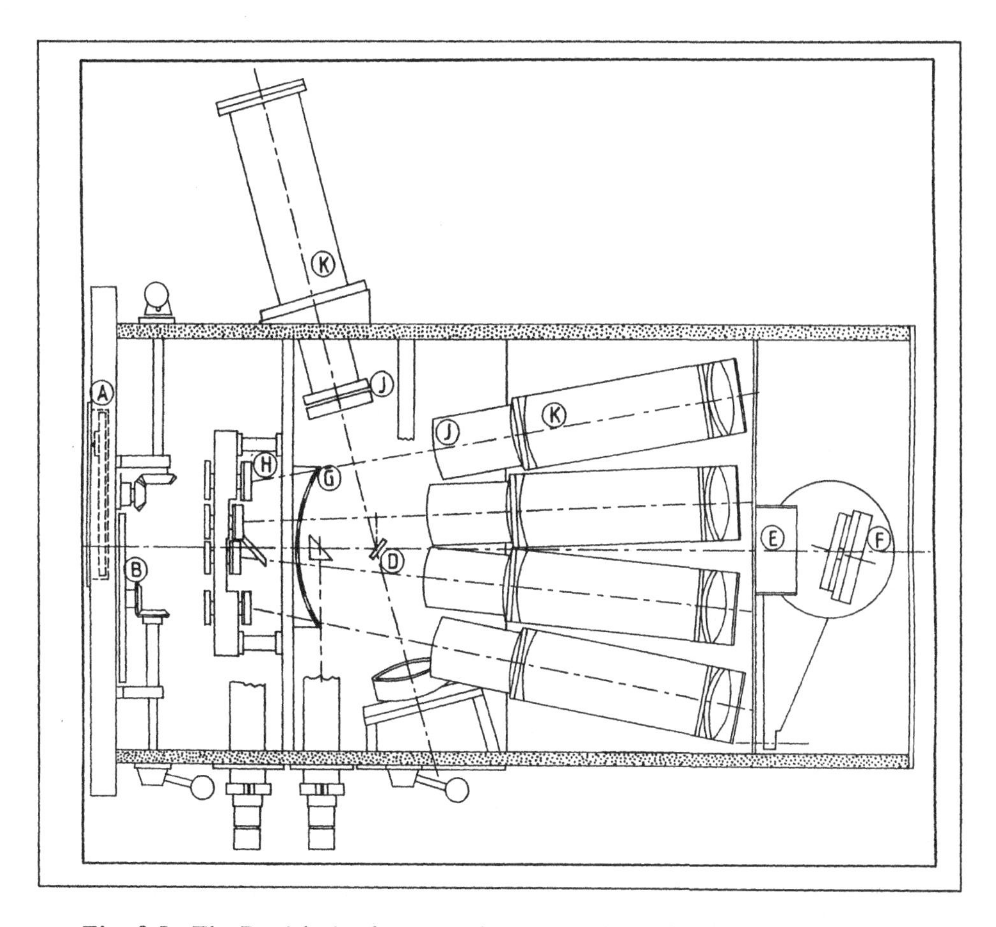

Fig. 3.5 *The Danish simultaneous photometer. Basic layout of the* uvby/Hβ *photometer (adapted from Florentin Nielsen 1983). A pivotable mirror D can be inserted into or extracted from the beam to select either the* Hβ *or four-color mode (photomultipliers K with Fabry lenses J). F is a reflectance grating which forms a spectrum on a curved surface (G) at which slots and (or) filters define the spectral bands. A is the diaphragm wheel.*

bands are purely filter-defined (see Fig. 3.7). A 85/15 percent beamsplitter ensures that an equal amount of light is directed to the two photomultipliers.

The simultaneous design is very efficient, since multi-color measurements are obtained in a really simultaneous way. Light losses in the higher-order spectra, however, make such instruments somewhat less efficient in terms of limiting magnitude (this relative inefficiency may be compensated by purely slot-defined passbands). The use of a grating has some detrimental effects on accuracy when stars are not properly centered (the spectrum moves relative to the position of the slots, which results in modified passbands, even when filters are added). Observers should thus frequently check the position

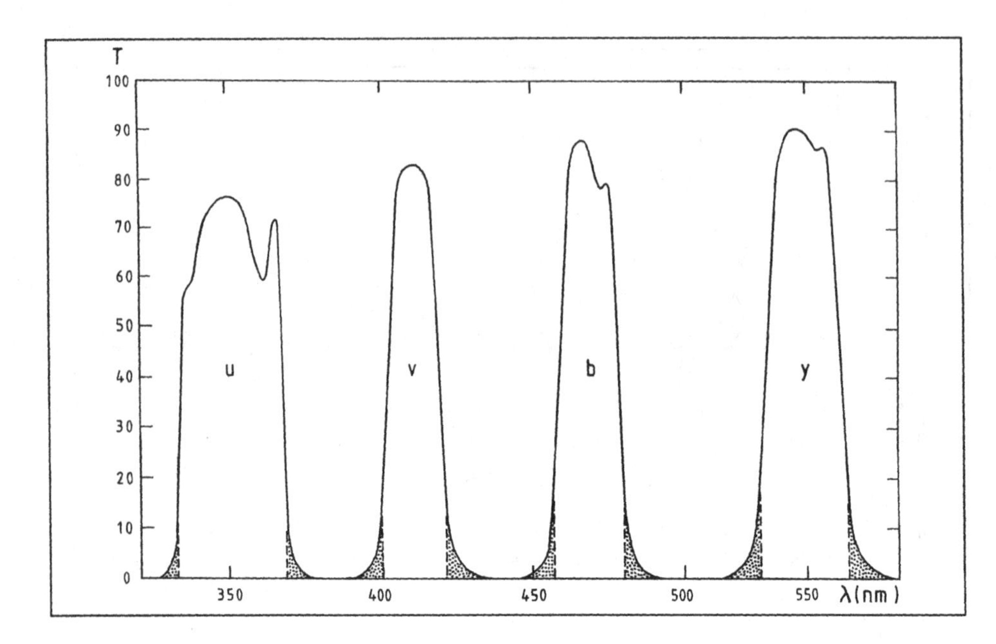

Fig. 3.6 *Transmission curves of the* uvby *filters. Shaded areas are the wings that are being cut-off by the exit slots in the simultaneous* uvby *photometer.*

angle of the grating with the use of a calibration lamp.

In photometers with a spectrograph design (slot-filter or slot configurations) thermal wavelength shifts may play an important role because the dispersive element (grating) and the mechanical structure undergo thermal expansion with changing temperatures. This wavelength shift could be of the order of several percents per degree centigrade (Young 1974b). Also, the presence of metal mirrors in the telescope-spectrometer system may cause a modified output. Such "spectrometric" photometers may have leakage problems too, especially if spectral orders are not well separated. Strong reflections between filter surfaces and diaphragm may occur; since this effect depends on the diameter of the diaphragm, systematic errors will show up.

Some observers use a classical single-beam simultaneous multi-color photometer to measure colors through *uniform* thin clouds, and it is common practice to use even single-beam mono-color photometers to do $H\beta$ photometry during atmospherically-poor nights (in the visible domain some types of clouds act almost like neutral filters; unfortunately this is not always the case). Non-uniform clouds add considerable noise to the data, but by increasing the number of measurements on each object, one can still arrive at valuable observations (see Fernie 1982).

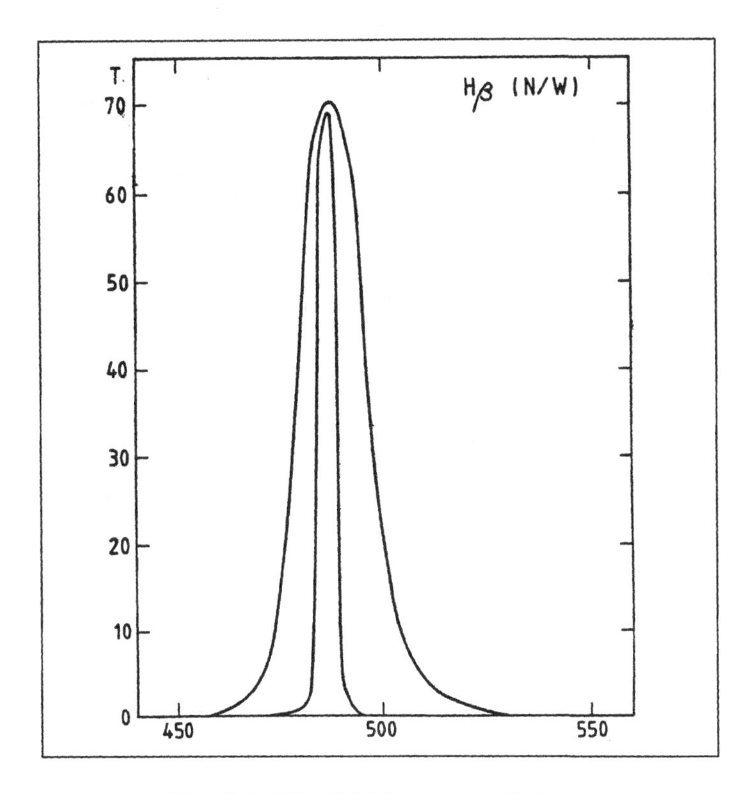

Fig. 3.7 The Hβ filters transmission.

3.4 The multi-star photometer

The idea of eliminating variations in atmospheric transparency, or even receiver response, by simultaneous monitoring of a variable star and one or several comparison stars has led to a number of prototypes of multi-star photometers. Some designers use two separate detectors; others use only one detector and a chopper to switch from one star to the other.

Guthnick (1913) already proposed the idea of measuring simultaneously the light from two nearby celestial objects. Several solutions are possible:

- the use of a twin telescope-photometer configuration on one mounting;
- the use of one telescope and a double-beam photometer.

The double-telescope design has been developed at Edinburgh observatory (especially in view of measuring sequences of magnitudes; Reddish 1966), but was followed by little imitation and development (mechanically duplicating the telescope introduces problems of its own).

Of the multiple-beam design some remarkable achievements have been made. Geyer & Hoffman (1975) describe a double-beam multichannel photoelectric photometer which

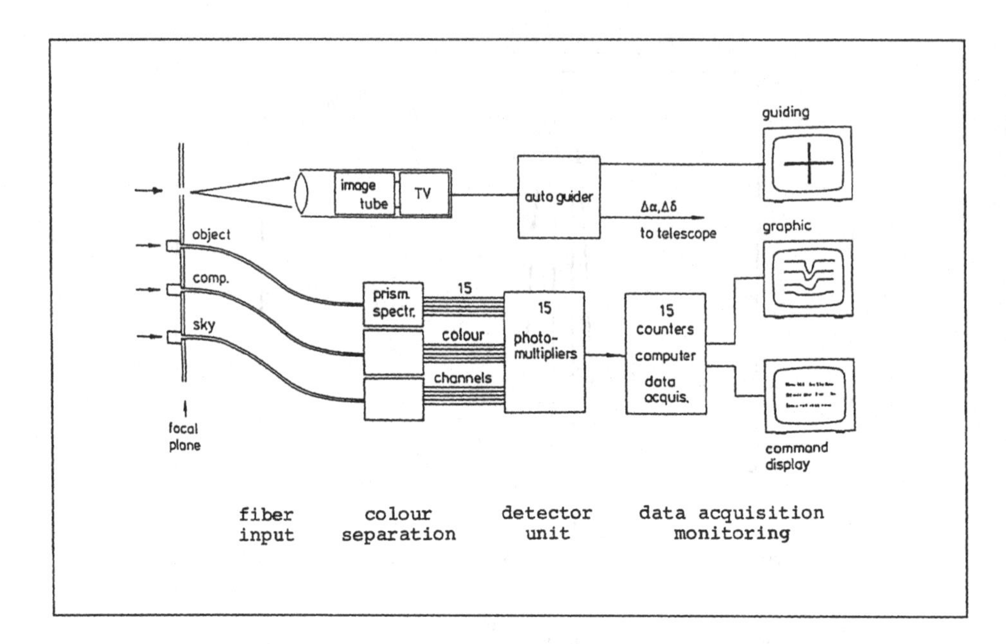

Fig. 3.8 Block diagram of the three-channel five-color MCCP photometer (from Barwig & Schoembs 1987).

allows two stars to be monitored within a small field (telescope of 106 cm of the Hoher List Observatory). The two coldboxes can move independently from each other, perpendicular to the optical axis of the telescope, and the whole unit can be rotated. This permits centering of any star in the telescope field. The photometric accuracy obtained meets the theoretical expectations: for a 12th mag star brightness differences with 1% accuracy have been measured during nights with transparency changes of 50%.

Barwig & Schoembs (1987) describe their Multi-Channel Multi-Color Photometer (MCCP, see also Barwig et al. 1987), that consists of three separate fiber-optic input channels, each splitting the light into five colors by means of highly efficient prism spectrographs. Fifteen cooled photomultiplier tubes allow the simultaneous measurement of an object, a nearby comparison star, and the sky background in 5 colors (see Fig. 3.8). The outcome of such observations during a non-photometric night is illustrated in Fig. 3.9. Note the high time-resolution of the measurements. According to the authors, atmospheric absorption up to about 80% would not degrade the accuracy of the reduced data (except for photon statistics). Note, however, that, unless program star and comparison star have identical spectral energy distributions, a residual error will remain because of extinction (see Chapter 6). The optical fiber design is also very sensitive to the exact centering of the stars, so that tracking errors and atmospheric instabilities (giving rise to image wandering or spreading) introduce additional noise.

Walker (1988) conceived an interesting design of a fiber-optic four-channel photometer. The light at four positions (determined by four X-Y arms) from the focal plane of the

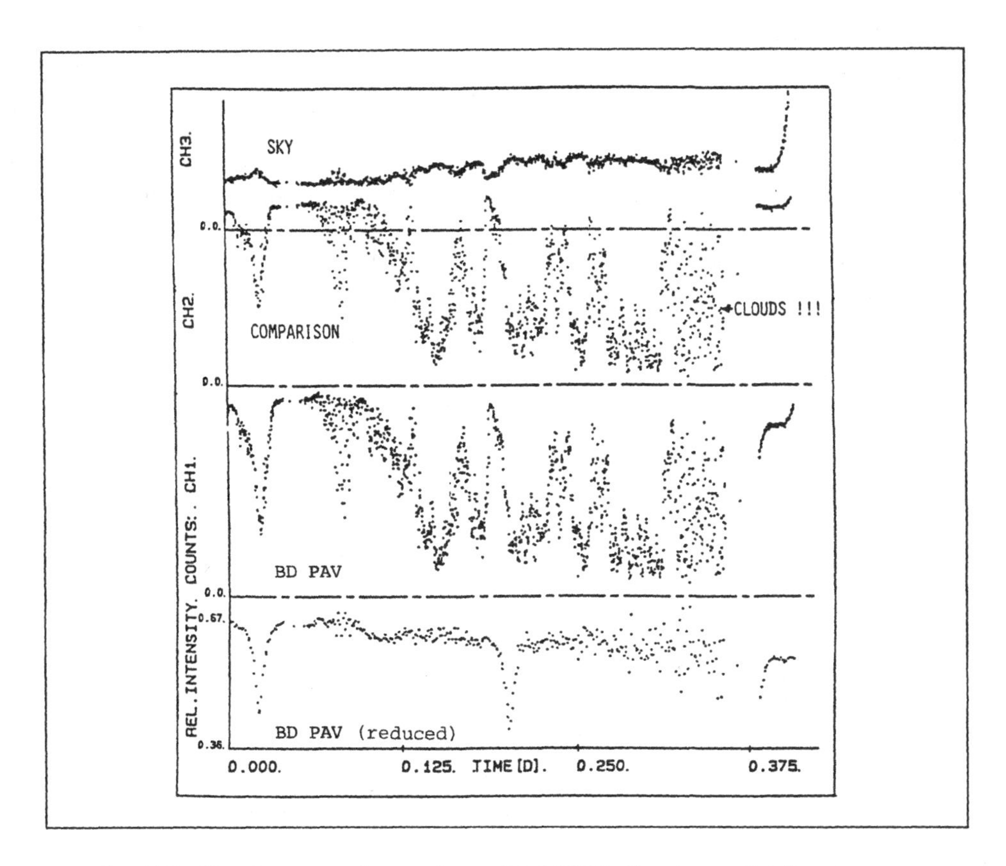

Fig. 3.9 *Simultaneous observations with MCCP. Results on BD Pav (an eclipsing cataclysmic binary), comparison star and sky during a non-photometric night. The reduced V-light curve (bottom) demonstrates the ability to compensate for variable atmospheric absorption (from Barwig & Schoembs 1987).*

telescope is brought with negligible losses to a single detector (photomultiplier) by use of optical fibers. The four positions (within a field of 1°) can be centered on four stars, or on three stars and one sky background location. The light is brought to a chopper box which uses a rotating perforated disk to allow sequential measurement of each of the fibers' light output. The light is then directed to a single photomultiplier tube. A second-generation photometer of this type has been built at Observatoire de la Côte d'Azur (Nice, France). In spite of the apparently different design, this photometer mimics the classical design in all aspects: each position has an aperture with a Fabry-lens assembly which delivers an image of the primary mirror on the end of every fiber optical bundle. Each arm can move so that the pattern of stars to be measured simultaneously can be changed according to the observer's needs, a situation which differs from the MCCP, which uses a metal mask to hold the three fibers and a guiding diaphragm at their precise position in the focal field. The apparatus seems to give very satisfactory results (see

Fig. 3.10). The *DQE* of this photometer is about 0.01 (due to chopping light losses, losses in the fiber optics and air-glass interfaces), whereas a conventional one-channel photometer should yield 0.05–0.10. This comparison is, however, unrealistic since, in variable star observations, one must also take into consideration the dead time of the telescope, when moving between the position of the variable and comparison stars and the sky. Taking into account that such multiple-channel photometers allow observation in less than perfect atmospheric conditions, one must admit that such designs deserve further development. It should be noted, however, that such instruments are of fairly sophisticated design and will work optimally only when they are permanently attached to a dedicated telescope.

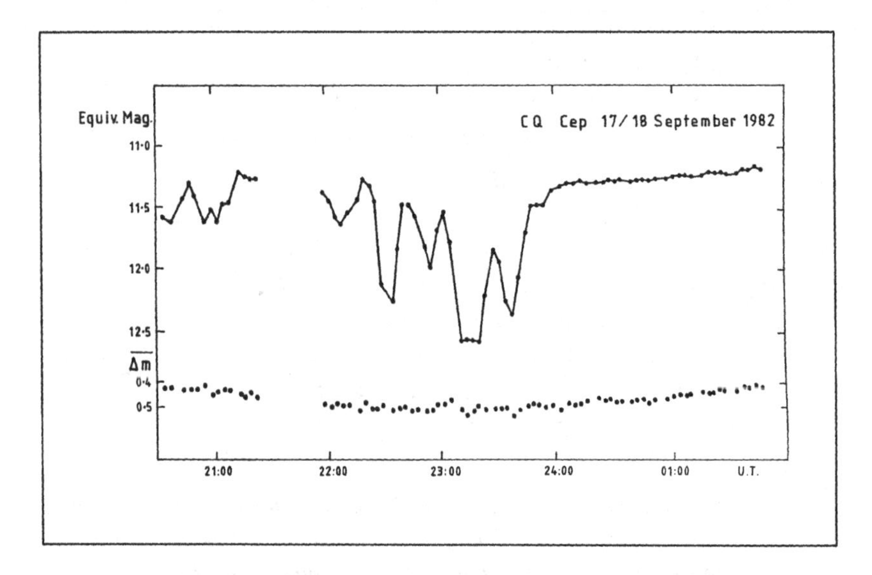

Fig. 3.10 *A four-star, fiber-optic photometer: example of results (Walker 1988). The upper diagram shows the variations in apparent brightness of CQ Cep due to transmission changes of the atmosphere, plus changes intrinsic to the star. The lower plot shows the difference between CQ Cep and two comparison stars on the same scale as the upper plot.*

Chapter 4 The photomultiplier

The *photomultiplier* is the essential part of most professional aperture photometers. It does not record an image, but only measures the amount of incident light. In summary the photomultiplier is *a light detector that produces pulses of charge proportional to the number of photons interacting.*

4.1 The photoelectric effect

The functioning of the photomultiplier tube is based on the same principle of functioning as the *photocell*, which converts light into electricity. The *photoelectric effect* was discovered in 1839 by Edmond Becquerel who observed that a pair of electrodes, immersed in an electrolyte, generated a current whose intensity increased with illumination, and depended on the wavelength of the infalling light. *Photoemission* (emission of electrons resulting from the action of light) was discovered by Hertz (1887), who found that ultraviolet light striking metallic electrodes lowered the voltage at which sparking occurs. For a historical review, we refer to Görlich (1962).

These phenomena were caused by a particular way light is absorbed by matter. An electron is pushed out of an atom or molecule by a photon; every material has a critical wavelength for this effect, and emission of electrons will occur only if the material is exposed to light of wavelength shorter than this critical wavelength. For a given material, the occurrence of the effect depends entirely on wavelength and is independent of the irradiance. However, only a few substances will release a significant number of electrons upon irradiation by visible light. The photoemissive materials having a high *photoelectric yield* are all semiconductors.

The number of electrons released is proportional to the irradiance over a large range of brightness. The maximum kinetic energy of the released electrons is approximately equal to the energy of the impinging photons minus the energy Φ (*photoelectric work function*), which is the amount of energy necessary to liberate an electron with zero kinetic energy at absolute zero temperature.

The maximum kinetic energy of an escaping electron is given by

$$\frac{1}{2}mv^2 = E = h\nu - \Phi \tag{4.1}$$

For each material there is a corresponding value of Φ (generally expressed in ev). The *threshold frequency* is determined by the case where the photon energy of infalling radiation is just sufficient to liberate an electron ($E = 0$),* and hence

$$h\nu_0 = \Phi \tag{4.2}$$

where ν_0 is the threshold frequency of the corresponding material, below which no photoemission can occur (note that the electron must be able to leave the surface, i.e., the threshold frequency is not the only parameter: the electron will lose energy due to collisions when it moves through the material, and if this energy loss is too large, it will not retain sufficient kinetic energy to overcome the electron affinity).

The major advantage of the photocell is not so much higher sensitivity, but its ability to measure irradiance differences much more precisely than visual or photographic techniques do. A big problem originally was the high noise level of amplifiers capable of dealing with such weak currents (a fraction of a microampere) as we deal with in astronomical observations. This situation was altered by the invention of the photomultiplier tube which allows the amplification of small photocurrents by the almost noiseless process of *secondary emission* previous to introduction of the signal into the circuit of a measuring device. Whitford & Kron (1937) were among the first to use a photomultiplier for astronomical purposes.

The photomultiplier in fact incorporates a photocell and a high gain current amplifier in one glass envelope. The basic radiation sensor is the *photocathode*. Light of sufficiently large energy incident on the photoemissive cathode releases photoelectrons which are accelerated by a positive potential of a few hundred volts towards a positively charged electrode (first *dynode*). Secondary electron emission is the property of some materials to emit several electrons when struck by an electron with sufficient energy. Every photoelectron impinging on the first dynode is capable of exciting electrons within the dynode to higher energy states; those with sufficient energy to overcome the work function of the dynode surface are emitted. These are accelerated onto the next dynode (at a still higher potential) where they, in turn, produce secondary emission. This process repeats itself for every dynode stage and yields cascade multiplication until the electrons leaving the last dynode are collected by the anode and produce an output electric current.

Electron multiplication takes place at each stage, and with a gain at each dynode of e.g., a factor of four, millionfold amplification can be achieved when ten or more dynodes are being used, and the resulting signal level is compatible with standard measuring equipment. The photosensitive surface, however, not only releases electrons when struck by light, but also does so due to heat, and this unwanted emission is amplified too (see Section 4.5.2).

The output signal is directly proportional to irradiance of the incident light. The measurement of irradiances is thus reduced to the measurement of electric currents.

Photomultipliers offer various interesting advantages, such as

- extremely fast response time: reaction times can be as short as a fraction of a nanosecond;

* This textbook explanation applies only at a temperature of 0 K.

- photomultipliers allow the measurement of very weak signals with signal-to-noise ratios limited by electron statistics;
- the nature of output of a photomultiplier enables application of pulse counting techniques.

They have also disadvantages:

- some (like S-1 *) have a relatively low quantum efficiency;
- they have a relatively limited wavelength range (from near-ultraviolet to near-infrared).

4.2 Types of photomultipliers

Figure 4.1 shows arrangements of the photocathode, the dynodes, and the anode inside photomultipliers of different brands and types. The photomultiplier tube at the top is the historical RCA 1P21, and is a so-called side-window design where the light enters laterally, and which has a cathode deposited on an opaque metal electrode: the photoelectrons are emitted from the side of the photocathode that is struck by incident photons (one also speaks of a reflective-mode photocathode, though this term is misleading). For the other model in Fig. 4.1 the photocathode is deposited on the inside of the window at a depth of 20 to 30 nm, and is semi-transparent. The head-on type is widely used, and has better uniformity than the side-on photomultiplier tube, but because of the limited escape depth of photoelectrons, the thickness of the semi-transparent photocathode is critical: much of the incident radiant energy is lost by transmission if the photocathode layer is too thin; in a too thick photocathode much of the incident energy is absorbed at a distance greater than the escape depth. The design of the photocathode-first dynode region is very important, since *all* photoelectrons from the photocathode must be focused on the first dynode.

The photomultiplier needs a metal housing to provide electric shielding, and magnetic shielding is needed to prevent the Earth's magnetic field from changing the multiplication of electrons as the telescopes moves. Photomultipliers must be rigidly mounted and may not have flexure. Because the applied voltages are high, the photomultiplier's base and socket must be clean and dry to avoid large leakage currents.

4.3 The essential parts of a photomultiplier tube

4.3.1 The glass housing

The housing serves to protect the inside parts and to preserve the vacuum inside the tube. The transmission of the glass determines the ultraviolet cutoff wavelength of

* The S-designation describes the spectral response curve of photomultipliers, including the effects of both window and cathode materials; it was standardized by the Electronic Industries Association. Many of these designations are now obsolete.

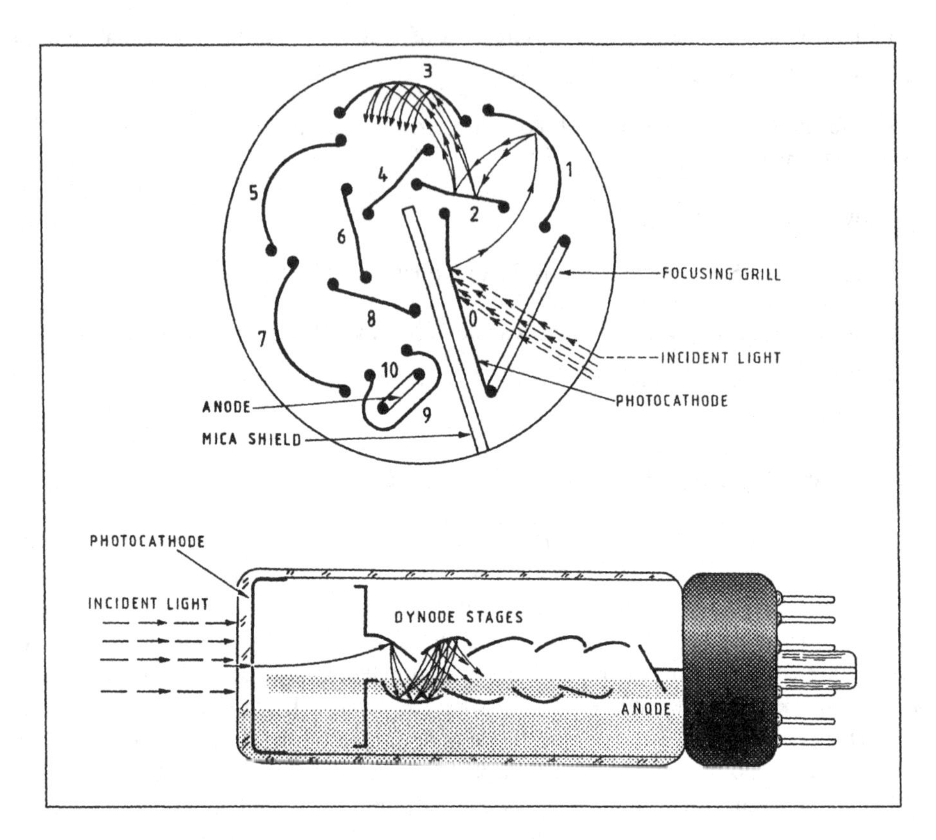

Fig. 4.1 *Photomultipliers. The top drawing shows a side-window design. In the bottom one the photoelectrons are emitted from the side opposite to the incident light (after Walker 1987).*

many photomultipliers. The most widely used material is lime or borosilicate crown-glass which transmits light with wavelength longer than about 350 nm. Ultraviolet transmitting glass cuts off at 185 nm, fused silica at 160 nm and MgF_2 at 115 nm, i.e., much below the atmospheric cutoff.

4.3.2 The photocathode

The photocathode is a most important element of a photomultiplier tube, because a good photomultiplier is primarily characterized by high photocathode quantum efficiency.

Pure metals are not suitable for building photocathodes because photoelectrons collide with the abundantly present free electrons in the metal, and so lose a lot of energy. Work functions of metals are higher too, so only highly energetic photons (ultraviolet radiation) are able to produce electronic emission. Metals also have high reflectivity, so less light can enter them. Photocathodes are semiconducting compounds, chosen in such a way that the desired spectral response and sensitivity are produced.

Whereas an ideal photocathode would release one photoelectron for every impinging photon (the quantum efficiency would then be 100%), in practice there are large losses in absorption of the incident photons, hence, existing photocathodes have low *QE* (when light falls upon a photocathode, only a few out of every 100 photons provokes ejection of an electron).

Photocathodes are significantly non-uniform, with sensitivity differences exceeding 25% (Rodman & Smith 1963). Sensitivity drops steeply towards the edge of the cathode, and the maximum sensitivity is not necessarily at the geometric centre.

The spectral-response range is determined on the long-wavelength side by the photocathode material and on the short-wavelength side by the composition and thickness of the window material. This combination of window material and photocathode type offers a wide variety of spectral response.

Photocathodes most commonly used are discussed below.

1. Ag-O-Cs (S-1) is sensitive from visible to infrared beyond 1 μm. It has a high dark emission, which can be reduced by cooling to -180°C. It is used where sensitivity beyond 850 nm is required.

2. Bi-alkali cesium-antimony (Sb-Rb-Cs or Sb-K-Cs) has high sensitivity and low noise. This photocathode is used when low dark current is important, and when high response in the red region is not required. The potassium alloy has improved blue sensitivity. The response is higher at dry-ice temperatures than at room temperature, but it drops again in liquid-nitrogen environment. These cathodes loose red-sensitivity when cooled.

3. Multi-alkali (Na-K-Sb-Cs, S-20) is widely used for its broad spectral response and high sensitivity. Multi-alkali cathodes behave similarly as bi-alkalis, except that the region of lowered response is located in the near infrared.

4. The historical RCA 1P21 photomultiplier has a blue-sensitive Cs-Sb (S-4) cathode and 9 dynodes. At a potential of 100 V per stage and an average secondary yield of 4, each photoelectron generates about 250,000 electrons at the anode.

Table 4.1 shows the main characteristics of a variety of photocathodes. S and O symbols (in the column "Type"), respectively, designate semi-transparent and opaque photocathodes. λ_{max} is the wavelength of maximum *QE*. The range of response of the S-8 photocathode is approximately that of the eye, but the shape of the response curve is quite different. Multi-alkalis have response curves which are more extended to the red side of the spectrum than is the case for bi-alkalis. This offers the interesting possibility of adding red filters to a blue filter-defined photometric system (but beware of red leaks which could then affect the blue filters, see Fig. 5.1). "Extended red multi-alkali" (ERMA) photocathodes are thicker at the expense of short wavelength response.

The superior photocathode surfaces are those with the largest width of spectral response. One should keep in mind that the photocathode color-sensitivity is dependent upon temperature, and that cooling of the photomultiplier shifts the sensitivity curve

Table 4.1 Some characteristics of photocathodes.

Response designation	Material	Type	λ_{max}	QE at λ_{max}	dark emission femto A cm^{-2}
S-1	Ag-O-Cs	O	800	0.40	900
S-3	Ag-O-Rb	O	420	0.55	-
S-4	Cs-Sb	O	400	13	0.2
S-5	Cs-Sb	O	340	18	0.3
S-8	Cs-Bi	O	365	0.77	0.1
S-9	Cs-Sb	S	480	5.3	0.3
S-10	Ag-Bi-O-Cs	S	450	5.6	70
S-11	Cs-Sb	S	440	15.7	3
S-13	Cs-Sb	S	440	14	4
S-17	Cs-Sb	O	490	21	1.2
S-19	Cs-Sb	O	330	24.4	0.3
S-20	Cs-Na-K-Sb	S	420	19	0.3
S-21	Cs-Sb	S	440	6.7	4
S-23	Rb-Te	S	240	2	0.001
S-24	Na-K-Sb	S	380	23	0.0003
S-25	Cs-Na-K-Sb	S	420	13	1
ERMA	Na-K-Sb-Cs	S	530	10.3	1.4

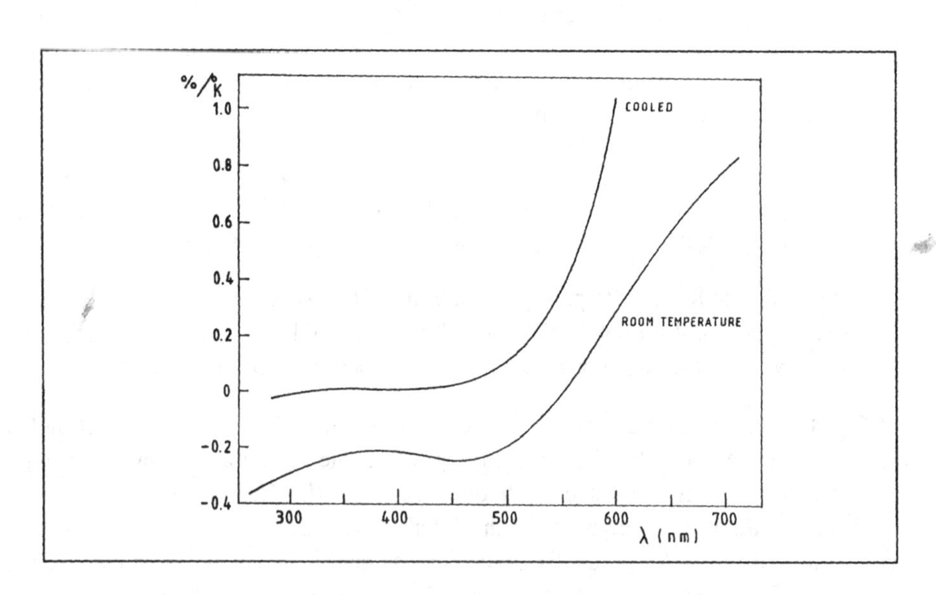

Fig. 4.2 *Temperature coefficient of S-4 photocathodes as a function of wavelength (after Lontie-Bailliez & Meessen 1959).*

bluewards, as is shown in Fig. 4.2. Note that in the blue spectral region, the temperature coefficient (relative change of cathode current per degree K) of a S-4 cathode is roughly independent of the wavelength, but steeply increases at wavelengths beyond 400 to 500 nm (see also Young 1963). Hence, if a photomultiplier cools during the night, blue and red color indices will be oppositely influenced.

Figure 4.3 gives the absolute spectral response, S, and quantum efficiency QE of various photomultiplier cathodes. It is immediately apparent that the red-sensitive cathodes (S-1 Cesium-oxide-silver) have an extremely low QE. This is due to a metal, silver, being the primary photoemitter.

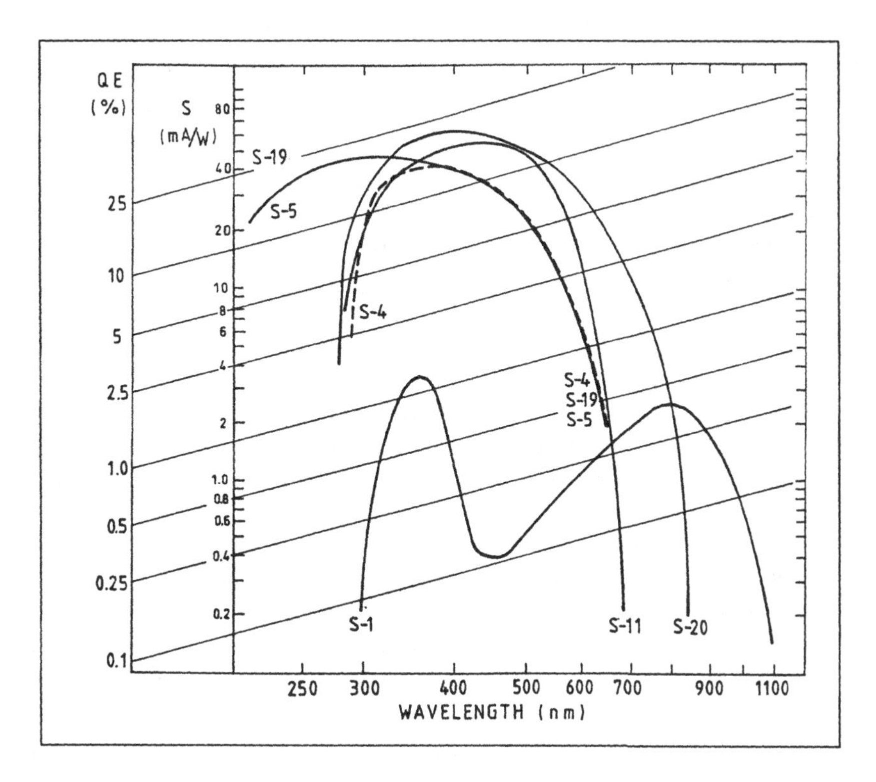

Fig. 4.3 Quantum efficiency. Spectral response S and quantum efficiency QE *of various photomultiplier cathodes. The broken line represents the spectral response of the S-4 photocathode of the RCA 1P21 tube.*

4.3.3 The dynodes

The number of secondary electrons, for a given photomultiplier, depends on the energy of the primary electrons. The average ratio of the number of secondary electrons leaving the surface of a dynode to the number of incident electrons, is called δ, the average secondary emission coefficient.

δ has a maximum which lies between 200 and 1000, and is approximately equal to

$$\delta = AE^{\alpha} \tag{4.3}$$

where A is a constant, E is the working interstage voltage, and α is a coefficient determined by the dynode material and the dynode geometry (usually around 0.7).

If n is the number of dynodes, the internal electron gain of the photomultiplier is, for the ideal case, given by

$$G = \delta^{n} \tag{4.4}$$

In practice, however, the cathode-first dynode collection efficiency f (about 90%), and the transfer efficiency g of electrons between dynodes intervene, so that

$$G = f(g\delta)^{n} \tag{4.5}$$

$g\delta$ usually is about 5. This means that a single photoelectron at the cathode of a 12-stage photomultiplier results in 5^{12}, or more than 200 million electrons being collected at the anode. Dynode emission is a strong function of dynode-to-dynode potential difference, thus the output signal of the photomultiplier is extremely susceptible to fluctuations in the power supply voltage; so in order to prevent the output current of the photomultiplier from being modulated by fluctuations in supply voltage, one needs a well-regulated voltage supply.

The manufacturer specifies maximum voltages to be applied between anode and cathode, between consecutive dynodes, and between anode and last dynode. Overstepping these voltages causes a strong increase in signal noise, and may cause damage to the emitting surfaces of the dynodes.

One expects a high degree of stability in current output with constant light flux and stable voltage per stage. Stability is increased when one operates the tube for at least half an hour at the intended level of irradiance.

The time delay in the secondary emission process is very short ($\sim 10^{-13}$ sec). *Transit time* is the time for the electrons to reach the anode. The spread in transit time affects pulse width and pulse resolution.

The dynode temperature coefficient is independent of wavelength, so temperature variations will not induce color effects at the stage of secondary emission.

4.4 Anode current and its measurement

The function of the anode is to collect the pulses of electrons emitted from the last dynode, and to deliver the charge to a measuring circuit.

Most photomultipliers have a maximum anode current rating of the order of 1 mA (some tubes, however, cannot take more than 0.1 mA), and it is recommended to keep the average anode current well below the specified value (usually below 1 μA) because extended operation at maximum average anode output current can lead to chemical changes at the surface of the dynodes, which would result in nonlinear effects in the anode current, increased fatigue effects, and permanent decrease in responsivity and tube life.

Anode current is proportional to irradiance of incident light within a couple of per cent. At high currents, a saturation effect is present. The photomultiplier exhibits temporary instability in anode output current for several seconds after voltage is switched on; this results in a hysteresis effect, the output response not being in step with the luminous excitation. It is therefore desirable to keep a photomultiplier tube switched on during standby periods.

The anode current can be measured in different ways. DC (for *direct current*) detection is the simplest: the output signal can be fed into a DC amplifier and into a strip-chart pen recorder. It can also be chopped and put into an AC amplifier. Or one can charge a condenser and measure the accumulated charge during equal time intervals (in that case the electrical signal produced by the anode is digitized by a voltage-to-frequency converter and delivered as a pulse train to the counter for summation). Or one can count the individual photon pulses per time unit. These methods not only vary in the complexity of the measuring electronics, but there are also fundamental differences in the properties of the output signal and of the associated noise component.

If the number of secondary electrons emitted per primary electron were constant, all pulses resulting from an electron leaving the cathode would be equal, and every photoelectron would be separately counted. All the information in the photocurrent would be recorded, and the *DQE* of the photomultiplier would equal the *QE* of the photocathode. Furthermore, the relative rms noise of anode current would be equal to the noise present in the photocurrent from the cathode. And all pulses would be of the same size.

Because of statistical dispersion in number and direction of emission of secondary electrons, the amount of multiplication is very different from one photoelectron to another, and the process of dynode multiplication produces a very broad statistical spread in the sizes of the output pulses (even at constant photocurrent) which causes a non-constant distribution of pulse-height. Thus, some pulses are much higher than the average, and some are much smaller.

Figure 4.4 shows the distribution of pulse heights for a tube in darkness and for an illuminated one. The variable pulse height of the dark current output and also of the amplified photoelectron count are caused by the statistical variation of secondary emission at each stage. These distributions are totally different because the photocurrent entirely originates at the cathode, while the dark current comes from both cathode and dynodes as well. Because the total dynode area is large in comparison with the cathode area, most of the thermoelectrons originate from the dynodes. As such, most of the dark pulses pass through fewer stages of multiplication than pulses coming from the cathode do, and the resulting electrons will cause shorter and generally lower pulses.

When output current is recorded, or collected at a condenser, all pulses (dark and signal) are being smeared together, and one cannot differentiate between the different pulses. By using an appropriate discriminator circuit one can reject all pulses smaller than a certain height by cutting off the asymptotic rise of the dark-pulse distribution and by passing the majority of pulses due to photoelectrons as square pulses all of standard size. So, each pulse passing through the discriminating window is counted with equal weight regardless of its size. This method is known as *pulse counting*, where one

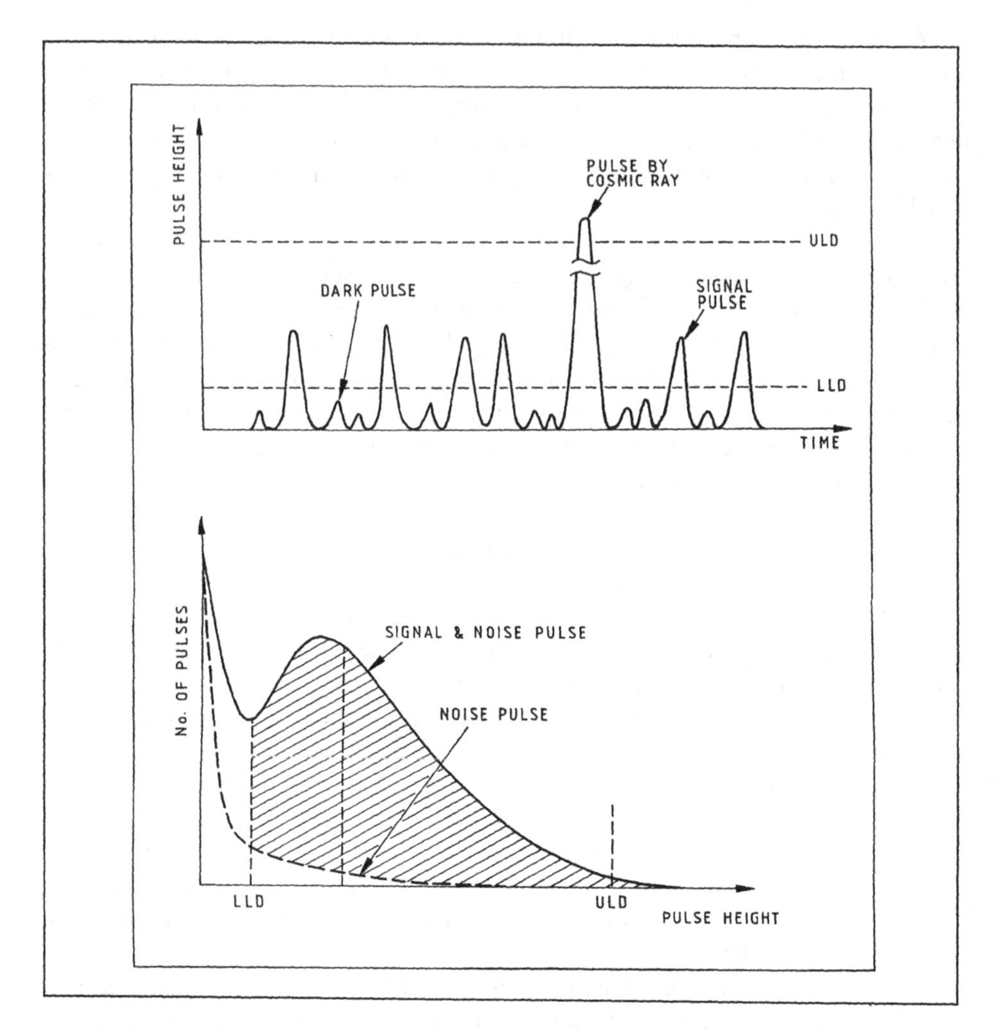

Fig. 4.4 *Distribution of pulse heights. See text. (Source: Hamamatsu catalog).*

count is the equivalent of one photoelectron. However, it is impossible to separate by discrimination small signal pulses from large dark pulses, and this introduces errors or noise.

This process is often not well understood by observers, who sometimes "play" with discriminator settings to increase the relative numbers of pulses whose heights exceed the preset level. This approach invariably leads to registration of spurious signals with a very high noise component. Another misconception is that thermal emission can be eliminated by discrimination: it is obvious that proper cooling only can reduce photocathode and dynode dark current!

Figure 4.4 (bottom) shows a typical pulse height distribution (PHD) of the output of a photomultiplier. One curve represents the number of counts per unit time due to dark current; the other curve is the sum of the signal and of the dark current. The horizontal axis gives pulse height, also called discriminator level. The lower level discriminator (LLD) and upper level discriminator (ULD) settings are indicated. Such a curve is obtained by viewing a weak steady light source or a faint star and recording signal, and signal plus noise. The dark-current curve descends asymptotically from high level at negligible pulse height to zero level at large pulse heights (this is due to the numerous small dark pulses arising in the ultimate stages of amplification). The optimal setting is the one which corresponds to the largest difference between both curves (somewhere midway between noise peak and signal peak, as indicated by a vertical dashed line).

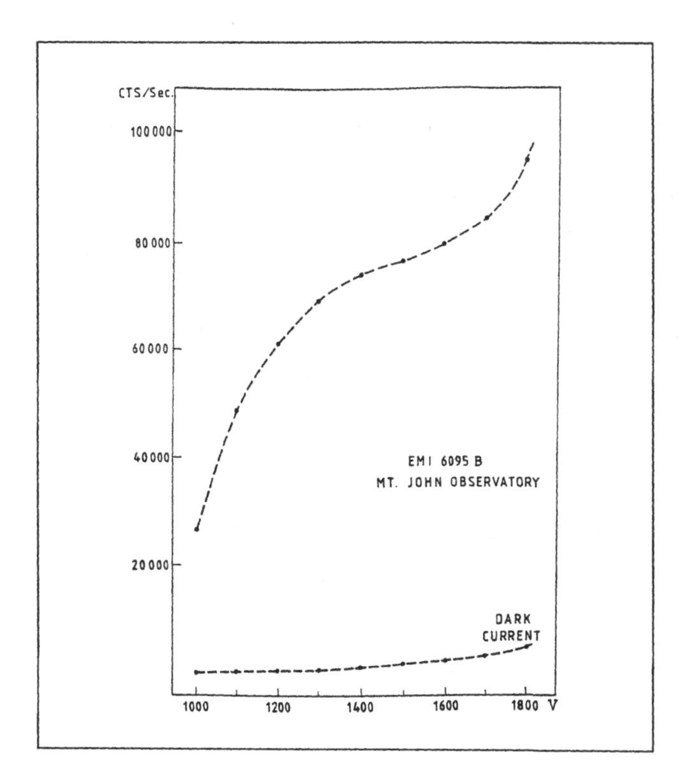

Fig. 4.5 Count rate as a function of high voltage.

Once the discriminator settings are fixed, one must still select the optimal voltage. This is done through the determination of the *count-rate plateau curve*, which gives counts against applied voltage, and is obtained by varying high voltage while monitoring output count-rate. Count rate (not pulse height) will initially increase rapidly with high voltage and then reach a pseudo-plateau region before resuming a rapid rise (see Fig. 4.5). Operating the tube within the voltage range of the plateau guarantees a signal deviation which is small, even with large voltage fluctuation.

In DC mode spurious pulses are integrated along with true pulses and so the signal-to-noise ratio degrades. Even in the absence of noise, all modes of measurement would not be equivalent: the pulse-height spread makes DC noisier than (ideal) pulse counting.

Pulse counting offers many advantages, such as exact integration times, direct digital output with on-line information about photon statistics and signal-to-noise ratio, little drift, discrimination against dark emission, insensitivity to leakage currents and high speed, but the equipment is complex and bulky, and is less "observer-proof".

4.5 Characteristics of photomultipliers

4.5.1 Sensitivity, amplification and drift

Sensitivity is a measure of the magnitude of the expected signal for a given light input and is generally given as the ratio of output current to input flux. Sometimes, for visual photometric applications, luminous sensitivity is defined on the basis of a tungsten lamp operated at a specific color temperature, and input flux is then given in lumens. In radiometry, radiant sensitivity is given as a function of wavelength.

Amplification is the ratio of anode output current to photocathode current at the applied voltage. Manufacturers provide luminous sensitivity (output anode current divided by the luminous flux incident on the photocathode) and cathode luminous sensitivity (photocurrent emitted per lumen of incident light flux on the photocathode at constant voltage) in A/lm. As stated previously, these concepts are somewhat inadequate, and it would be more appropriate to use parameters based on radiant power (radiant sensitivity and cathode radiant sensitivity). Sensitivity may vary with time, an effect that manifests itself as a slow *drift*.

QE and spectral radiant sensitivity S have the following relation:

$$QE = 1240S/\lambda \tag{4.6}$$

S (A/W) and λ in nm (see Fig. 4.3).

Note that a photomultiplier can detect a single photon, hence the detection threshold is zero.

4.5.2 Dark current

Dark current is the anode current which flows in the anode circuit when the tube is operated in complete darkness. Dark current is the combined contribution of thermal emission, electrical leakage and background radiation (see Figs. 4.6 and 4.7). The dark current decreases with time after the tube is placed in darkness. Dark current varies from one tube to the other, even if they are of the same type.

Thermionic emission is due to electrons gaining enough thermal energy to escape spontaneously from the cathode or dynodes and mimic photoelectrons. Thermionic emission shows an exponential dependency on absolute temperature. It is also proportional to the area of the photocathode and dynodes; therefore, astronomy applications benefit from photomultipliers where the photocathode surface is adapted to the size of the illuminated spot. In many tubes (e.g., those with alkali-antimonide cathodes), the thermal component of dark current is not of thermionic origin, but is caused by residual gas ions (mainly H^+ and H_2^+) striking the photocathode.

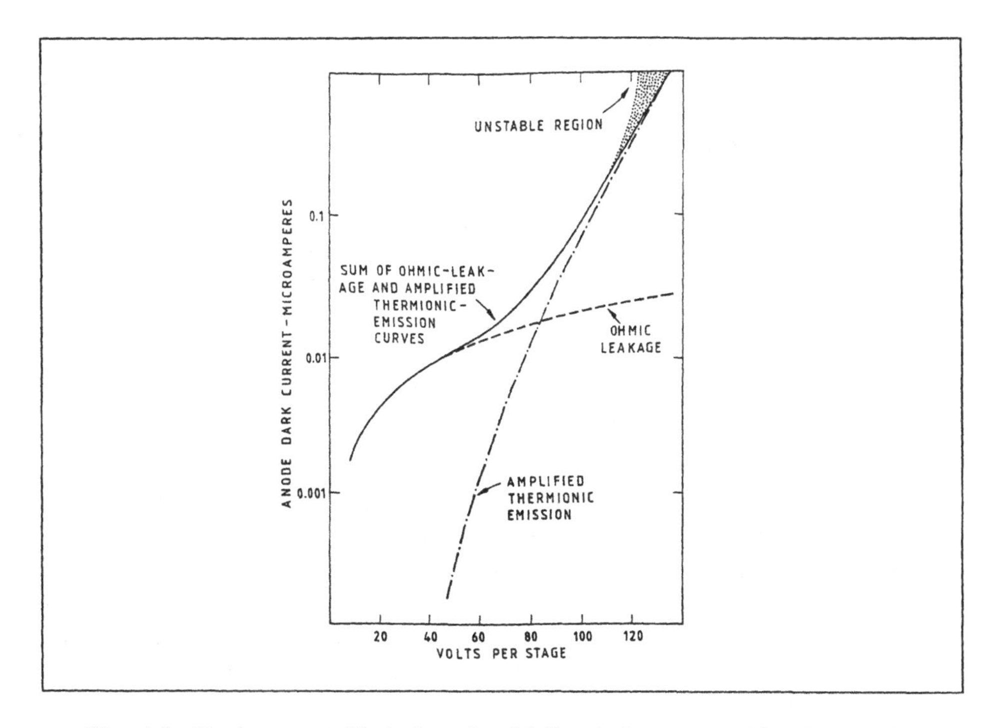

Fig. 4.6 *Dark current. Variation of multiplier dark current with voltage (after Engstrom 1980).*

Cooling the photomultiplier decreases the intensity of the dark current, but one should also keep the temperature constant, since temperature variations cause changes in the spectral response characteristics of the photocathode. Figure 4.7 illustrates the variation of cathode dark current with temperature. Red sensitivity increases most with temperature. There is also a lower limit to cooling: the conductivity of the photocathode material decreases with decreasing temperature, and excessive cooling would result in nonlinear effects. A good working temperature should remain above −100°C. Since most of the dark-current reduction is achieved at temperatures above −40°C, cooling with dry ice (−60° to −80°C) is sufficient. Some difficulties may arise when improper cooling results in condensation of water vapor which fogs the window and causes leakage at electrical contacts.

Inter-electrode *Ohmic leakage* is due to imperfect insulation at the photomultiplier's base (intrinsic leakage between two closely spaced socket pins with a potential difference

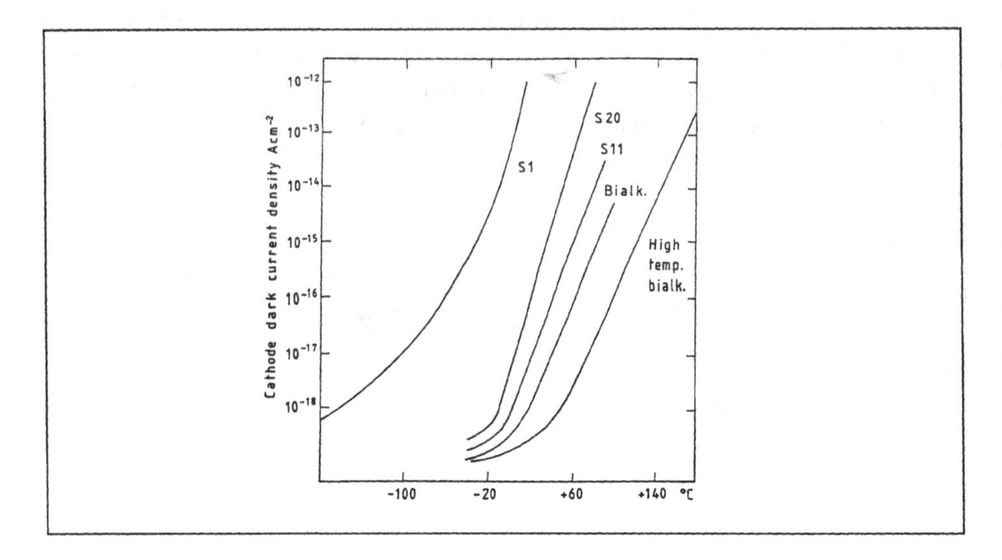

Fig. 4.7 Dark current. Variation of cathode dark current with temperature (after Sharpe 1970).

of several hundred volts, accentuated by the presence of dirt, grease and humidity). This leakage is roughly proportional to the voltage per stage.

Some noise cannot be reduced by cooling: at very low photon rates one can measure pulses originating from cosmic rays and decay of radioactive isotopes in the photomultiplier window. Quartz windows contain less radioactive materials, therefore this residual noise might be lower. Cerenkov light flashes produced by cosmic rays traversing the window cause large dark-current pulses with a low rate of production (Young 1966), especially in UV transmitting windows. These pulses, because of their sizes, contribute a lot to the mean square fluctuation. This is another advantage of pulse counting: not only are the numerous small dark current pulses overlooked, but the big pulses cannot contribute to deterioration of the S/N level.

Dark current and dark noise are quite different things and they should not be confused. Dark current has its own noise component. The presence of this current and the associated noise are critical factors in determining the lower level of light detection.

With increasing voltages the dark current increases proportionally with the gain characteristics of the tube. At still higher voltages the dark current fluctuates violently and produces large bursts of pulsed noise. In such operational conditions, the life of the tube is shortened and optical feedback of dynode glow to the cathode occurs.

Excessive working temperature may cause permanent damage to the tube. This can happen when high temperature causes an increase in dark current beyond maximum anode current. Since some photocathode materials are very volatile (e.g., Cesium) they may in such conditions evaporate and condense on cooler parts of the tube, causing reduced gain and less sensitivity.

4.5.3 Dead time

At high count rates, the recorded count rate n (in events per second, i.e., in Hz) underestimates the true count rate N because some pulses are too close to each other to be resolved. In a first approximation one can estimate that every time an amplifier/discriminator detects a pulse, it must wait a time τ before it can detect another pulse. This dead time typically is of the order of 30 nanoseconds or more, and is the theoretical pulse resolution of the system. During τ, $N\tau$ events occur, so for each event, $N\tau$ events are missed, and for N events, $N^2\tau$ are missed. Thus the apparent rate is $n = N - N^2\tau = N(1 - N\tau)$, hence $N = n/(1 - N\tau)$ or, approximately,

$$N = n/(1 - n\tau) \tag{4.7}$$

With increasing pulse rates, a greater percentage of output pulses will fail to be resolved, so accurate knowledge of dead time is important when measuring very bright stars. τ must be determined for every amplifier-discriminator and pulse-counting channel. The approximation of a Poisson statistical distribution of the pulses (see Section 1.1) fails at high rate: photons obey Bose-Einstein statistics, so there is an excess of quasi-simultaneous events. Because the coherence time of light is very short, this effect becomes important at very high count rates and does not affect ordinary photometry (it would be much more important for monochromatic emissions such as laser light). Fernie (1976) discusses the validity of Eq. (4.7) for count rates exceeding 500 000 Hz.

Chapter 5 Photometric filters

5.1 Fundamental characteristics

Filters select specific regions from the ultraviolet, visible and infrared region of the electromagnetic spectrum. They are fully described by their thickness and diameter and by their spectral transmission and reflection curves. Categories useful in astronomical photometry are:

- short- and long-wavelength *cutoff* filters,
- filters for *isolating* (*band-pass* filters) or *rejecting* a spectral region,
- *neutral-density* filters.

In most astronomical cases, filters isolate a range of wavelength. Cutoff filters may be used when another component of the detection equipment (including the atmosphere) completes the definition of the passband. Rejection and cutoff filters are useful in rejecting unwanted wavelength ranges leaked by other filters. Neutral filters are used to decrease the level of irradiance (e.g., to avoid detector saturation, or for calibration purposes).

Band-pass filters are sometimes characterized by a central wavelength and their FWHM, i.e., the spectral separation between 50% transmission cut-on and cut-off points of the passband curve (definition of the bandwidth of a filter is complicated because the transmission curve is not rectangular, but bell-shaped). Filters can be broad-band (bandwidth of about 30 to 100 nm, such as in the *UBV* system), intermediate-band (width from 9 to 30 nm, such as in the Strömgren *uvby* system), or narrow-band (width below 9 nm). We refer to Chapter 16 for a description of photometric filter systems.

There is a fundamental difference between filters which derive their effects from absorption (or scattering) of light inside the material (gelatin or glass filters) and those where selective transmission is produced by interference (interference filters).

The optical thickness (i.e., the physical thickness times the refractive index) of all the filters used in a photometer should be equal.

Filters are defined by the amount of light that they block. The following quantities are used to describe their properties:

- *Transmittance* is the ratio of the transmitted radiant flux F^{t} to the flux incident on the filter surface F^{i}
$$T = F^{\mathrm{t}}/F^{\mathrm{i}} \tag{5.1}$$
The notion of *opacitance* (= $1 - T$) is sometimes used.

- It is often convenient to use the logarithmic density scale: the *optical density* is the common logarithm of the reciprocal of transmittance
$$D = \log(1/T) \tag{5.2}$$
The advantage is that optical densities are additive (as long as reflection can be ignored). Optical density is dimensionless. The notation 0.5D simply means a density of 0.50.

- *Reflectance* ρ is the ratio of the reflected radiant flux F^{r} to the incident flux
$$\rho = F^{\mathrm{r}}/F^{\mathrm{i}} \tag{5.3}$$

- *Absorptance* α is the ratio of the radiant flux lost by absorption, F^{a}, to the incident flux
$$\alpha = F^{\mathrm{a}}/F^{\mathrm{i}} \tag{5.4}$$

One has
$$F^{\mathrm{i}} = F^{\mathrm{a}} + F^{\mathrm{r}} + F^{\mathrm{t}} \tag{5.5}$$
and
$$1 = \alpha + \rho + T \tag{5.6}$$
Transmittance, reflectance and absorptance are wavelength dependent so that one defines T_λ, ρ_λ and α_λ. They may also be dependent on the incidence angle of the radiation.

The transmittance of a series of stacked filters with individual transmittances T_i is the result of a complex calculation because of interreflections between the various surfaces. When those reflections between filters are negligible, one may write
$$T = \prod_i T_i \tag{5.7}$$
The optical density is then
$$D = \sum_i D_i \tag{5.8}$$
Equation (5.7) (or 5.8) is valid for air-spaced absorption filters, which is a common combination in astronomical equipments. Interference filters, because of the substantial reflection effects at their surfaces, do not follow this law, unless they are slightly inclined to one another so that the reflected beams do not reach the detector (this, however, modifies the passband, see Section 5.4). Polarization effects also cause deviations from this rule.

5.2 Gelatin filters

Gelatin filters (of the kind used in many photographic applications) are very cheap, and are easy to cut and to replace. But they are delicate to handle, and they are very sensitive to deterioration by moisture. The only cleaning possible is the removal of dust. They have a very high thermal expansion (may produce wrinkles), they deteriorate easily, most of them do not resist prolonged exposure to bright light, they exist in a rather limited variety of passbands, and they may not be uniform. Strength and durability may be improved by mounting the gelatin between glass plates. Gelatin filters are rarely used in photoelectric photometers.

5.3 Glass filters

Glass filters are relatively cheap, they are very stable over long time intervals and have a moderate thermal expansion. But they break easily if mounted under tension.

Good-quality glass filters have uniform transmission properties over their entire aperture. The spectral properties are relatively insensitive to the angle of incidence of the light, but since tilting the filter increases the optical path length, transmission is reduced.

Some filters have parasite spectral leaks, such as the *U* filter of the *UBV* system which has an additional passband in the near infrared. This is called a *red leak.* See for example Fig. 5.1. Similarly, a *blue leak* can also occur: photons which are absorbed by the filter excite its fluorescence at wavelengths to which the detector may be sensitive. The response will then depend on the distance between the filter and the detector. Leak signals can be substantial, especially if the spectral energy distribution of the source and the detector response happen to peak in that specific spectral region. Blue measurements of red stars made with a multi-alkali photocathode may be very much affected by a red leak. Such defects can be compensated by the cutoff of the detector itself, by the inclusion of an additional filter which blocks the red transmission or the fluorescent radiation, or by observing objects through the red leak separately, with a specific filter, in order to establish an empirical correction.

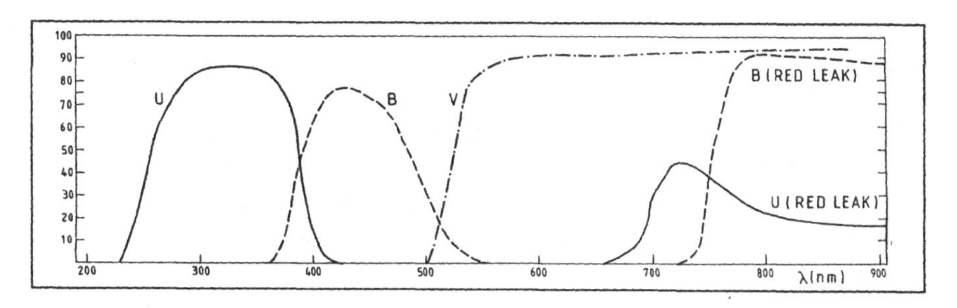

Fig. 5.1 *Passband of a* UBV *system filter set.* U *and* B *filters have a "red leak". The* V *filter is a short-wavelength cutoff filter.*

The temperature effect on the transmission of a glass filter was described by Young (1967a, 1974b). Increasing temperature produces a broadening of all absorptions produced by the glass, a decrease in transmission, and a shift of the absorption spectrum to longer wavelengths.

The magnitude of the effects partly depends on the expansion coefficient of the glass and on the thickness, and can amount to several tenths of nm/°C. Figure 5.2 illustrates the variations of the cutoff wavelength. At the short wavelength side of the passband, the effects reinforce each other; at the red side of the spectrum they are opposed. These temperature effects show up in astronomical studies. For example, seasonal effects have been noticed in *UBV* photometry. They have opposite sign in both hemispheres, and they may add up to about 0.04 mag (Young 1968).

One may combine several absorption filters of various composition in a "sandwich" construction to further limit the bandpass, but, as was stressed before, caution should be exercised that no multiple reflections occur. This problem can be cured by slightly tilting the individual filters (this causes polarization), or by cementing the components together.

5.4 Interference filters

Interference filters are based on the phenomenon of optical interference and are like very thin Fabry-Perot interferometers. When incident light undergoes multiple reflections between two plane-parallel semi-reflecting surfaces separated by a transmitting medium, each transmitted wavefront undergoes an even number of reflections which, depending on the optical path difference, emerge in phase or out of phase, and thus produce constructive or destructive interference. For wavefronts to emerge in phase the optical path difference between them must be an integral number of whole wavelengths. If we consider a beam perpendicular to the filter, the optical thickness of the spacer must be $nd = m\lambda/2$, where m is an integer, n is the refractive index and d is the physical thickness. For $m = 1$ (the simplest case for interference filters) the optical thickness is exactly a half wave. Contrarily to Fabry-Perot interferometers, such a thin space is not an air gap but a deposited layer of dielectric material. The terminology of interferometry is, however, conserved and one generally speaks of a *cavity* for this layer.

Similar interferences are produced on the reflected wavefronts which, for their part, undergo odd numbers of reflections (natural manifestations of this phenomenon are the familiar iridescent patterns of color reflected by soap films and oyster shells). Constructive interferences by reflections on a $\lambda/4$ layer of a material with high refractive index, deposited on a glass substrate, provide highly reflective mirrors with broadband reflectance (note that a $\lambda/2$ phase shift at the reflection on the surface between the high-n layer and the glass substrate is added to the $\lambda/2$ optical path to yield an optical path difference of a full wavelength). Such quarterwave reflectors separated by the halfwave cavity form the simplest interference filters.

The performance of interference filters is described by several parameters which are not well standardized. Figure 5.3 shows the main characteristics of narrow-bandpass filters.

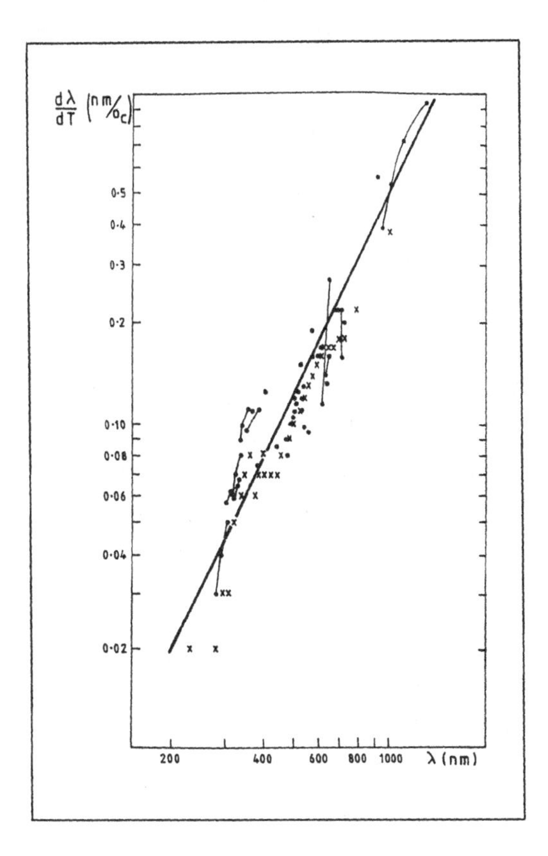

Fig. 5.2 Temperature coefficients of the cutoff wavelength for various optical and filter glasses. The straight line represents the linear relation between log $d\lambda/dT$ and log λ for glass of 2 mm thickness. Thicker filters lie above this line, and thinner below it (from Young 1974b).

The transmittance of a Fabry-Perot filter is approximately

$$T = \frac{T_1 T_2}{(1 - \sqrt{\rho_1 \rho_2})^2 + 4\sqrt{\rho_1 \rho_2} \sin^2 \delta} \tag{5.9}$$

where ρ_1, ρ_2 and T_1, T_2 are, respectively, the reflectances and transmittances of the first and second reflectors (seen from within the spacer), and

$$\delta = \frac{2\pi}{\lambda} n d \cos\phi + \frac{\epsilon_1 + \epsilon_2}{2} \tag{5.10}$$

ϵ_1 and ϵ_2 are the phase changes on reflection at both reflectors (from the spacer side); n is the refractive index of the spacer, d its thickness, and ϕ the refraction angle. Maxima and minima of transmission, respectively, occur for

$$\lambda_{\max} = \frac{2\pi d \cos\phi}{i - (\epsilon_1 + \epsilon_2)/2\pi} \qquad (i = 0, 1, \ldots) \tag{5.11}$$

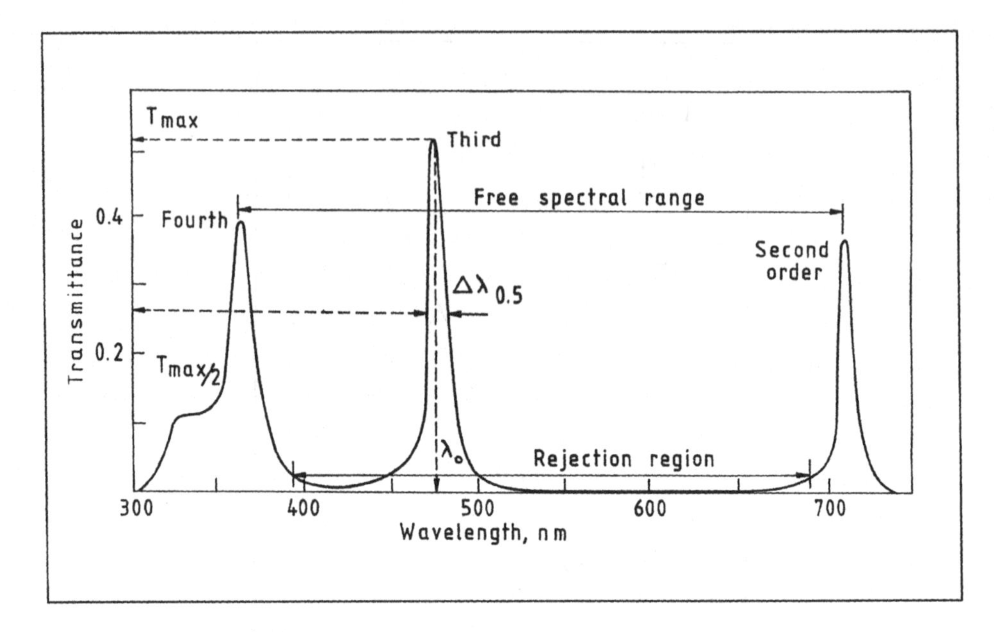

Fig. 5.3 *Parameters used to describe narrow-bandpass interference filters. The curve represents an unblocked second-order Fabry-Perot filter (from Dobrowolski 1978).*

$$\lambda_{\min} = \frac{2\pi d \cos\phi}{i + (\epsilon_1 + \epsilon_2)/2\pi} \qquad (i = 1, 2, \ldots) \tag{5.12}$$

Neglecting variations of transmittance and reflectance of the reflectors between both wavelengths, the maximum and minimum transmissions can be written

$$T_{\max} = \frac{T_1 T_2}{(1 - \sqrt{\rho_1 \rho_2})^2} \tag{5.13}$$

and

$$T_{\min} = \frac{T_1 T_2}{(1 + \sqrt{\rho_1 \rho_2})^2} \tag{5.14}$$

The rejection ratio $T_{\min}/T_{\max}$ is then

$$\frac{T_{\min}}{T_{\max}} = \left(\frac{1 - \sqrt{\rho_1 \rho_2}}{1 + \sqrt{\rho_1 \rho_2}}\right)^2 \tag{5.15}$$

When the angle of incidence is not 0° but some angle θ ($\leq 20°$) the change in wavelength of the transmission peak $\delta\lambda_{\max,\theta}$ and the half width $\Delta\lambda_{0.5}$ can be expressed as a function of an *effective index* $\mu^\star$, as

$$\frac{\delta\lambda_{\max,\theta}}{\lambda_{\max}} = -\frac{\theta^2}{2\mu^{\star 2}} \tag{5.16}$$

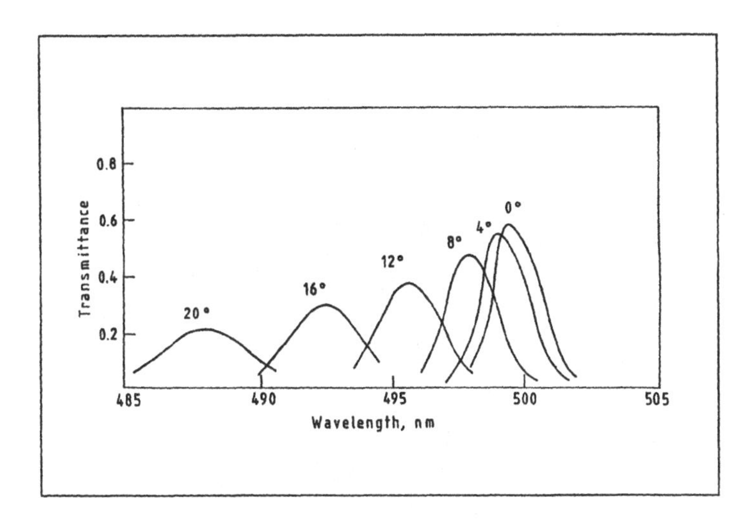

Fig. 5.4 *Transmittance of inclined interference filters. The transmitance of a dielectric interference filter is represented as a function of the angle of incidence (from Dobrowolski 1978).*

and

$$\frac{\Delta\lambda_{0.5,\theta}}{\Delta\lambda_{0,5}} = \left[1 + \left(\frac{\theta^2 \lambda_{\max}}{\mu^{\star 2} \Delta\lambda_{0.5}}\right)^2\right]^{1/2} \tag{5.17}$$

In many applications the incident beam is not parallel but has some semi-angle that we shall designate by α. Expressions (5.16) and (5.17) become

$$\frac{\delta\lambda_{\max,\alpha}}{\lambda_{\max}} = -\frac{\alpha^2}{4\mu^{\star 2}} \tag{5.18}$$

and

$$\frac{\Delta\lambda_{0.5,\alpha}}{\Delta\lambda_{0,5'}} = \left[1 + \left(\frac{\alpha^2 \lambda_{\max}}{2\mu^{\star 2} \Delta\lambda_{0.5}}\right)^2\right]^{1/2} \tag{5.19}$$

Multiple-cavity filters which are obtained by depositing several stacks one on top of the other yield high transmission in the passband spectral region, and a large rejection outside this region. These filters permit isolation of wavelength intervals down to a few nanometers; the wavelength at which they function is controlled by making the films thinner or thicker. While single-cavity filters have shallow slopes of the transmission curve, multiple-cavity filters have steep passband slopes and almost square tops. Figure 5.5 gives the transmission curves of band-pass interference filters as a function of the number of cavities used in the filter construction.

As temperature changes, all layer thicknesses change, and so do the layer refractive indices. Still, interference filters have a temperature dependence which is an order of magnitude smaller than is the case for glass and gelatin filters. The central wavelength increases linearly with increasing temperature, and vice versa; the magnitude of

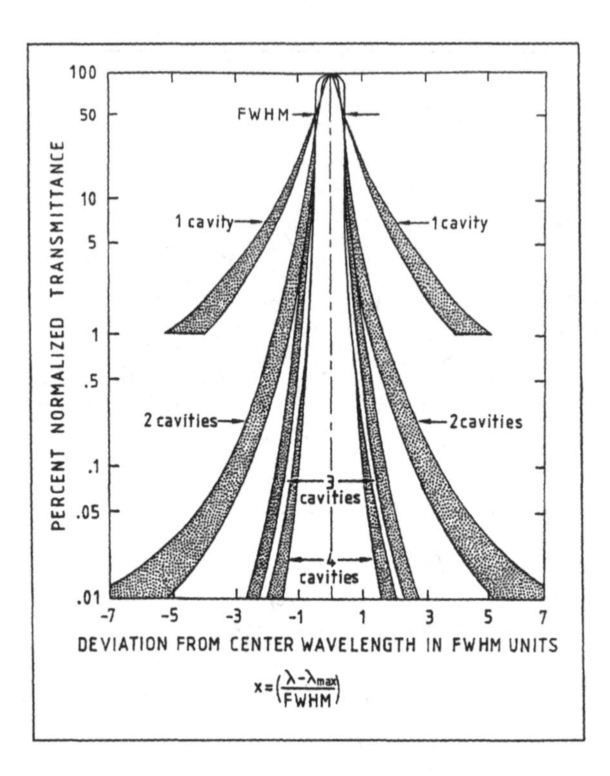

Fig. 5.5 *Interference filters. Transmission curve as a function of the number of cavities (adapted from Melles Griot Optics Guide 1985).*

the change is generally about 0.01 to 0.02 nm/°C, but this shift factor is wavelength-dependent. One should, however, keep in mind that some interference filters have very narrow bandwidths, so that even small shifts in central wavelength may cause drastic changes in measured transmission at a given wavelength.

These filters are usually designed for use at ambient temperatures of 20 to 25°C, but they may be operated at temperatures as low as −50°C. Interference filters used in infrared photometers at cryogenic temperatures will perform differently because of the large deviation from the "design temperature". This is clearly illustrated in Fig. 5.6 which shows, for infrared filters, the wavelength shift as a function of temperature, and in Fig. 5.7 which shows the variation with temperature of the response curve of a typical filter. Note that the rate of temperature change should be kept below 5°C/min in order to prevent rupture due to the thermal stress between the different composing materials. Repeated cooling to cryogenic temperatures may also deteriorate interference filters.

As a rule, one side of the filter is given a high reflectance toward the outside. It should face the source of radiation, so that the reflective coating serves as a heat shield that reduces thermal load on the filter, thereby extending the lifetime of the filter. Few astronomical applications, however, pose the risk of overheating, with the exception of solar research. Filter orientation may also alter the fluorescence characteristics. More-

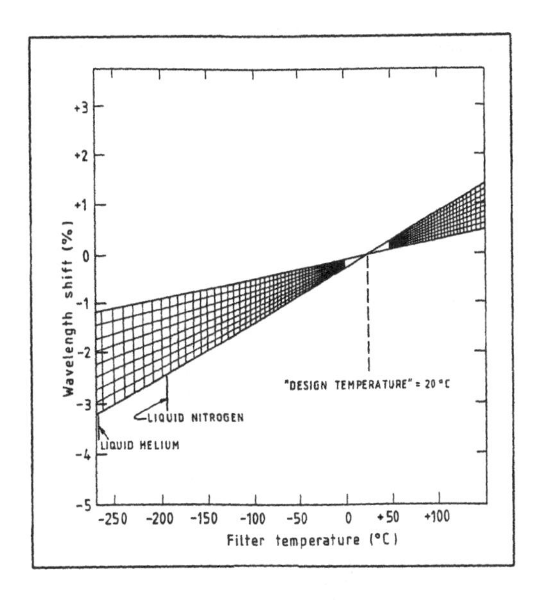

Fig. 5.6 *Temperature effect on the wavelength of an infrared filter (source: OCLI Technical report 1980).*

over, the high reflectance of these filters causes a problem when they are used in a stacked configuration. Reflections cannot always be eliminated by a slight tilting, since the latter causes broadening of the passband.

Filters usually have a square or circular form, with dimensions of 1 inch or more. Manufacturers usually hermetically seal their circular filters in anodized aluminium rings to provide maximal protection.

Interference filters have several severe handicaps:

- They deteriorate with time due to moisture, because the layers are porous and adsorb water in the cracks. Some show drift of peak wavelength with age. Deterioration occurs at the edges first, hence these filters should be made large enough so that only the central regions must be used.
- They have fixed size and cannot be cut to smaller pieces.
- They are normally designed to be used with collimated light at normal incidence; both peak transmittance and peak reflectance shift towards shorter wavelength as the angle of incident light is increased (see Eqs. (5.16–5.19) and Fig. 5.4), so one is bound to a fixed and limited range of F-ratios. The peak transmittance is strongly reduced with decreasing F-number, and the narrower the bandpass, the more pronounced this effect is. This restricts the applicable F-ratios to values of at least 12 to 15.
- They are expensive.

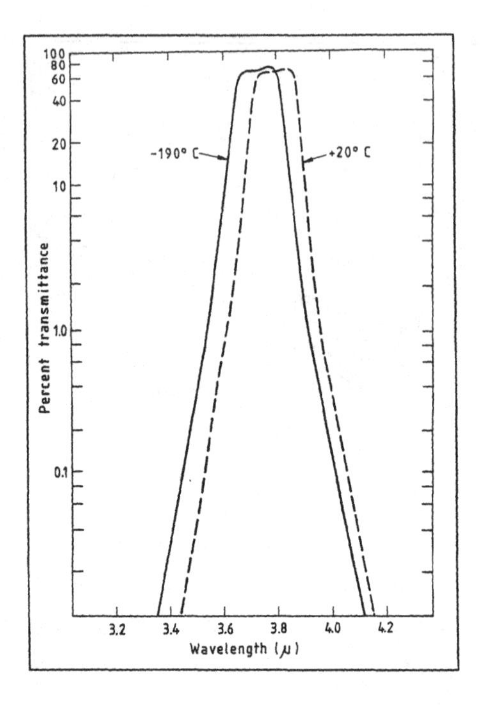

Fig. 5.7 Temperature effect on the response curve of an infrared filter. Compare with Fig.5.4 (source: OCLI Technical report 1980).

- They are very difficult to duplicate, which is a severe handicap for the reproduction of photometric systems.
- Their passbands may strongly shift (by as much as 5 nm between dry and vapor-saturated air at 20°C) with changes in humidity.

5.5 Custom-design filters

For some specific applications, manufacturing according to the customer's specifications may be necessary. In those cases the manufacturer will ask for adequate and precise specifications, for instance:

- Size and thickness.
- Center wavelength and allowed tolerance. If filters are to be used for systematic observations of objects with high radial-velocities, such as stars belonging to another galaxy, the Doppler shift of the central wavelength may be substantial, and the central wavelength must be adjusted accordingly.

- FWHM and corresponding tolerance.

- Out-of-band attenuation level according to wavelength (this depends on the spectral response of the detector used).

- Angle of incident radiation.

- Operating temperature range, since the specifications of the filters can only be matched at one specific temperature. The specification of operating temperature is especially important for filters which work in a cooled environment.

If necessary, some tolerances must be relaxed, since the above-mentioned requirements are not always independently realizable.

5.6 Image-quality filters

Image-quality filters are needed for some applications, e.g., photometry with imaging devices such as CCDs. Normal interference or glass filters do not always have the optical quality necessary for imaging. The manufacturing tolerances for image-quality filters (polishing and flatness of the glasses) are tighter, and multiple images are avoided by depositing additional anti-reflection coatings.

5.7 Neutral-density filters

A special kind of filter is the neutral-density filter. These filters have a flat wavelength response and are used to attenuate light over a broad wavelength range in a calibrated way (e.g., to dim the light of stars too bright to be observed with the available detector). They are categorized by their optical density. Some neutral-density filters are glass filters, others are metallic alloy coatings deposited on a glass substrate. Some neutral-density filters are available as graded strips or disks.

The optical density given by the manufacturer may differ significantly from its real value, so careful calibration is crucial. This can be done by observing a star that is bright enough to be measured through a neutral-density filter and simultaneously not too bright to be measured without attenuation. This method, however, assumes that one can correct any non-linearity of the measuring system. Accurate calibration, if done in different passbands, will also reveal any wavelength dependence (deviation from neutrality, see Section 10.3.6).

The lack of good-quality neutral-density filters in a photometer is often compensated by the use of a diaphragm in front of the telescope, but this is only a worst-case solution (see Section 15.6.5).

5.8 Circular variable filters

Circular variable filters are a very useful variation of the interference filter. The coating of those circular interference filters is deposited close to the edge, and its thickness varies with position. Once properly positioned in the light beam, the central wavelength varies linearly when the disk is rotated. The size of the light spot is an important parameter in fixing the effective passband, which can be as small as one percent of the central wavelength. These filters provide a way to perform sequential spectrophotometry in a broad wavelength interval, and are frequently used for observation in the infrared spectral region.

Chapter 6 Atmospheric extinction

6.1 Absorption, scattering and dispersion of light

Atmospheric extinction, in its broadest sense, is the reduction of the intensity of radiation as a result of absorption and scattering by the Earth's atmosphere (see Fig. 6.1). Both processes reduce the radiant flux from a given beam of light (about one sixth of the amount of perpendicularly incident light is extinguished in the visible domain).

Absorption of radiant energy by an atmospheric molecule is a destructive process: a photon is annihilated, and its energy is transferred to the absorbing molecule, which undergoes a transition from a lower to a higher state of energy. Absorption may lead to subsequent emission. It may also lead to destruction of the molecule.

Scattering involves the collision of a photon with a particle, with a consequent change in the direction of motion and of the energy content of the photon (and of the particle). Atmospheric scattering is caused by particles of different sizes, i.e., air molecules ($10^{-4}\mu$m), haze particles (10^{-2} to 1 μm), fog and cloud droplets (1 to 10 μm) and dust grains. Generally, the smaller these particles are, the higher is their concentration. Scattering by air molecules is approximately inversely proportional to the fourth power of λ (close to $\lambda^{-4.1}$), and is called *Rayleigh scattering*. Scattering by dust grains (i.e., cosmic dust, volcanic ash, combustion products, organic particles released by trees and plants, and even sea salts injected into the air as small droplets of seawater by the breaking of ocean waves), and by liquid water (films of water around hygroscopic particles, or water droplets carrying the hygroscopic particles in solution) is called *aerosol* scattering and depends on the size distribution of the particles. Scattering by small *spherical* particles (size of the order of the wavelength of light) is inversely proportional to λ and is known as *Mie scattering*. The larger particles cause scattering which is virtually independent of wavelength. The actual size distribution of the particles of various shapes roughly results in a λ^{-1} scattering. The decomposition of atmospheric extinction in its components due to Rayleigh and aerosol scattering, and due to absorption by molecular bands (O_3, H_2O,

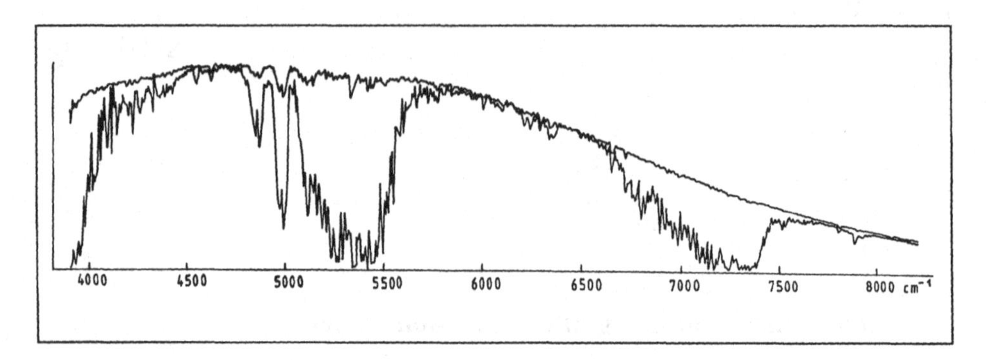

Fig. 6.1 *Near-infrared (1.2 to 2.6 μm) spectrum of the Moon.*
Upper curve: *the spectrum of the Moon from a jet aircraft at an altitude of 12500 m. The spectrum still contains the atmospheric absorption features due to the atmosphere above this altitude.*
Lower curve: *spectrum of the Moon from Catalina Observatory obtained at a zenith angle of 58°. This figure is a combination of Figs. 1 and 3 of Johnson et al. (1968).*

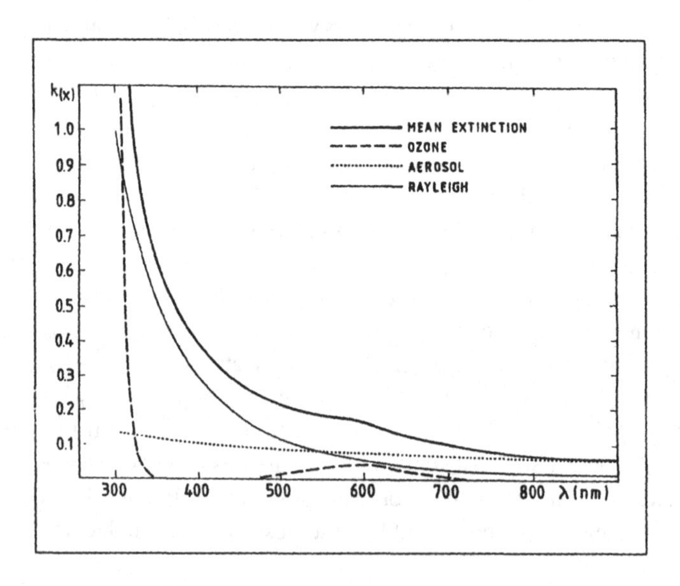

Fig. 6.2 *Components of atmospheric scattering (after Rufener 1986).*

O_2,...), depends on the adopted atmospheric model and on the numerical values of its parameters. Figure 6.2 shows such a decomposition.

All scattering depends on wavelength in the sense that extinction in the red part of the visible spectrum is much less troublesome for photometrists than it is at the ultraviolet side (see Fig. 6.3 for the wavelength-dependence of extinction).* Near the zenith, the extinction at sea level is more than 15% in the visual region. It strongly increases towards the blue and becomes 100% below 290 nm. Towards the red end of the visual spectrum extinction drops below 10%, except in molecular bands.

This wavelength dependence is also the reason why the daytime clear sky is blue, and not the color of sunlight (white-looking sky is caused by high cirrus that fill the air with ice crystals and act as a diffusor of sunlight). A small amount of Rayleigh scattering by a few particles conserves the direction of the dominant light beam, but when large particles are densely packed, multiple scattering occurs, and in the extreme case (fog) such a situation may lead to complete loss of sense of direction of the initial beam.

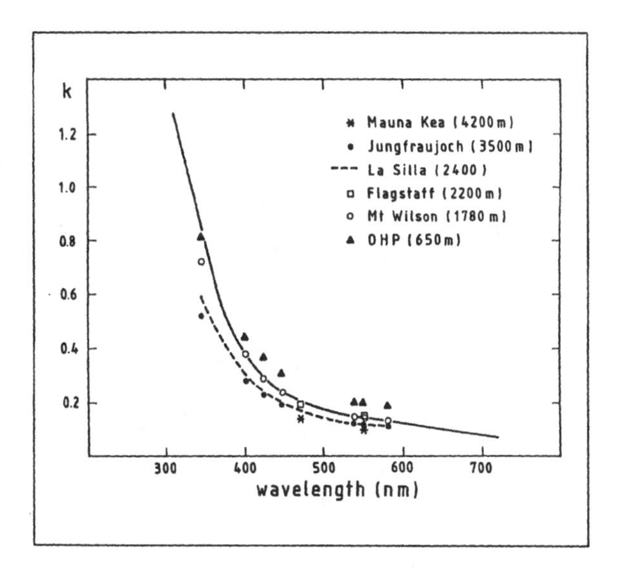

Fig. 6.3 Atmospheric extinction in function of wavelength for different sites. The full line covers data from Melbourne (1960); the dashed line is based on the 7-color data from Rufener (1986). Other data are from Hall et al. 1975 (□,✳), Cramer 1991 (•) and Golay 1974 (▲, ○). Note how close the dashed curve lies with respect to the data for the world's highest photometric sites (•,✳).

For evidence on long-term variability of extinction, and on its relation to volcanic eruptions, see Hall et al. (1975), Taylor et al. (1977), Sterken et al. (1986), and Rufener (1986). Seasonal variations of extinction are discussed by Laulainen et al. (1977); see also Roosen et al. (1973). Rufener's retrospective analysis, covering almost 10 years of extinction observations, led him to conclude that—even during clear weather conditions—the atmospheric extinction is intrinsically variable, with typical peak-to-peak variations

* For a short description of the complexities of extinction in the infrared spectral region, see Section 12.1

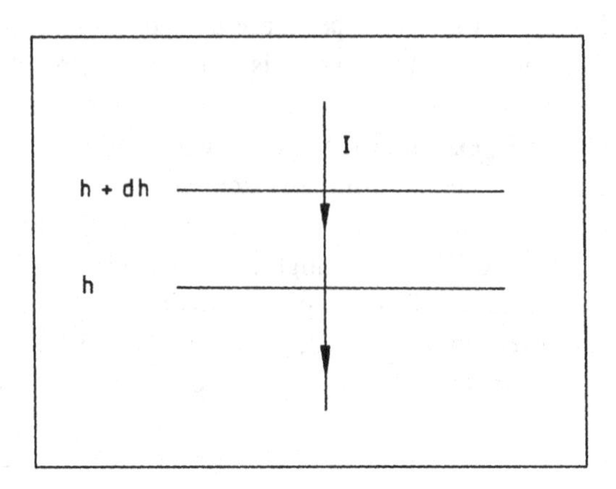

Fig. 6.4 Vertical light beam.

of 0.02 to 0.03 mag per unit air mass for a given night. The observed variations are systematically larger in the ultraviolet than in the more redward part of the spectrum. The amplitude of variation is, of course, lowest in good-quality sites (see also Fig. 6.3). The principal contribution to the variation comes from scattering. Much more prominent are variations due to particles brought into the atmosphere by the regularly occurring volcanic eruptions; timescales of 1 or 2 years are to be expected for all the particles to fall through the atmosphere, causing a slow downward trend in the value of the extinction.

Atmospheric refraction is the bending of light rays, making an object appear higher in the sky than it really is. The error on zenith distance goes from 0 at zenith up to half a degree at horizon. It is, however, non-linear (approximately proportional to the tangent of zenith distance) and only exceeds 1 arc minute at zenith distances larger than 45°. *Atmospheric dispersion* is due to the atmosphere refracting light of shorter wavelength more than it does with light of longer wavelength. All stellar images, in fact, are short, vertical spectra, with the red end pointing down. The effect is not present for stars in the zenith, but the distance from red to blue may amount to several seconds of arc at zenith distances exceeding 65°.

6.2 Monochromatic extinction

In this and in the following sections, we shall mainly use the monochromatic definitions in the wavelength domain.

The variation of the spectral radiance (or specific intensity) of a vertical light ray between altitudes h and $h + dh$ (see Fig. 6.4) is given by

$$dL(\lambda, h) = \kappa(\lambda, h)\rho(h)L(\lambda, h)dh \tag{6.1}$$

where $\kappa(\lambda)$ is the absorption coefficient per unit mass and $\rho(h)$ is the specific mass of air (since light is going downward, a positive dh yields an increase of radiance).

The absorption coefficient includes contributions from various processes and it varies with the composition of the atmosphere. Absorption by atoms and molecules (H_2O, O_3...) occurs in limited spectral ranges, while scattering by molecules and by aerosols in suspension in the atmosphere is responsible for the continuous extinction (see Section 6.1).

6.3 Definition of the air mass

Equation (6.1) integrates as

$$L(\lambda, h_0) = L(\lambda, h_1)e^{-\int_{h_0}^{h_1} \kappa(\lambda,h)\rho(h)dh} \tag{6.2}$$

with h_1 denoting the outer boundary of the atmosphere, and h_0 the altitude of the observing site. Of course, stars are not observed at the zenith only. Let z be the zenith distance of the direction of the incoming light, i.e., its angle with the vertical (see Fig. 6.5). Between altitudes $h + dh$ and h, the light rays will have traveled through a thickness $dh \sec z$ of air and Eqs. (6.1) and (6.2) become

$$dL(\lambda, h) = \kappa(\lambda, h)\rho(h)L(\lambda, h) \sec z dh \tag{6.3}$$

and

$$L(\lambda, h_0) = L(\lambda, h_1)e^{-\int_{h_0}^{h_1} \kappa(\lambda,h)\rho(h) \sec z dh} \tag{6.4}$$

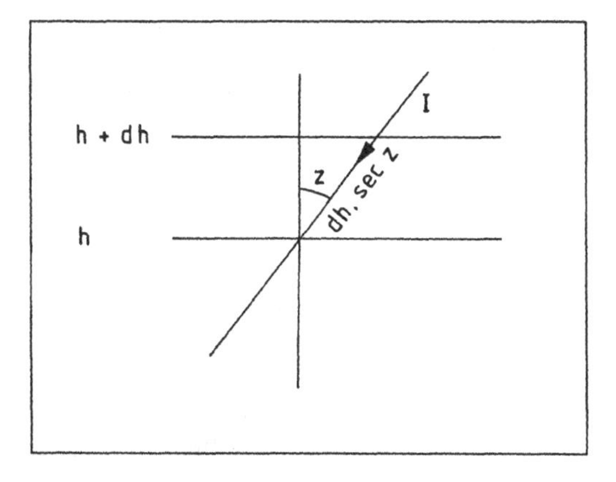

Fig. 6.5 Inclined light beam.

But in reality the angle $z = z(h)$ varies with the altitude because of the refraction of the light in the atmosphere and also because of the curvature of the Earth (see Fig. 6.6). It can be shown (Newcomb 1906; see also Young 1969a) that for a horizontal ray, the curvature due to refraction is about one sixth of the curvature of the atmosphere. In a first

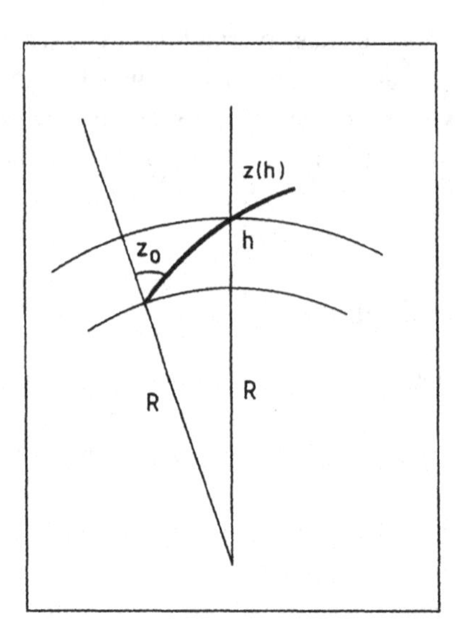

Fig. 6.6 *Path of a ray of light. The angle made by a ray with the vertical varies because of the curvature of the Earth and because of the atmospheric refraction.*

approximation we assume that the absorption coefficient $\kappa(\lambda, h)$ is constant throughout the atmosphere. This is more or less true for the major atmospheric constituents, but not for those which have variable distributions (ozone, dust, water droplets...). Equation (6.4) takes the form

$$L(\lambda, h_0) = L(\lambda, h_1)e^{-\kappa(\lambda)\int_{h_0}^{h_1} \rho(h)\sec z(h)dh} \tag{6.5}$$

The integral represents the mass of air contained in a column parallel to the light ray and of cross-section unity. However, this is *not* what photometrists call the *air mass*. Instead, they define the (relative) air mass X as

$$X = \frac{\int_{h_0}^{h_1} \rho(h)\sec z(h)dh}{\int_{h_0}^{h_1} \rho(h)dh} \tag{6.6}$$

which is a dimensionless quantity. The air mass X is defined as the number of times one has, along the line of sight, the quantity of air seen in the direction of the zenith. X depends on the path of the light ray, which at any altitude h is determined by the direction z. Hence we shall write $X = X(z_0)$, expressing the air mass as a function of the *apparent* zenith angle ($z = z_0$) at the observer's level h_0. For small zenith distances, the atmosphere may be considered to be a flat slab of constant density, and in this approximation, X simply is the secant of the apparent zenith distance. Using the definition of air mass, Eq. (6.5) becomes

$$L(\lambda, h_0) = L(\lambda, h_1)e^{-\kappa(\lambda)X(z)\int_{h_0}^{h_1} \rho(h)dh}$$

$$= L(\lambda, h_1)e^{-\kappa(\lambda)X(z)\alpha} \tag{6.7}$$

α being the true quantity of air measured vertically above h_0. The advantage of this definition is more apparent when using the logarithmic form. Equation (6.7) becomes

$$\log L(\lambda, h_0) = \log L(\lambda, h_1) - (\log e)\kappa(\lambda)X(z)\alpha \tag{6.8}$$

In our simplified model, the light beam does conserve its geometry when z changes so that the irradiance is proportional to the radiance. In fact, the extent of the beam falling in the entrance pupil of the telescope is proportional to the cosine of the atmospheric refraction angle (at the horizon it barely exceeds half a degree), and the resulting systematic error is always negligible. On the other hand, the solid angle subtended by the object—even a star is not really pointlike—is reduced by a factor of approximately $(n-1)\sec^2 z$, where n is the refractive index of the atmosphere (Young, 1974b). The resulting error is about 0.01 mag at air mass 6, and much smaller at lesser air masses. This is also negligible. Hence we can write (6.8) as

$$\log E(\lambda, h_0) = \log E(\lambda, h_1) - (\log e)\kappa(\lambda)X(z)\alpha \tag{6.9}$$

or, in magnitude form,

$$\begin{aligned} m(\lambda, h_0) &= m(\lambda, h_1) + 2.5(\log e)\kappa(\lambda)X(z)\alpha \\ &= m(\lambda, h_1) + k(\lambda)X(z) \end{aligned} \tag{6.10}$$

with

$$k(\lambda) = 2.5(\log e)\kappa(\lambda)\alpha \approx 1.086\kappa(\lambda)\alpha \tag{6.11}$$

The extinction affecting the monochromatic magnitude is proportional to the air mass. $k(\lambda)$ is called the *extinction coefficient* at wavelength λ. Relation (6.10) is known as *Bouguer's law.* It is important to keep in mind the basic asumption made in deriving this simple and very frequently used law, namely that *the absorption coefficient* $\kappa(\lambda, h)$ *per unit mass is considered to be constant throughout the atmosphere.*

6.4 Calculation of the air mass

At large zenith distances, the curvature of the atmosphere, the atmospheric refraction, and the variation of atmospheric density with height complicate the exact calculation of the air mass.

Bemporad (1904) has calculated the air mass for a spherical atmosphere, taking into account the bending due to the refraction, and assuming that the absorption coefficient κ per unit mass is constant. As we have seen, the latter hypothesis allows the definition of a quantity $X(z)$ independent of λ. Based on Bemporad's tables, interpolation formulae have been derived which are commonly used in photometry. Among the most widely used formulae, Hardie (1962) gives the following third-degree polynomial:

$$X = \sec z - 0.0018167(\sec z - 1) - 0.002875(\sec z - 1)^2 - 0.0008083(\sec z - 1)^3 \tag{6.12}$$

Hardie clearly uses the *true* zenith angle, i.e., the value the zenith angle would assume in the absence of atmospheric refraction, though Bemporad's tables use the *apparent* zenith angle as argument. This desire for a high precision on the air mass was guided by the hope that it would reflect on the resulting photometric reductions. However, many factors contribute to make this hope unjustified.

First it should be mentioned that Bemporad worked out corrections for temperature and pressure which are not included in the above formula. Those corrections can exceed the 1% level and must lead to systematic errors of this order in any formula.

The constancy of κ throughout the atmosphere is not well verified. The aeorosol component is mainly concentrated in the lower layers of the atmosphere, with a scale height of the order of one kilometer. On the other hand, the ozone component is important at altitudes of around 30 km. The resulting air masses for such dissimilar distributions are quite different, and are not in agreement with Bemporad's tables. Those contributions have opposite effects—aerosols, for example, increase the air mass above Bemporad's value, while ozone yields lower air masses. Hence on the average Bemporad's tables are rather acceptable, but in specific wavelength ranges they will be either above or below the correct air mass. This is particularly important in the Chappuis bands (around 580 nm) of O_3 and below 340 nm where this molecule is also an important absorber. The errors can reach 0.01 at air masses of the order of 2.5–3, thus making deceptive any impression of high precision coming from any interpolation formula. There is not much to do about that, except to avoid large zenith angles. In principle, however, it would be quite possible to use different air masses at different wavelengths, $X = X(\lambda, z)$, as long as an appropriate model of the extinction is available. But we shall see later that the reduction of such data, especially in *heterochromatic* photometry, is a complicated matter.

Young (1974b) pointed out a common misuse of Bemporad's tables that consists in calculating true zenith angle instead of apparent zenith angle, and using it as the argument (this was already noticed for the interpolation formula (6.12)). Computer programmes are more likely to yield the true zenith angle z_t rather than the apparent zenith angle z_a (affected by the refraction) because the former is based exclusively on readily available astronomical data: time and coordinates.

As a formula using true zenith distance, and giving satisfactory agreement with Bemporad's results up to an air mass of 4, Young & Irvine (1967) propose

$$X(z_t) = \sec z_t \left(1 - 0.0012(\sec^2 z_t - 1)\right) \tag{6.13}$$

For the error caused by using z_t instead of z_a, we refer to Fig. 12 of Young (1974b). For a calculation based on modern atmospheric data, see Kasten & Young (1989).

6.5 Heterochromatic extinction

The continuum part of the monochromatic extinction increases monotonically towards the blue and UV parts of the spectrum. When heterochromatic photometry is carried out, measurements at the blue side of the passband are thus more affected than those at the red side. The resulting effect is that when a star is observed at larger zenith angles, it becomes redder: the effective wavelength is displaced toward the red and the effective passband gets narrower. In other words: *at large air masses the monochromaticity is increased because of the progressive removal of the extinction at shorter wavelengths.* Because of this change in the equivalent passband, a progression of air-mass increments yields smaller and smaller extinction increments. The phenomenon, known as *Forbes effect* (Forbes 1842), is illustrated in Fig. 6.7. Stars of different colors yield different results, and blue stars are more affected than red ones. This alone shows that wide-band photometry is intrinsically a much more complicated technique than is its monochromatic counterpart.

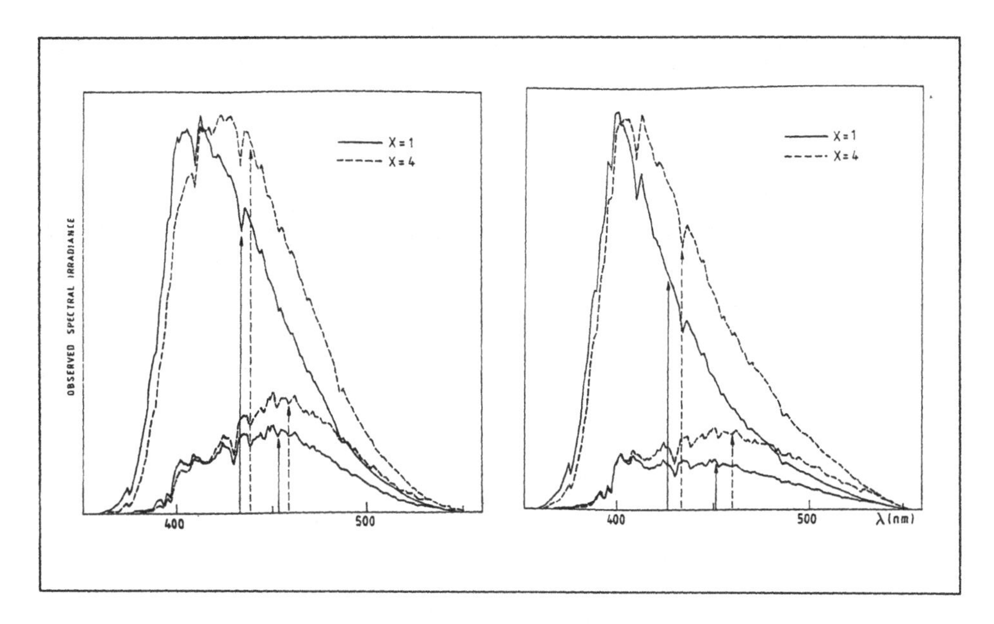

Fig. 6.7 B *response functions for an O star and an M star.* Left: *"observed" energy distribution (bottom panel of Fig. 1.17, corrected by the mean monochromatic extinction, taken from Fig. 6.3) at $X = 1$.* Right: *the same data corrected for atmospheric extinction at $X = 4$. The vertical arrows indicate the positions of the effective wavelengths (Eq. 1.41) of the bandpasses for each observed spectral energy distribution (see Fig. 1.17). It is obvious that the effective extinction is much larger for the early-type star than it is for the late-type star.*

Let us now go back to the expression (1.16) of the spectral irradiance measured by a photometer with an instrumental response $s(\lambda)$.* We write it—using somewhat lighter

* $s(\lambda)$ includes the spectral sensitivity of the photocathode, the filter transmission, and also the transmission properties of the telescope optical system.

notations

$$E_{\mathrm{m}} = \int E\, s_{\mathrm{e}}\, s\, d\lambda \tag{1.16}$$

The atmospheric transmission function s_{e} is given by (6.7) and (6.11)

$$\begin{aligned} s_{\mathrm{e}} &= e^{-\kappa X \alpha} \\ &= e^{-kX/(2.5\log e)} \end{aligned} \tag{6.14}$$

The Strömgren-King theory considers Eq. (1.36) resulting from the Taylor expansion of (1.16),

$$E_{\mathrm{m}} = \left(S_0 + \frac{1}{2}\mu_2^2 S_0''\right) \int s d\lambda \tag{6.15}$$

where the higher-order terms have been omitted, and with $S_0 = S(\lambda_0)$. Substituting $S = E s_{\mathrm{e}}$, we obtain (with $E_0 = E(\lambda_0)$ and $s_{\mathrm{e},0} = s_{\mathrm{e}}(\lambda_0)$)

$$E_{\mathrm{m}} = E_0 s_{\mathrm{e},0}\left(1 + \frac{1}{2}\mu_2^2 \frac{S_0''}{S_0}\right) \int s d\lambda \tag{6.16}$$

The term between brackets can be considered as the heterochromatic correction to be applied to the monochromatic transmission s_{e} calculated at the effective wavelength λ_0. The observed heterochromatic magnitude is

$$\begin{aligned} m(h_0) &= -2.5\log\left[E_0 s_{\mathrm{e},0}\left(1 + \frac{1}{2}\mu_2^2 \frac{S_0''}{S_0}\right) \int s d\lambda\right] + C \\ &= -2.5\log\left[E_0 s_{\mathrm{e},0}\left(1 + \frac{1}{2}\mu_2^2 \frac{S_0''}{S_0}\right)\right] + C' \end{aligned} \tag{6.17}$$

(C, C' are constants) while the extra-atmospheric heterochromatic magnitude is given by the corresponding expression without extinction correction,

$$\begin{aligned} m(h_1) &= -2.5\log\left[E_0\left(1 + \frac{1}{2}\mu_2^2 \frac{E_0''}{E_0}\right) \int s d\lambda\right] + C \\ &= -2.5\log\left[E_0\left(1 + \frac{1}{2}\mu_2^2 \frac{E_0''}{E_0}\right)\right] + C' \end{aligned} \tag{6.18}$$

(which is *not* directly related to the extra-atmospheric monochromatic magnitude at λ_0, $m(\lambda_0, h_1) = -2.5\log E_0 + C''$). The total extinction correction (extra-atmospheric minus observed value), expressed in magnitudes, is

$$\begin{aligned} \Delta m &= m(h_1) - m(h_0) \\ &= 2.5\log s_{\mathrm{e},0} + 2.5\log\left(1 + \frac{1}{2}\mu_2^2 \frac{S_0''}{S_0}\right) - 2.5\log\left(1 + \frac{1}{2}\mu_2^2 \frac{E_0''}{E_0}\right) \\ &= -k_0 X + 2.5(\log e)\left[\ln\left(1 + \frac{1}{2}\mu_2^2 \frac{S_0''}{S_0}\right) - \ln\left(1 + \frac{1}{2}\mu_2^2 \frac{E_0''}{E_0}\right)\right] \end{aligned} \tag{6.19}$$

The second terms in the arguments of the logarithmic functions in Eq. (6.19) are usually small and one can use the approximation $\ln(1+x) = x$. Hence

$$\begin{aligned} \Delta m &= -k_0 X + 1.25(\log e)\mu_2^2\Big(\frac{S_0''}{S_0} - \frac{E_0''}{E_0}\Big) \\ &= -k_0 X + 1.25(\log e)\mu_2^2\Big(\frac{2E_0' s_{e,0}'}{E_0 s_{e,0}} - \frac{s_{e,0}''}{s_{e,0}}\Big) \end{aligned} \tag{6.20}$$

The first term is the monochromatic extinction (Eqs. 6.10 and 6.11) at wavelength λ_0. The second term contains the first and second derivatives of the atmospheric extinction, and the first derivative of the spectral distribution, i.e., a local spectral index. Those derivatives need to be evaluated.

Equation (6.20) is, in fact, a *color transformation* between the extra-atmospheric heterochromatic magnitude $m(h_1)$ at mean wavelength λ_0, and the measured heterochromatic magnitude $m(h_0)$, but it is usually called extinction "correction" rather than transformation. King (1952) remarked that "it is easiest to regard the photometer as afflicted with an altitude-dependent color equation". Equation (6.20) simply states that *the extinction is dependent on the spectral type, and that the dependency is a function of air mass.*

The atmospheric transmission curve is smooth except around the wavelengths of molecular absorption bands. Astronomical spectra on the other hand are generally crowded with absorption and emission features which differ from one object to the other, so that evaluating the derivative of E is no trivial matter. Various approaches can be made to deal with this problem.

6.5.1 Second-order coefficient

The first approach is to ignore King's theory, or to misread it (this is a commonly adopted position; see for instance Hardie, 1962, Henden and Kaitchuk, 1982). The overall reasoning is very simple: since using an average extinction correction gives an error which depends on the color of the star, one can modify Eq. (6.10) by introducing a color-dependent term

$$m(h_1) = m(h_0) - (k + \hat{k}\,CI)X(z) \tag{6.21}$$

The coefficient $\hat{k}$ is called the *second-order extinction coefficient* while k is called the *principal extinction coefficient.* The color is described by *some* color index* CI which is given by the measurements. k is the extinction coefficient for a star of color $CI = 0$ while $\hat{k}$ represents the increase of that coefficient for a star of color index $CI = 1$. Notice that this equation does not include the Forbes effect (the increasing reddenening and monochromaticity of the radiation as the air mass increases) if the color index is extra-atmospheric. Obviously Eqs. (6.20) and (6.21) are not compatible, since the

* In the early fifties only one color index was available, i.e., the International System color index (see Section 16.3), and—outside the Balmer discontinuity—a single gradient would well describe the spectral energy distribution over a substantially wide wavelength interval.

extra-atmospheric color index is directly related to E' and does not depend on s_e or its derivatives. A better agreement can be obtained when CI includes a contribution from the atmospheric extinction. This is the reason why CI is often taken as the instrumental color index (not corrected for extinction, and hence depending on the air mass). We shall see that this choice is also incorrect. Unfortunately, the method got wide use because of its simplicity, not because of its results: it can be used relatively easily without computers, and it is possible to set up simple systematic observing schemes allowing the calculation of the coefficients without too much work. Although those arguments no longer have any weight—see below—this is still the way most, if not all, wide-band photometry is being reduced.

Young & Irvine (1967), and Young (1974b) have developed King's discussion to show that, under rather general conditions, the heterochromatic extinction can be approximated by an expression similar to Eq. (6.21)

$$\Delta m = -(k + \hat{k}\,\overline{CI})X \tag{6.22}$$

Instead of the extra-atmospheric color index CI or the instrumental index CI_m, they have shown that the *arithmetic mean* $\overline{CI} = \frac{1}{2}(CI + CI_m)$ must be used. This supposes that one has measurements at a second wavelength λ_1 in the vicinity of λ_0 ($|\lambda_0 - \lambda_1| \ll \lambda_0$). The color indices are

$$\begin{aligned} CI_m &\simeq CI + X\Delta k \\ \overline{CI} &= \frac{1}{2}(CI + CI_m) \simeq CI + \frac{1}{2}X\Delta k \end{aligned} \tag{6.23}$$

with

$$\Delta k = k_1 - k_0$$

Hence the heterochromatic extinction has a contribution in X^2 which did not appear in Eq. (6.21), and which involves the gradient of the atmospheric extinction. The term of the first order in X includes the monochromatic extinction at the effective wavelength, and an additional term involving the gradient of the stellar spectral distribution.

To summarize, Eq. (6.22) is justified and correctly handles the Forbes effect when

- the color index between neighboring wavelengths λ_0 and λ_1 (with the condition $|\lambda_0 - \lambda_1| \ll \lambda_0$) is used;
- the color index $\overline{CI}$ used for the second-order term is the arithmetic mean between the extra-atmospheric index and the measured index.

But most of the time, the first condition imposed on the effective wavelengths is not fulfilled, so a suitable color index cannot be calculated. In other words, the derivatives at λ_0 are not available and an exact extinction correction is not possible. King developed his theory for black bodies (see Section 8.2.1), with the idea that the spectrum of a stellar photosphere is sometimes not very different from that of an ideal radiator. This is valid for some stars in adequate wavelength ranges, but not around the Balmer jump or close to important absorption and emission features. For instance, no one would approximate the spectrum of a carbon star, of a Wolf-Rayet or of an emission nebula with a black-body energy distribution.

The prospect for obtaining valid extra-atmospheric magnitudes in heterochromatic photometry is thus very bleak. In general the extinction correction will not yield those exact values but something else, more or less related to them. Under those conditions, one could wonder about the usefulness of heterochromatic photometry. There is, however, some homogeneity in all that business. Everybody uses more or less the same equations, with the same corrections for approximately the same instrumental response in a given system—even if they do not yield what they were initially expected to yield. Hence data obtained with different equipment and reduced by different photometrists can be compared with a better precision than the above discussion would predict. The point is that the magnitudes will not be the quantities defined by Eq. (1.17) but simply those defined by the correction model. This makes illusory any attempt to obtain accurate values of the relative irradiances—a fortiori of the absolute irradiances—from the "magnitudes" (see also Rufener & Nicolet 1988). Consequently, if someone could perform accurate heterochromatic extinction corrections he would face the problem of producing data that finally cannot be related to those of others working with the same photometric system. This discussion raises the important item of homogeneity in photometry: not only should the equipment be compatible (we will come back to that in Chapter 11) but the reduction techniques have to be compatible too. This is somewhat analogous to the problem of quantifying the solar activity: if one wishes to compare modern data with those obtained one century ago, one must follow the same protocol, even if it looks somewhat flawed. Changing the extinction-correction scheme in wide-band photometry is bound to lead to disaster, but if one wishes to obtain milli-magnitude accuracy in differential photometry of variable stars, a complete treatment of the extinction correction will certainly pay off.

6.5.2 A rigorous second-order method

Further refining of King's theory could help set up newer and better photometric systems, giving a really accurate estimate of the extra-atmospheric irradiance and more likely allowing exact reproducibility of the results. From the theoretical point of view, such attempts have been eagerly pursued by Young (1988, 1992).

Young (1992) carried the Strömgren-King series expansion to third- and fourth-order terms, and concluded that these terms are by no means negligible, because photometric response profiles are unlike any low-order polynomial, and are not well-represented by a Taylor series. As an illustration, Fig. 6.8 gives an example of a profile with its associated high-order moments (the profile is representative for a typical asymmetric profile like the Johnson-Morgan B filter). The importance of the far-red tail of this moderately asymmetric response function is evident: where symmetric profiles have odd moments which are zero, in asymmetric profiles all odd moments exist. Some asymmetric filter profiles (for example, the two-cavity interference filters) even have infinite moments.

Young finally stresses the point that the principal factor contributing to convergence of the Taylor expansion (1.16) is by no means the vanishing of the nth derivative $d^n S/d\lambda^n$,

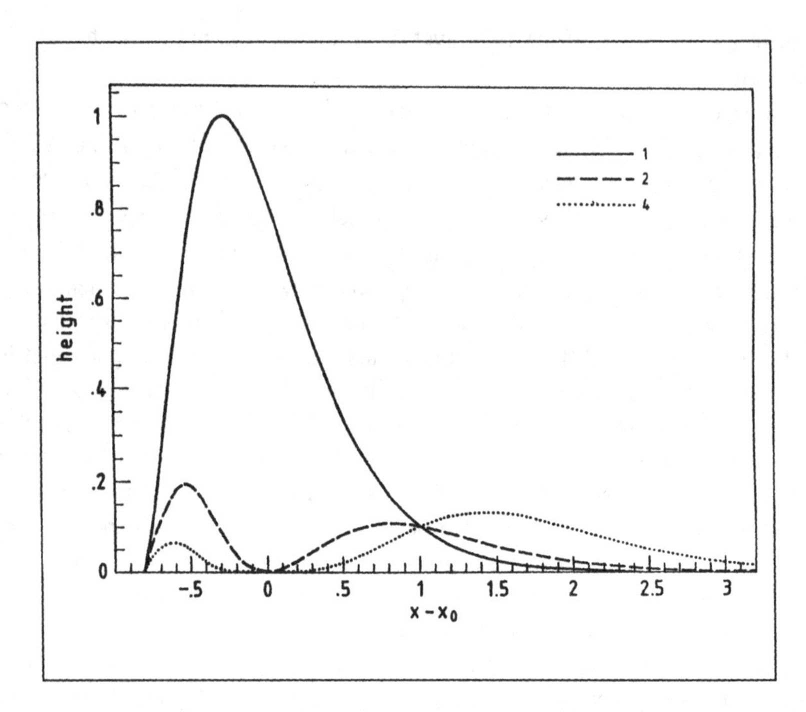

Fig. 6.8 *Realistic approximation to an asymmetric filter function like Johnson-Morgan* B *(Young 1992). The full curve represents the normalized function* $x^2\, e^{-x}$*, and the dashed and dotted lines, respectively, are the integrands of its second and fourth moments (the origin of the x-axis is at the centroid of the passband).*

nor μ_n^n, but rather the quickly growing factor $n!$. This is specially important as the derivatives grow with the presence of stellar features. For what concerns the extinction correction, the nth derivative of k more than compensates this effect.

As we have noticed, and as one would expect from a theory based on a Taylor series expansion, the key problem is the evaluation of the derivatives of the involved functions, i.e., E and k (in a rigorous method, one should not make the assumption that k is a pure power law). Consequently, Young (1988) shows that both first and second derivatives can be determined accurately, and that such estimates lead to much more accurate extinction and transformation corrections.

The previous section dealt with a crude linear approximation where two nearby passbands provided some information on the behaviour of the stellar spectra as well as of the atmospheric transmission. It is obvious that a good knowledge of the second derivatives would bring additional strength to the theory. To obtain them one needs a third filter so that the relevant differences can be computed.

What is exactly meant by derivatives? Stellar spectra are very irregular with many discrete features which simply cannot be represented by a limited Taylor's expansion.

As stated by Young (1988):

> "Evidently, the only sensible physical interpretation of these derivatives is that they refer not to the true stellar energy distribution, which fluctuates wildly, but to some smooth function that approximates it, and that has well-behaved derivatives".

It is clear that if one would not take this approach, one would end up with excessively high values of E' if λ_0 were to fall close to the center of even a small spectral line. Therefore, Young describes the true stellar irradiance as a smoothed part—a parabolic function of λ—and a remainder:

$$E(\lambda) = E_s(\lambda) + E_r(\lambda)$$

The smoothing is chosen such that the equations are valid when E_s and its derivatives are used, without any involvement of E_r. Hence from Eq. (1.16) we get the condition

$$\int E_r\, s_e\, s\, d\lambda = 0 \tag{6.24}$$

The atmospheric transmission is generally quite smooth, except around telluric molecular features—which are usually avoided by passbands of photometric systems. Hence we can expand s_e in a Taylor series and we get for condition (6.24)

$$s_{e,0} \int E_r\, s\, d\lambda + s'_{e,0} \int (\lambda - \lambda_0) E_r\, s\, d\lambda + \frac{1}{2} s''_{e,0} \int (\lambda - \lambda_0)^2 E_r\, s\, d\lambda = 0 \tag{6.25}$$

Because for a given star $s_e(\lambda_0)$, $s'_e(\lambda_0)$, $s''_e(\lambda_0)$ depend on air mass, all three integrals should vanish in Eq. (6.25). Those three conditions on E_r allow to determine the three coefficients of the parabolic function E_s. We refer the reader to Young's paper (1988) for the mathematical aspects of the actual calculation of the derivatives of E_s and s_e from the photometric measurements. The resulting equations describing the extinction correction involve three magnitudes for each star and three extinction coefficients for each night (or for each observing period, as long as the atmospheric conditions remain stable). Moreover, six additional parameters describe the instrumental system (bandwidths, band spacings and zero points), and these should remain constant for a given instrumental setup. Although more complex than the classical analysis, this formulation is easily manageable with the computing facilities presently available.

How close need the wavelengths be to give a reasonable accuracy? Numerical experiments by Young (1988) show that a spacing equal to half the bandwidth (FWHM) is about adequate, the errors increasing rapidly for wider spacings. It is interesting to note that Young (1974b) had already reached similar conclusions from a theoretical analysis based on the "sampling theorem". This theorem, named by Shannon (1949), requires the sampling of the function $\ln S(\lambda)$ to be dense enough to resolve all significant structure, i.e., the passbands must overlap enough so that they yield a good approximation of the smoothed energy distribution (Young 1974b, 1988, 1992). The overlapping bands are by no means redundant: they add the necessary information to allow accurate transformation. Unfortunately, no existing photometric system does respect the sampling theorem. Considering that the most widely used system (Johnson-Morgan's *UBV*) violates this criterion by a factor of 5, it is easy to understand why it is so inaccurate. The Geneva seven-color system fares better (moreover, its stability and the homogeneous reduction methods ensure a good degree of reproducibility), but a narrow-band system, such as Strömgren's *uvby*, is badly undersampled (by a factor of 10). This does not show up in the extinction determination (except in *u*), because of the quasi-monochromaticity. We shall see later how important this is for the color transformation problem.

6.6 Conclusions

The monochromatic atmospheric extinction is easily calculated—even with sophisticated multi-component atmospheric models—so that photometric systems using narrow bandpass filters can yield realistic values of the extra-atmospheric irradiance. This is the case of the Strömgren system, for instance. When heterochromatic photometry is considered, the common practice of using color indices defined over a wide spectral domain to correct for the Forbes effect is grossly inadequate. However, in order to retain a minimum of homogeneity in a widely accepted system such as Johnson-Morgan's *UBV*, it is almost mandatory to stick to the old practices.

The work by King (1952), Young & Irvine (1967) and Young (1974b) shows the best way to handle heterochromatic photometry within the limits of present systems—i.e., systems largely undersampled with considerable spacings between bandpasses.

If and when correctly sampled systems are used, it becomes possible to calculate accurate extinction corrections. The complete description of a better second-order technique has been presented by Young (1988), the key requirement being that three filters be available in a relatively small spectral range, so that accurate determination of the derivatives becomes feasible. More rigorous higher-order solutions have also been proposed (Young 1992), with more stringent requirements on the sampling of the spectrum. This is unfortunately almost contrary to the design of all existing photometric systems since the wavelengths were chosen to sample different aspects of the spectrum and are often widely apart, and when two passbands are close to one another, one can be assured that they are meant to study an important feature of the spectrum (e.g., one band will measure the continuum, and the other an emission line, or they will be on two sides of a discontinuity). In such cases the numerical evaluation of the derivatives will have to be done on a discontinuous function and it is likely to yield poor results. The only way of having three close passbands is to design an entirely new system. Anyway, as we have already explained, the better treatment of the extinction would make it incompatible with previous results.

A good example of the problems involved in the calculation of the atmospheric extinction is found in cometary photometry and more specifically in the OH band at the wavelength of 308 nm (see Section 16.4.8). The absorption is quite large, and an appreciable fraction of it is due to the ozone layer so that the air mass model has to be different from the usual one. The filter bandwith is relatively large compared to the OH cometary feature but, since the cometary continuum is much lower than the emission, the "effective" passband is quite narrow—and the monochromatic approximation, at the OH wavelength, is well suited. However, the parameters have to be calculated from standard stars (there are unfortunately no standard comets to give the right values). The color effect is quite important on the stellar continuum so that the monochromatic approximation is not valid. Estimating the OH extinction from the stellar observation has to be done rather empirically.

Chapter 7 Atmospheric turbulence: scintillation and seeing

The atmosphere of the Earth changes both the irradiance and the direction of the light that passes through. On a very short timescale, both effects may be approximately considered to consist of a constant term, and of a variable quantity. The constant part of attenuation of light is the extinction, whereas the constant term of the deflection of light is the refraction. The rapid variations are random fluctuations—respectively called *scintillation* and *seeing*—which are caused by atmospheric turbulence.*

Turbulent motion occurs in a very wide range of spatial and temporal scales. The largest turbulent elements have sizes ranging from kilometers to hundreds or thousands of kilometers (large meteorological systems). Their kinetic energy is progressively transferred to smaller eddies, and the associated large-scale thermal variations are broken down into smaller and smaller scale fluctuations until viscous dissipation makes the temperature uniform over distances of millimeters. The corresponding turbulent motions cover time scales of days and weeks down to seconds and less. Turbulence as a rule has a multilayered structure, and is mostly confined to thin horizontal layers with high wind gradient, where small-scale wavelike undulations appear. Turbulent layers are separated by less turbulent regions.

Small-scale turbulent motions in the troposphere and in the lower stratosphere are of major concern in photometry. They are associated with temperature variations from place to place of a few hundredths of a degree, and these bring along refractive-index fluctuations which are roughly proportional to the linear sizes of the eddies. Such eddies (the smallest of which are of the order of a few mm) act as weak lenses. Hence, the atmosphere constitutes a complex optical system which distorts the plane-parallel wavefronts arriving from a star. When such a wavefront enters a turbulent layer, the small-scale variations in the refractive index produce wrinkles of up to several wavelengths amplitude in the wavefront. The total effect of the different turbulent layers prevents the telescope

* For a summary, see Stock & Keller, 1960; for a detailed discussion of the effects of turbulence on optical astronomy, see Van Isacker (1953), Reiger (1963), Young (1967b, 1969b, 1974a), Roddier (1981) and Coulman (1985).

optics from forming a diffraction pattern appropriate to a single point source. At large zenith distances one looks through more air, and effects get worse.

Turbulence degrades the image of a point source in two ways:

- Local convergence or divergence of the wavefront increases or decreases the irradiance. These random fluctuations in irradiance constitute the *scintillation* which is most evident to the naked eye as *twinkling* of point sources, particularly of the brightest stars. Twinkling consists of variations up to a couple of magnitudes, and is strongest for stars close to the horizon. Only the atmospherical waves with length comparable to the telescope aperture or larger contribute to scintillation. Reiger (1963), using a model that accounts for turbulent motion and refractive-index fluctuations in the entire atmosphere, derived analytical expressions for the correlation function (average product of the relative irradiance fluctuation for two parallel rays), for the spectrum of irradiance fluctuations, and for the rms angular fluctuations of starlight. Young (1967b, 1969b, 1974a) experimentally confirmed Reiger's work, and extended it.

- The random variations of the local direction of the light rays (normal to the wavefront) result in random motion of the image. This is called *seeing*. Irregular motions associated with seeing cannot be detected with the naked eye, but with small telescopes one can already see dancing of the image with an amplitude of several arc seconds around the mean position, the instantaneous image being no larger than the Airy disk (see Eq. 2.1). At greater aperture, the dancing decreases (for a qualitative explanation, see Fig. 7.1), and a blurred image is seen, which is, in fact, a changing pattern of speckles, each of which is comparable to an Airy disk, with a lifetime of a few hundredths of a second. The aperture of transition from one effect to the other depends on atmospheric conditions, and is generally found between 10 cm and 1 m. Image motion, however, would not cease completely until the aperture is much larger than the outer scale of the turbulence (which could mean hundreds of meters or more!). Theory predicts, and observation confirms (Young 1971), that the area of the seeing disk is nearly proportional to the air mass.

Since the refractive power of the atmosphere decreases with increasing altitude, and because the deviation increases linearly with distance, the source of seeing is likely situated nearer to the ground, while scintillation is produced at higher elevations above the ground.

Scintillation and seeing show little correlation: it is possible to have poor seeing together with negligible scintillation, but the reverse situation cannot occur.

The following paragraphs contain a brief, qualitative, description of the effects of scintillation and seeing on image quality.

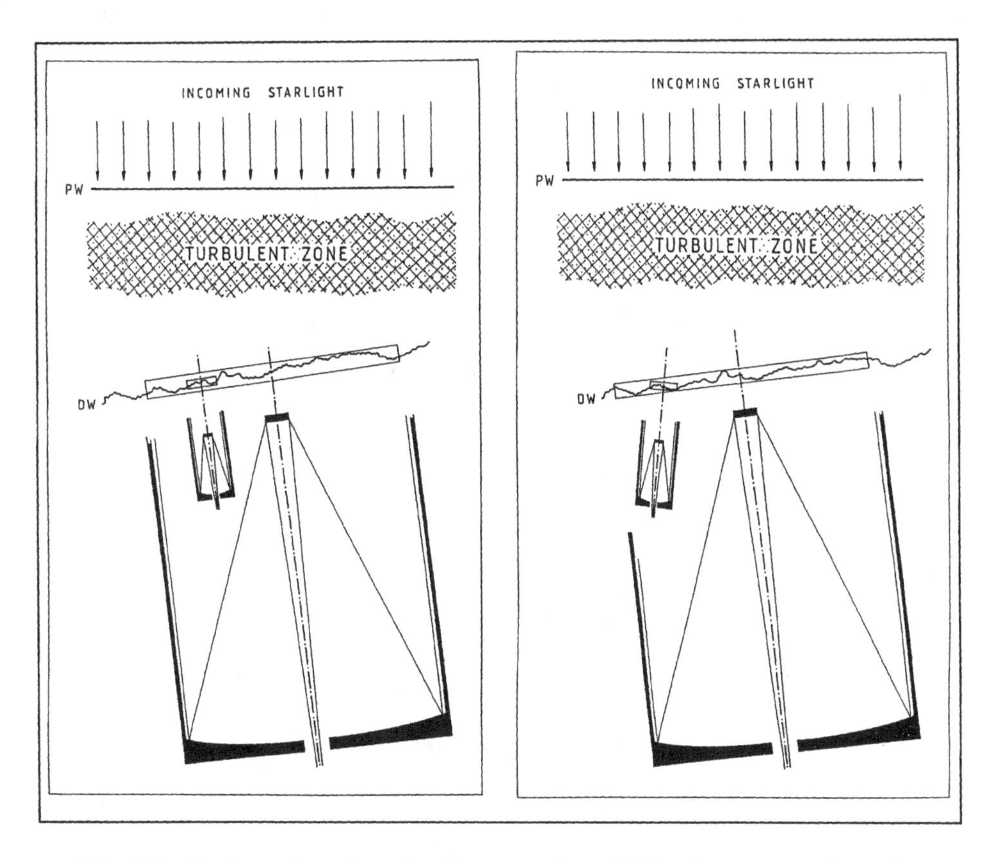

Fig. 7.1 *Seeing: distortion of a plane wavefront (PW) after passage through a zone of turbulence. Left and right diagrams describe two situations separated by a short time interval during which the disturbed wavefront (DW) has traveled a short distance. The average direction of the wavefront entering the telescope determines the apparent direction of a star. The optical axes of both large and small telescopes point in the direction where the star, centered in the focal-plane diaphragm, can be observed. In practice, of course, the telescope is pointed at a fixed position, and it is the stellar image that is meandering due to the continuously varying tilt of the distorted wavefront. The angular displacement is much larger in the case of a small telescope than it is for a larger one, because a telescope of small aperture sees only a small portion of the disturbed wavefront.*

7.1 Scintillation

Scintillation produces a (rapidly varying) *shadow pattern* at the ground, the individual dark and light bands being several centimeters across. When one looks at the entrance pupil of a telescope, one sees a moving pattern of bright and dark patches, with characteristic spacing of the order of 10 cm. Again, we find that the mean amplitude of scintillation has a larger disturbing effect for small telescopes than for large-aperture instruments that merge many light and dark patches.

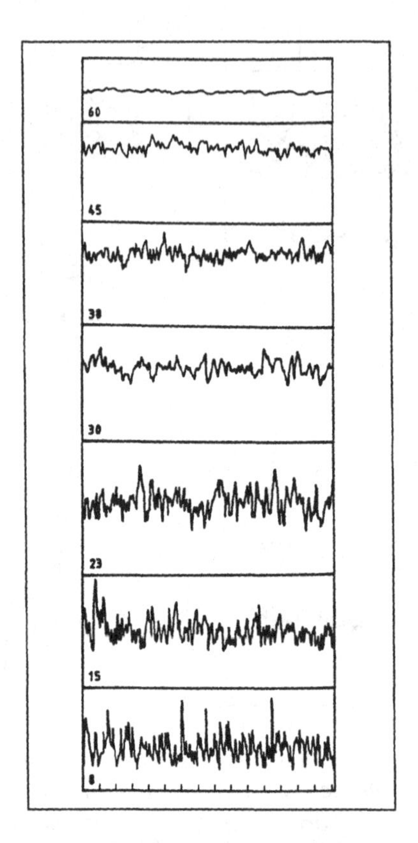

Fig. 7.2 *Scintillation for various telescopic apertures (expressed in cm). The x-axis scale is 0.1 sec (after Warner 1962).*

Planets twinkle less than stars, because planets show an apparent diameter, hence their light rays are not parallel, and as long as the changes in the direction of the light paths are smaller than the apparent diameter, their twinkling will be hardly noticeable. Scintillation of a planet becomes visible as soon as the changes in direction of the light rays are of the same order of magnitude as the apparent diameter.

Warner (1962) photometrically observed Deneb with a 24-inch telescope using stops of 18, 15, 12, 9, 6 and 3 inches in decreasing order during a transparent night of good quality. Typical tracings, obtained with the various apertures, are shown in Fig. 7.2. The amplitude of light variability is only 10% of the mean light level for the 24-inch aperture, and increased to about 130% at the 3-inch aperture, with a dramatic increase in amplitude when stopping down from 24 to 18 inch. Warner finds that the amplitude

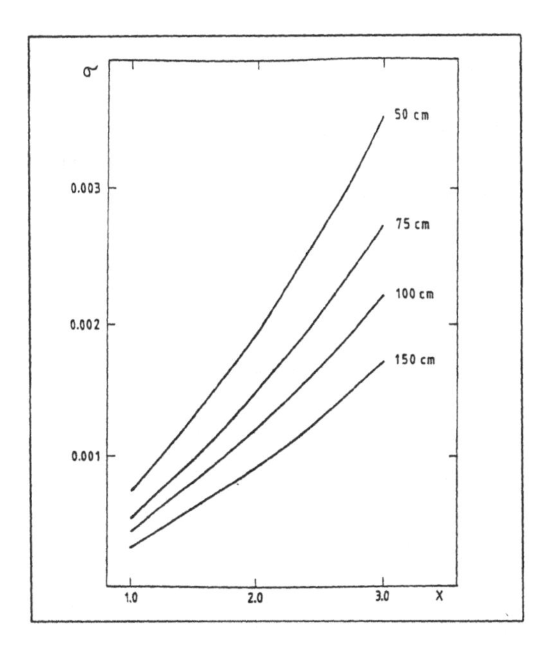

Fig. 7.3 *Scintillation. rms scintillation error σ (mag) to be expected under typical conditions as a function of air mass X for telescope apertures of different sizes (integration time 10 seconds).*

of variation is inversely proportional to the 3/5 power of aperture. Note that with the changes in light amplitude there are associated changes in temporal frequency.

Figure 7.3, based on Table II of Young (1967b), gives the rms scintillation error to be expected under typical conditions as a function of air mass for telescope apertures of different sizes and integration time 10 seconds.

Scintillation effects can only be compensated using a two-star photometer at one single telescope, since the amplitude of the scintillation depends critically upon the closeness of the two light paths.

Scintillation, just as photon noise, is a random error. The amplitude of scintillation noise is nearly independent of wavelength, but depends on air mass (Rufener 1964, Stock 1968, Young 1974b): it is roughly proportional to the square of air mass in the wind azimuth and to the $3/2$ power of it across the wind. It is also proportional to the $-2/3$ power of telescope aperture,* directly proportional to the wind speed, and inversely proportional to the square root of the integration time.

Young (1967b) gives a table of apparent magnitudes at which scintillation and photon noise are equal for various sizes of telescopes, altitudes and zenith distances (see also Young 1974b).

* Consequently, the weight of a photometric observation, when photon noise is negligible, varies as the 4/3 power of telescope aperture.

7.2 Seeing

Because refraction depends on air density and on the degree of turbulence, the effect is largest in the lower regions of the atmosphere. In clear weather conditions the seeing effect is maximum around noon and is much less around sunset and sunrise. Amplitude is a couple of arc sec; the largest amplitudes occur at frequencies around a few Hz. In autoguided exposures, this image motion sets a lower limit to the time constant of the telescope guiding-system: the guiding should be such that high-frequency atmospheric image motion does not interfere. In aperture photometry, the observer may be forced to adapt the size of the focal-plane diaphragm.

A large contribution to seeing is to be found in the interface layer between the troposphere and the stratosphere, the tropopause (at altitudes of 7 to 15 km) where waves easily develop. Above the tropopause, the reduced density of air is unable to give a substantial effect. Consequently, as far as the *free atmosphere* is concerned, the improvement in the seeing condition is regular, but relatively slow, up to the altitude of the highest mountain peaks. However, seeing is strongly affected by the local topography and microclimate and by the immediate surroundings, so the free-atmosphere conditions are only part of the problem:

- The Sun's thermal radiation is absorbed by the Earth's surface during daytime and is released at night. Hence, the air in contact with the ground is heated in daytime and cooled at night, in a larger proportion than is the free atmosphere. The ground layer is thus affected by bad seeing. This effect strongly depends on the nature of the local surfaces: it is small for lawn, or aluminum- and white-painted surfaces, and very high for asphalt covers.

- Valleys and low plains get a specially thick ground layer because the air cooled at night on the surrounding hills and mountains flows down the slope into the lower areas, creating the *inversion layer* discussed below.

- A temperature inversion (a warmer layer of air lying above a colder layer) above the observing site often gives awful seeing because waves form at the interface layer where the temperature gradient is highest. However, above the inversion layer, the conditions are very stable and provide excellent seeing.

- The ground layer on mountain sites is very thin, but it is surmounted by a turbulent zone (a few tens of meters thick) where the wind directly interacts with the ground layer, and which causes severe seeing problems. On top of this layer, another layer (thickness of 10 to 20% of the mountain height) constitutes a transition zone to the free atmosphere. Because of their respective thickness and activity, the windlofting region, the transition zone and the free atmosphere contribute more or less equally to the total seeing effect.

- The upwind side of a mountain suffers less from turbulence than does the down-wind side.

- Proximity to the ocean and isolation from other summits in a mountain range have positive influence on seeing.

- Temperature gradient and, hence, turbulent conditions often originate inside the dome and even inside the telescope. Turbulence is also created by the wind flow around the dome or telescope shelter.

European astronomers generally use a seeing scale where 1 corresponds to excellent conditions, 2 and 3 mean good to fair, and 4 and 5, respectively, are poor to awful. American astronomers prefer a scale going up to 10, but with 10 meaning perfect conditions. Modern measurement techniques allow more objective estimates of the seeing in terms of the actual profile of the images.

Chapter 8 Color transformation

8.1 Introduction

The *color transformation* is a crucial step in the analysis of photometric data. Once the data have been reduced to outside the Earth's atmosphere they have to be compared with standard data obtained at other sites with other telescopes, other filters and other detectors. Even when one works in a well-established system, such as the Johnson-Morgan *UBV* system (see Chapter 16), the instrumental responses of various equipments are seldom identical, i.e., the ratio of the functions $s(\lambda)$ is not a constant (this ratio is rarely unity because of differences in the collecting areas, in the filter transparency, and in the detector efficiency, but this is of secondary importance, since it translates directly into a scaling factor or a magnitude zero-point).

Two major color transformations are usually encountered. Most often one has to transform data to the standard system for which the equipment has been designed; such transformations we call *normal* transformations. But sometimes one wants to estimate the transformation between two completely different systems, for instance between *UBV* and *uvby*; such transformations we name *special* transformations.

There are also two practical situations according to whether the filter bandwidth is narrow or wide. In the first instance the importance of spectral features grows; widening the passband, on the other hand, increases the importance of broad spectral features, i.e., of the continuum (including the atmospheric continuum). Evidently, narrow bands will lead to more intractable problems. Those difficulties are not often realized. Let us consider strictly monochromatic photometry. Our measurements are done at wavelength λ_1, while the standard system is defined at λ_2. How could we perform a transformation between magnitudes measured at different wavelengths, without knowing the spectrum of every star? It is clear that this is impossible with single-wavelength data. Multi-filter photometry is necessary but, of course, one has to limit the number of passbands to a minimum to keep the distinction between photometry and spectrophotometry, and to preserve the feasibility of performing measurements.

8.2 Narrow-band photometry

8.2.1 Blackbody radiation

It is interesting to evaluate the color transformation for blackbodies, since many stellar sources have spectra resembling this ideal emitter.

Blackbodies are ideal thermal radiators in the sense that they emit a continuous spectrum whose characteristics are completely specified by the temperature. That is the case for any emitter (whatever its color) which is at a uniform temperature and in thermodynamic equilibrium with its own radiation field.

The spectral radiance of a blackbody at temperature T is given by Planck's equation

$$L(\lambda) = \frac{2hc^2}{\lambda^5}\left[\exp\left(\frac{hc}{\lambda kT}\right) - 1\right]^{-1} \tag{8.1}$$

where h is Planck's constant, and $k = 1.38062 \times 10^{-23}$ JK^{-1} is Boltzmann's constant.

Figure 8.1 illustrates the spectral distribution of blackbody radiation; the curves have the following characteristics:

- $L(\lambda)$ increases at all wavelengths with increasing temperature;
- the peak of the curve shifts toward shorter wavelength for higher temperature;
- $\Delta L/\Delta T$ is greater at the blueward side of the wavelength of maximum radiance λ_m;
- in a $\log\lambda - \log L$ diagram the shape of the curve is invariant with T. In other words: the curve at any temperature can be obtained by shifting one curve up along the straight line connecting the maxima.

Two well-known approximations to Planck's law follow from Eq. (8.1), i.e.,

a) if $hc/\lambda kT \ll 1$, Eq. (8.1) becomes

$$L(\lambda) \simeq 2ckT\lambda^{-4} \tag{8.2}$$

This relation is known as the Rayleigh-Jeans approximation, valid at longer wavelengths.

b) if $hc/\lambda kT \gg 1$, Eq. (8.1) becomes

$$L(\lambda) \simeq 2hc^2(\lambda)^{-5}\exp(-hc/\lambda kT) \tag{8.3}$$

which is known as Wien's approximation, valid at shorter wavelengths.

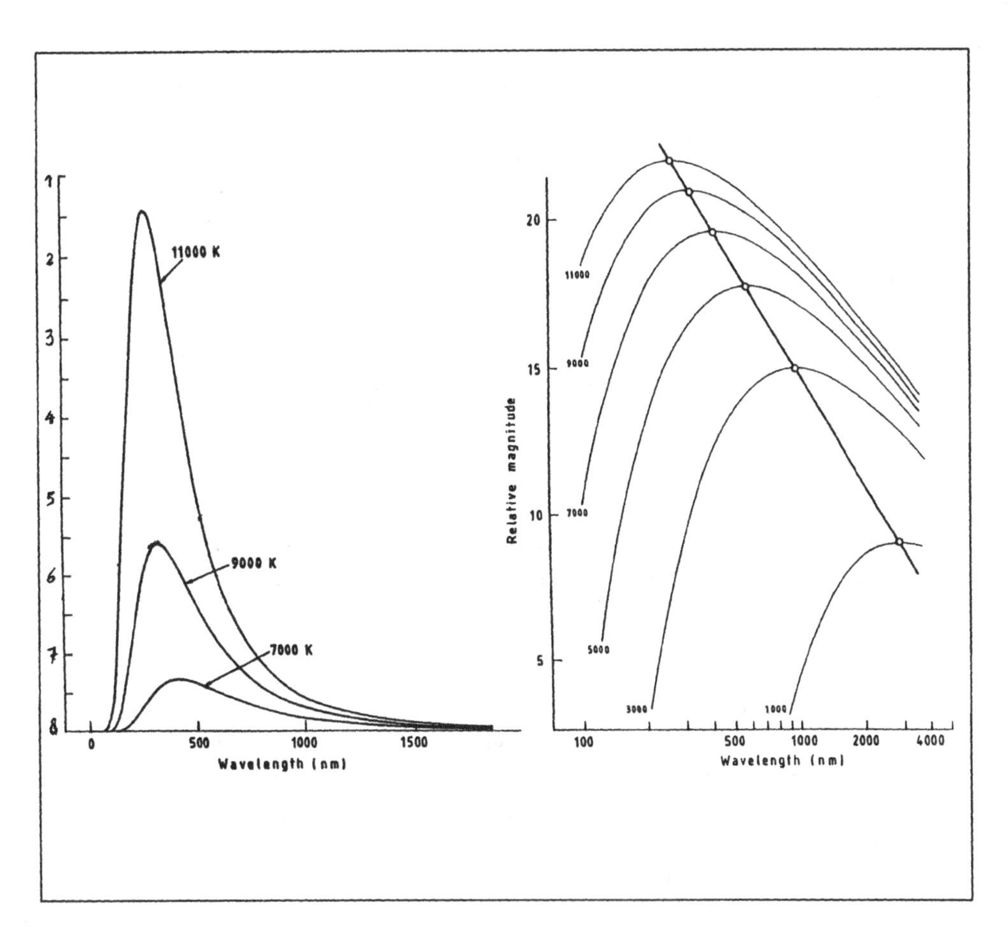

Fig. 8.1 *Blackbody radiance for different temperatures. Relative radiance (in decimal log and in magnitude). The straight line connecting the maxima represents Wien's displacement law, which is the curve* $\lambda_m = 2898/T$ *(*λ *in* μm*) obtained by differentiation of Eq. (8.1).*

In the absence of any extinction, the monochromatic magnitude of a blackbody at temperature T is given by

$$m(\lambda) = 12.5 \log \lambda + 2.5 \log \left[\exp\left(\frac{hc}{\lambda kT}\right) - 1\right] + C(\lambda) \tag{8.4}$$

The slowly varying $\log \lambda$ term is essentially a *zero-point* contribution since it depends only on the wavelength, and not on characteristics of the energy source or of the atmosphere. Therefore it can be incorporated into C, together with the sensitivity of the equipment which is also a function of λ. At the wavelengths of interest we often are in the conditions of the Wien approximation ($hc/\lambda kT \gg 1$) so that Eq. (8.4) becomes

$$m(\lambda) = 2.5(\log e)\left(\frac{hc}{\lambda kT}\right) + C(\lambda) \tag{8.5}$$

Hence the magnitude includes a linear function of $1/\lambda$ with a proportionality factor depending on the temperature of the blackbody.

Similarly, a color index defined between wavelengths λ_1 and λ_2 includes a linear term in $(1/\lambda_2 - 1/\lambda_1)$,

$$m(\lambda_2) - m(\lambda_1) = 2.5(\log e)\frac{hc}{kT}\left(\frac{1}{\lambda_2} - \frac{1}{\lambda_1}\right) + C(\lambda_2) - C(\lambda_1) \qquad (8.6)$$

Let us assume that a standard system is established, with wavelengths λ_3 and λ_4, and that the wavelengths of our instrumental system with which we measure blackbodies are λ_1 and λ_2. Presumably, we have $\lambda_1 \simeq \lambda_3$ and $\lambda_2 \simeq \lambda_4$, expressing the fact that we tried to reproduce the standard system with our equipment. In the case of the Wien approximation, however, this restriction is not necessary as we shall see now.

A relation giving $m(\lambda_3)$ is derived from Eqs. (8.5) and (8.6):

$$m(\lambda_3) - m(\lambda_1) = \frac{\lambda_3^{-1} - \lambda_1^{-1}}{\lambda_2^{-1} - \lambda_1^{-1}}[m(\lambda_2) - m(\lambda_1)] + C \qquad (8.7)$$

Again all zero-point terms have been grouped in C which is a function of all λ_1, λ_2 and λ_3. The color tranformation necessary to bring the magnitude $m(\lambda_1)$ of a blackbody to the $m(\lambda_3)$ system may thus be written

$$m(\lambda_3) = m(\lambda_1) + A_3\,[m(\lambda_2) - m(\lambda_1)] + C_3 \qquad (8.8)$$

Constant A_3 can be calculated from the values of the wavelengths, while the zero-point constant C_3 would require a more thorough knowledge of the observing equipment. The best way is to determine both constants empirically from observations of at least two blackbodies of different temperatures (blackbodies at same temperature would yield identical indices, as indicated by Eq. (8.6)).

We can write in the same manner

$$m(\lambda_4) = m(\lambda_2) + A_4\,[m(\lambda_2) - m(\lambda_1)] + C_4 \qquad (8.9)$$

where

$$A_4 = \frac{\lambda_4^{-1} - \lambda_2^{-1}}{\lambda_2^{-1} - \lambda_1^{-1}} \qquad (8.10)$$

Equations (8.8) and (8.9) lead to the color-transformation formula for indices

$$\begin{aligned} m(\lambda_4) - m(\lambda_3) &= (1 + A_4 - A_3)\,[m(\lambda_2) - m(\lambda_1)] + C_4 - C_3 \\ &= A_{43}\,[m(\lambda_2) - m(\lambda_1)] + C_{43} \end{aligned} \qquad (8.11)$$

where

$$A_{43} = \frac{\lambda_1\lambda_2(\lambda_4 - \lambda_3)}{\lambda_3\lambda_4(\lambda_2 - \lambda_1)} \qquad (8.12)$$

For small wavelength departures between both sets of filters ($\lambda_1 \simeq \lambda_3$, $\lambda_2 \simeq \lambda_4$), and for photometric systems with rather closely spaced bands ($|\lambda_2 - \lambda_1| \ll \lambda_1$), one has

$$A_{43} = \frac{\lambda_4 - \lambda_3}{\lambda_2 - \lambda_1} \qquad (8.13)$$

Large color corrections are most likely for very closely spaced bands.

8.2.2 Other spectral distributions

Let us emphasize that, except for the last point, the preceding results did not involve any assumption on the spacing between any of the wavelengths involved. They were only based on Wien's law for blackbody sources. The same reasoning applies equally well to any spectral distribution satisfying equations similar to (8.5), i.e., when the magnitude can be written as

$$m(\lambda) = f_1(\lambda)g(T, \ldots) + f_2(\lambda) \tag{8.14}$$

where the f's are functions of λ only, and g depends on the physical parameters of the emitters (temperature, etc.). Color indices can then be written as

$$m(\lambda_2) - m(\lambda_1) = [f_1(\lambda_2) - f_1(\lambda_1)]g(T, \ldots) + f_2(\lambda_2) - f_2(\lambda_1) \tag{8.15}$$

which is a generalization of (8.6).

It is obvious that Eq. (8.14) does not adequately represent the spectral distribution of most classes of astrophysical objects. Over a wide wavelength interval (Wien's approximation) it is satisfactory for blackbodies, as we noted above. One might think of other spectral distributions which could fit Eq. (8.14) over a more restricted domain. Using a limited Taylor expansion around $\lambda_0 \in [\lambda_a, \lambda_b]$, $(\lambda_b - \lambda_a \ll \lambda_0)$, one obtains the following equation

$$m(\lambda) = \left(\frac{dm}{d\lambda}\right)_{\lambda_0} (\lambda - \lambda_0) + m(\lambda_0) \tag{8.16}$$

which has the required form (note that $\left(\frac{dm}{d\lambda}\right)_{\lambda_0}$ depends on the physical parameters characterizing the sources). One can draw an important conclusion:

> *as long as* **all** *the functions we are dealing with in actual photometry (spectral distribution, sensitivity curves, extinction) satisfy the conditions of Taylor expansion, and as long as we stay within the domain where the expansion can be limited to the first-order terms, Eqs. (8.8) or (8.11) are valid.*

Unfortunately these requirements are rarely met in practice. The crucial point is that a single Eq. (8.14) (or a single Taylor development) should be valid at both sides of the interval—i.e., generally, but not necessarily, over the entire interval $[\lambda_a, \lambda_b]$. There should not be two different equations, one locally around (λ_1, λ_2) and another one around (λ_3, λ_4). If independent Taylor developments are performed around λ_1 and λ_2, with unrelated coefficients $\left(\frac{dm}{d\lambda}\right)_{\lambda_1}$, $m(\lambda_1)$, $\left(\frac{dm}{d\lambda}\right)_{\lambda_2}$ and $m(\lambda_2)$, then the parameters describing the stellar energy distribution cannot be eliminated and they enter the relation between the color indices.

Generally the energy distributions are not smooth. Seen at the spectral resolution of narrow-band photometry, spectral lines may represent abrupt discontinuities that do not permit limited Taylor expansion (the true bandwidth of the filter-detector response curve has to be convolved with the spectral distribution in order to exactly appreciate the importance of those lines). The wavelength intervals are generally (we might even say "always") much larger than what even a less careful mathematician would admit for a first-order Taylor series expansion. Photometric systems are designed with filters often separated by tens or hundreds of nanometers, leading to $|\lambda_b - \lambda_a|/\lambda_0 \simeq 0.1$. In a normal color transformation problem (e.g., between an instrumental system and the standard system), the instrumental wavelengths (λ_1, λ_2) are, hopefully, very close to the standard ones (λ_3, λ_4). However, λ_1, λ_2 are always too far apart and Eq. (8.8) thus becomes quite unfounded. In the special kind of color transformation that we mentioned in the introduction—a transformation between different systems—the wavelengths (λ_1, $\lambda_2, \ldots$) are often separated from (λ_3, $\lambda_4, \ldots$) by several tens of nanometers, and the basic conditions are even more violated.

Hence, more complex relationships between the color indices can be considered. They might be non-linear or/and they might involve three, or more, color indices. All this depends on how the magnitude $m(\lambda)$ can be related to the physical parameters of the energy source. The example discussed above shows that the correct implementation of general analytical relations can only be a matter of coincidence, and one should carefully check the applicability in the photometric system used. It is also clear that different classes of spectra (or astrophysical sources) would yield different relations. For instance quasi-black-body emitters and emission-line objects should certainly be treated with separate transformations.

The only safe way to deal with the color transformation problem is to comply with the conditions of the limited Taylor expansion, i.e., to consider closely spaced filters (relative to the bandwith). We arrive at the same kind of problems as with the extinction determination: *there actually exists no photometric system that provides such apparently redundant data.*

8.3 The general color-transformation problem

The distinction between monochromatic (narrow-band) and heterochromatic (wide-band) photometry is not as important for color tranformation as it is for atmospheric extinction. Contrarily to the case of atmospheric reddening, bandwidth effects are always present in color transformation problems: color transformation in narrow-band photometry can be affected by individual spectral lines, while broad-band photometry is affected by wider spectral features. The fundamental parameter to consider is the ratio between the bandwidth and the spacing in wavelength of the filters, i.e., *the relative wavelength sampling of the system.*

The discussion we gave of the monochromatic case is still valid here. There is, however, a practical difference: the relative sampling of broad-band photometry is higher than that of a monochromatic system. Most photometric systems have more or less equally spaced filters, while the bandpasses vary by a factor of ten or twenty, and so does the sampling. Consequently, errors due to the failure of the Taylor expansion over a given wavelength interval are much more likely to occur in monochromatic photometry than in heterochromatic photometry.

For heterochromatic photometry the problem is complicated by the integration over a wide range of wavelengths. Most often, the largest variations between various filters used to measure the same band in a given system are due to different shapes of the bandpasses, not to differing central wavelengths. Instrumental functions, such as the sensitivity and the filter transmission-curves, are often smooth, but this is not the case for high-transmission multiple-cavity interference filters, nor for slot-defined spectrophotometers. In order to gain transmission at the central wavelength—and consequently to make the systems more efficient—modern filter transmission curves are now steep-sided, but unfortunately, ripple-topped (see for example system 3 in Fig. 11.1), and this adds enormously to the complexity of the color-transformation problem. Consequently, the discussion relative to the monochromatic case should be replaced by a theory similar to that of the atmospheric extinction where the whole effect is due to profile variations.

We have already noted that out-of-the-atmosphere data are very difficult to evaluate in wide-band photometry. Assuming that the atmospheric extinction component is well understood and that it has been removed from the data, the basic color transformation can be carried out between extra-atmospheric values. The more logical way could be to go via the monochromatic magnitudes. The relation between heterochromatic magnitudes (effective wavelength λ_0) and the corresponding monochromatic magnitudes can be estimated from Eq. (6.18) (with slightly different notations),

$$\begin{aligned} m_{\text{hetero}}(h_1) - m_{\text{mono}}(\lambda_0, h_1) &= -2.5\log\left[E_0\left(1 + \frac{1}{2}\mu_2^2\frac{E_0''}{E_0}\right)\right] + 2.5\log E_0 + C \\ &\simeq -1.25(\log e)\mu_2^2\frac{E_0''}{E_0} + C \qquad (8.17) \end{aligned}$$

with all zero-point contributions grouped in the C terms.

If this equation can be solved, we obtain the monochromatic magnitude at λ_0. The same should be done for the other system (presumably the standard system, with which we calibrate our measurements) at a nearby wavelength λ_1. Both monochromatic magnitudes should then be compared according to the scheme discussed above, and this is a lengthy and cumbersome process. We already mentioned many difficulties arising in the monochromatic transformations, and we concluded that empirical recipes are required with the existing undersampled systems. We can only go further in that direction.

Relation (8.17) involves the second derivative E'' of the stellar spectrum and depends on the square of the instrumental bandwidth. Just as for the calculation of the heterochromatic extinction, this expression can only be used when a Taylor expansion is valid all over. The derivative should be evaluated locally, inside the bandpass. Once again, this is impossible unless a dense sampling of the spectrum is realized, which is not the case for any of the existing photometric systems.

Consequently, in practice there is no alternative to using empirical relations, involving color indices (which do not give correct representation of the local derivatives), over wide spectral ranges (which certainly cannot admit Taylor expansion all over).

Chapter 9 Interstellar extinction

9.1 Monochromatic extinction

Star counts reveal that the observed space density of stars decreases with increasing distance from the Sun. Since it is clear that the Sun is not at the center of a concentration of stars, the only explanation for the phenomenon is the assumption that light is being absorbed by an interstellar medium. The most convincing argument in support of that assumption came from Trumpler (1930) who found that the diameter of open clusters, calculated from the angular diameter of the cluster and from the distance moduli of the member stars, seemed to increase with distance. Obviously, the clusters are not larger, but they are closer than what the derived distances suggest, and the excess in inferred distance is caused by interstellar absorption not being taken into account. Trumpler derived a mean interstellar extinction coefficient of 0.7 magnitude per kpc in the photographic domain.

Halm (1917) already noted that the extinction in the photographic domain expressed in magnitudes was 22% larger than that in the visual domain: interstellar reddening is not uniform with respect to wavelength, but is selective. In other words, the energy distribution of a reddened star and of an unreddened star of the same spectral type and luminosity class is not simply put off by a translation (if it were, it would mean that interstellar absorption is neutral), but differs by an amount that grows with decreasing wavelength. It was soon apparent that the extinction obeyed a (now called) λ^{-1} law. Since the Wien approximation of stellar energy distribution roughly goes with λ^{-1}, it follows that stars reddened by interstellar extinction have approximately the same spectral intensity distribution as have unreddened stars of a lower temperature (the color temperature of a reddened star is lower than the temperature indicated by the degree of ionization in their spectra). Hence, observed color indices must be "de-reddened" before any analysis can be made.

There are widely varying amounts of reddening (distant globular clusters lying outside the galactic plane, for instance, are little reddened because of the concentration towards the galactic plane of the interstellar material).

The photometrist faces three distinct problems: (i) establishing the reddening curve, (ii) determining the possible variations in this curve for different parts of the sky that may result from variations in the optical properties of the interstellar medium, and (iii) dealing with the reddening effect in the process of data reduction.

The light emitted by a star is partly scattered and partly absorbed by interstellar dust before it reaches the upper atmosphere of the Earth. The phenomenon is quite analogous to atmospheric extinction by dust and aerosol particles. The most obvious differences for the photometrist arise from the fact that we do have some control on the atmospheric extinction: we know its average properties (height, density, composition, etc.) as a function of wavelength, we can observe from sites at various altitudes—eventually from above the atmosphere—and we can observe the same objects at various air masses from a same site in order to evaluate the amount of extinction. None of this is possible in the case of interstellar extinction. Study of interstellar extinction requires accurate photometric measurements in several bands of a large number of stars of known spectral type. Not only the amount of extinction must be determined, but also the dependence on galactic latitude, galactic longitude, and distance from the Sun.

The interstellar medium consists of two components: gas (largely atomic and molecular hydrogen), and dust (probably silicate mantles around graphite cores, or bare graphite particles, with sizes ranging from nm to μm). The interstellar absorption and scattering processes are mostly due to the dust component. Dust and gas are thoroughly mixed with a dust-to-gas mass ratio of the order of 10^{-2}. The structure of the interstellar medium is very clumpy, with clouds of various sizes (from parsecs to hundreds of parsecs) and densities (from 10^6 to 10^{12} atoms or molecules per cubic meter) in an intercloud medium of about 10^5 atoms per cubic meter. The densest clouds are concentrated in a thin sheet in the galactic plane. Absorption and scattering by dust limits observation to only a few kpc in the plane.

The degree of extinction (expressed in magnitudes) is proportional to the quantity of material along the line of sight, and in a homogeneous medium the extinction would be proportional to the distance traveled by the light. The extinction is commonly designated by A and is sometimes normalized per kiloparsec.

An equation like (6.1), written for the Earth's atmosphere, is also valid for the interstellar extinction, i.e.,

$$dL(\lambda, h) = \kappa(\lambda, h)\rho(h)L(\lambda, h)dh \tag{9.1}$$

h is the distance to the Earth, dh is the thickness of a slab of interstellar material perpendicular to the line of sight, $\kappa(\lambda, h)$ is the monochromatic absorption coefficient per unit mass and $\rho(h)$ is the specific mass of the absorbing medium.

By integrating (9.1) over the whole line of sight (ls) we obtain

$$L(\lambda, \text{Earth}) = L(\lambda, \text{star})e^{-\int_{\text{ls}} \kappa(\lambda,h)\rho(h)dh} \tag{9.2}$$

Function $\kappa(\lambda)$ is quasi universal. It is known to vary little from cloud to cloud, although some notable exceptions are known. Hence Eq. (9.2) may be rewritten as

$$L(\lambda, \text{Earth}) = L(\lambda, \text{star})e^{-\kappa(\lambda)\int_{\text{ls}} \rho(h)dh} \tag{9.3}$$

and,

$$\log L(\lambda, \text{Earth}) = \log L(\lambda, \text{star}) - (\log e)\kappa(\lambda) \int_{\text{ls}} \rho(h)dh \tag{9.4}$$

Let us call m_{obs} the observed magnitude of a star and m_{int} its "intrinsic" magnitude, i.e., corrected for the interstellar extinction. From Eq. (9.4) we obtain

$$\begin{aligned} m_{\text{obs}}(\lambda) &= m_{\text{int}}(\lambda) + 2.5(\log e)\kappa(\lambda) \int_{\text{ls}} \rho(h)dh \\ &= m_{\text{int}}(\lambda) + A(\lambda, \text{ls}) \end{aligned} \tag{9.5}$$

The apparent magnitude is thus increased by an amount $A(\lambda, \text{ls})$, which is called *total absorption* (one commonly writes, e.g., A_V for total absorption in the V-band of the Johnson-Morgan system). Note that we do not normalize A to 1 kpc.

The amount of interstellar absorption A is expressed as the product of the absorption coefficient $\kappa(\lambda)$, which is only a function of wavelength, and of a factor depending on the quantity of absorbing material distributed along the line of sight. Let us define the function

$$\zeta(\lambda) = \kappa(\lambda)/\kappa(\lambda = 550\text{nm}) \tag{9.6}$$

which describes the interstellar absorption law. The normalization is such that $\zeta(\lambda) = 1$ in the "visible" domain, i.e., in the Johnson-Morgan V or an equivalent band. For reasons readily apparent, and which are detailed below, it is wise to stay with monochromatic bands, but the use of Johnson-Morgan's V band is deeply entrenched for historical reasons.

We write the extinction correction as

$$A(\lambda, \text{ls}) = \zeta(\lambda) A_1(\text{ls}) \tag{9.7}$$

with

$$A_1(\text{ls}) = 2.5(\log e)\kappa(\lambda = 550\text{nm}) \int_{\text{ls}} \rho(h)dh \tag{9.8}$$

$A_1(\text{ls})$ is a function of the location of the observed star. Simple models of the distribution of the absorbing medium in the Galaxy allow derivation of an average law (see, e.g., Jaschek & Jaschek 1987).

There exist maps giving the absorption as a function of l (the galactic longitude), b (the galactic latitude) and the distance r to the star (see, e.g., Neckel 1966, Neckel & Klare 1980). They are approximations of the three-dimensional cloud distribution in the galactic disk, in the vicinity of the Sun. Fig. 9.1 gives the galactic distribution of dust up to 3 kpc. Big complexes of dust extending up to 1 kpc are apparent. The complexes are separated by areas which are nearly dust-free.

The reddening curve $\zeta(\lambda)$, in different parts of the sky, is obtained from comparison of spectrophotometric measurements of reddened and unreddened (i.e., nearby with known distance) pairs of stars of the same spectral type, which are in the same direction. The curve is quasi-linear in $1/\lambda$ over the visible part of the spectrum (Fig. 9.2).

The function $\zeta(\lambda)$ is tabulated in Table 9.1. The wavelength dependency is very near $1/\lambda$ from the UV to the IR. This is shown in Fig. 9.2. Note, however, that this law

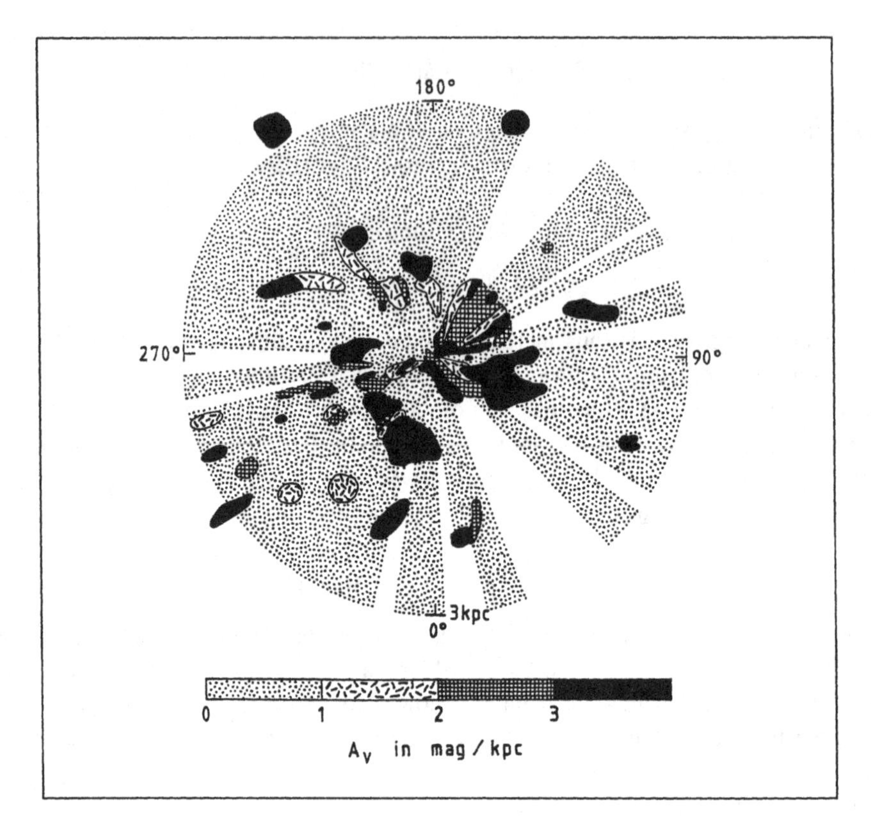

Fig. 9.1 *Map of galactic dust distribution up to 3 kpc. Shaded areas cover four levels of interstellar absorption per kpc (A_V/kpc) (after Neckel & Klare 1980).*

does not hold exactly, and that there are significant deviations from linearity at long and especially at short wavelengths, as can be seen in Fig. 9.2.

Equations (9.5) and (9.7) give

$$m_{\rm obs}(\lambda) = m_{\rm int}(\lambda) + \zeta(\lambda) A_1(\rm ls) \tag{9.9}$$

which is analogous to Eq. (6.10) obtained for the atmospheric extinction. ζ plays the role of the extinction coefficient, and A_1 that of the air mass.

9.1.1 Reddening vector

Consider now a set of wavelengths $\{\lambda_i\}$ and the corresponding monochromatic magnitudes $\mathbf{m} = \{m(\lambda_i)\}$. From Eq. (9.9) we have

$$\mathbf{m}_{\rm obs} = \mathbf{m}_{\rm int} + \mathbf{Z} A_1(\rm ls) \tag{9.10}$$

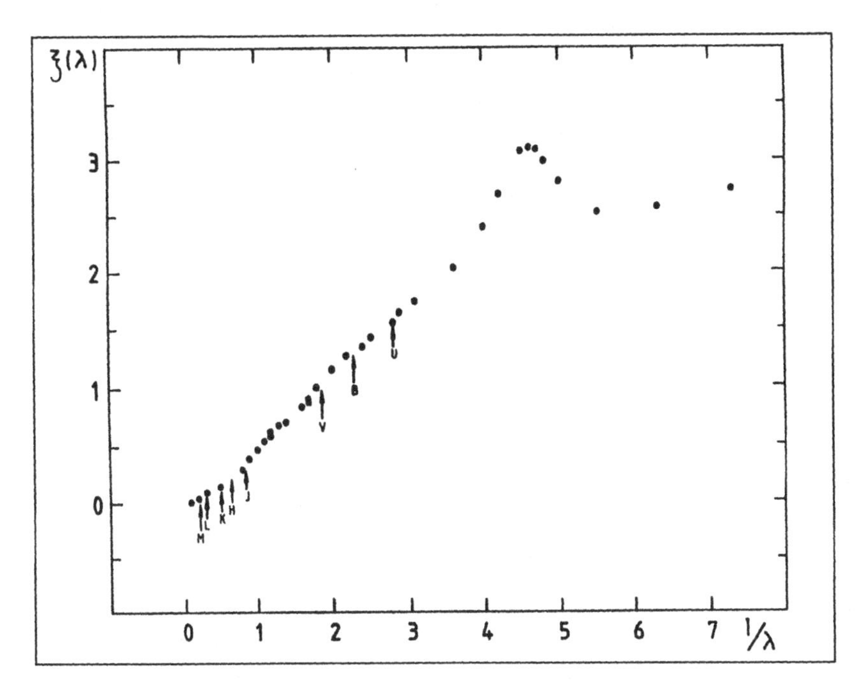

Fig. 9.2 *Interstellar extinction curve. A plot of ζ as a function of $1/\lambda$. It shows the wavelength dependence of the extinction produced by interstellar matter. The positions of the* UBVJHKLM *passbands are indicated.*

where $\mathbf{Z} = \{\zeta(\lambda_i)\}$, for $i = 1$ to n, is a vector depending only on the wavelengths $\{\lambda_i\}$. For any star, the reddening correction will be a vector parallel to $\mathbf{Z}$, and its norm will be proportional to $A_1(\text{ls})$

$$|\mathbf{m}_{\text{obs}} - \mathbf{m}_{\text{int}}| = |\mathbf{Z}| A_1(\text{ls}) \tag{9.11}$$

Instead of magnitudes one may also consider color indices (see Section 1.15)

$$\mathbf{c} = \{c_i\} = \{\sum_j a_{ij} m(\lambda_j)\} \qquad (i = 1 \text{ to } n) \tag{1.30}$$

We get

$$\begin{aligned} \mathbf{c}_{\text{obs}} &= \mathbf{c}_{\text{int}} + \mathbf{Z}^\star A_1(\text{ls}) \\ &= \mathbf{c}_{\text{int}} + \mathbf{E} \end{aligned} \tag{9.12}$$

where

$$\mathbf{Z}^\star = \{\sum_j a_{ij} \zeta(\lambda_j)\} \qquad (i = 1 \text{ to } n) \tag{9.13}$$

Again we have a reddening direction which is the same for any star in the n-dimension space of color indices.

Table 9.1 The interstellar absorption law $\zeta(\lambda)$ (from Scheffler 1982, and Schild 1977).

λ (nm)	$\zeta(\lambda)$	λ (nm)	$\zeta(\lambda)$	λ (nm)	$\zeta(\lambda)$
137.5	2.73	340.0	1.64	809.0	0.62
159.0	2.57	363.6	1.56	844.6	0.58
183.0	2.53	403.6	1.43	871.0	0.54
200.0	2.80	425.5	1.35	970.0	0.47
208.0	2.98	456.6	1.26	1061	0.40
214.0	3.09	500.0	1.16	1087	0.38
219.0	3.10	548.0	1.00	1250	0.30
223.0	3.07	584.0	0.91	2200	0.15
236.0	2.68	605.0	0.88	3400	0.10
250.0	2.39	643.6	0.83	4900	0.05
274.0	2.03	710.0	0.71	8700	0.01
320.0	1.74	755.0	0.68	10000	0.01

This simple relation leads us to a way of determining the amount of interstellar extinction. The intrinsic—unreddened—color indices of astronomical objects are not evenly distributed in the $\{c_i\}$ space. They form subgroups or sequences, for instance a line in a 2-D space (e.g., the main sequence in the HR diagram) or a surface in a 3-D space. Reddening moves the representative points out of such sequences, according to the amount of absorbing material through which the objects are seen. Hence the intrinsic colors can be found at the intersection of the area of the subgroup and the line drawn from the reddened position parallel to the reddening vector and the subgroup (see Fig. 9.3).

The intersection may consist of several values or even a whole range when the vector is tangential to the subgroup, or when the width of the latter is substantial (see Fig. 9.4). A quality of a photometric system is to ensure that such situations do not occur for the classes of objects for which the system is designed.

9.1.2 Color excess

The quantity $\mathbf{m}_{\text{obs}} - \mathbf{m}_{\text{int}}$ is the *magnitude excess* produced by the interstellar absorption, and we called it $\mathbf{A}$ (cf. 9.10). Similarly, the quantity $\mathbf{E} = \mathbf{Z}^{\star} A_1(\text{ls})$ appearing in Eq. (9.12) is the *color excess*.

$$\begin{aligned}\mathbf{E} = \mathbf{Z}^{\star} A_1(\text{ls}) &= \{\sum_j a_{ij}\zeta(\lambda_j)\} A_1(\text{ls}) \\ &= \{\sum_j a_{ij} A(\lambda_j)\} \end{aligned} \tag{9.14}$$

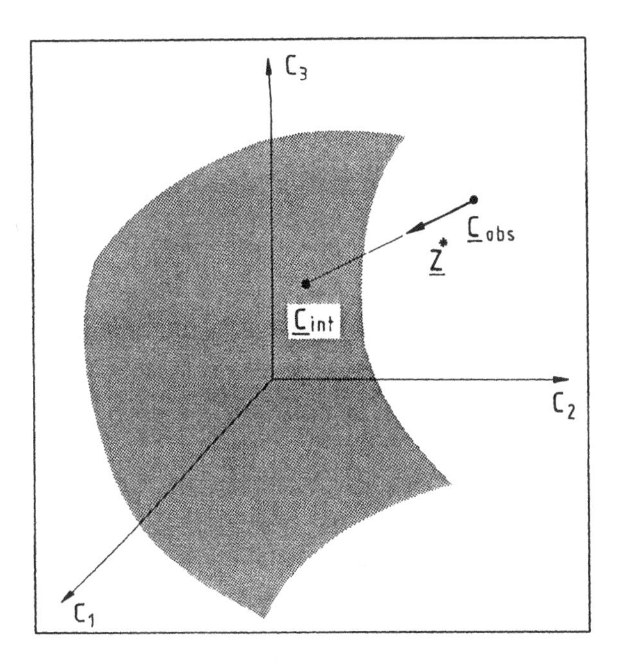

Fig. 9.3 *Measuring the interstellar reddening. In a three-color index diagram the intersection of the reddening vector* $\mathbf{Z}^\star$ *and of the surface representative of the unreddened stars gives the intrinsic colors* $\mathbf{c}_{\mathrm{int}}$ *of a star observed at* $\mathbf{c}_{\mathrm{obs}}$.

For example in the Strömgren system (see Section 16.4.5) one would write relations like

$$A_u = u_{\mathrm{obs}} - u_{\mathrm{int}} \tag{9.15}$$

and

$$A_u = A_1(\mathrm{ls})\zeta_u \tag{9.16}$$

Similarly, for two-color indices there exists a color excess which is denoted by E,

$$\begin{aligned} E_{b-y} &= (b-y)_{\mathrm{obs}} - (b-y)_{\mathrm{int}} \\ &= A_1(\mathrm{ls})\zeta^\star_{b-y} \\ &= A_b - A_y \end{aligned} \tag{9.17}$$

It is exactly the color excess that will allow us to determine the total interstellar absorption: when the optical properties of the interstellar matter are everywhere alike, the interstellar absorption will only depend on the differences in amounts of absorbing material, and the ratio of total-to-selective absorption (see Eq. (9.29)) will be constant.

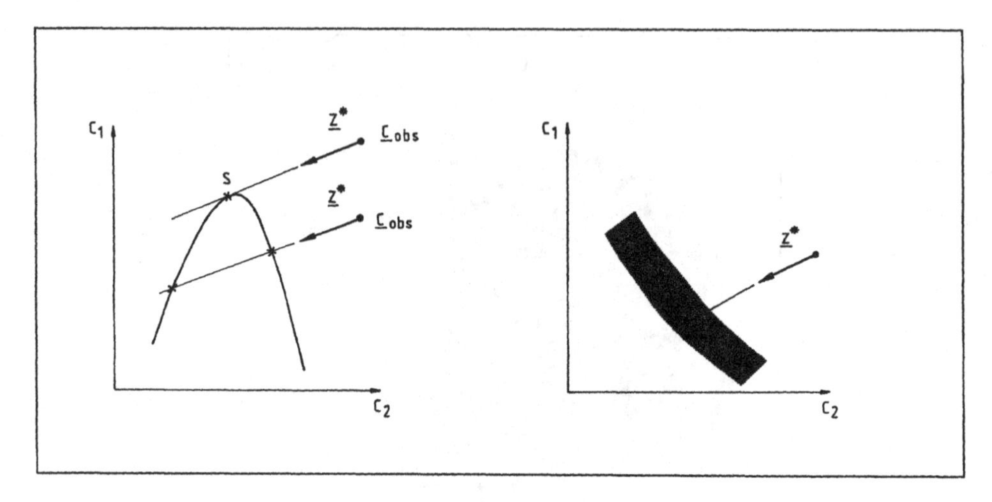

*Fig. 9.4 Measuring the interstellar reddening. In problematic cases the intersection of the reddening vector and the unreddened stellar sequence may be almost tangential or may even not yield a unique solution (*left part*); the intersection or the sequence may also have considerable scatter (*right*) (illustrated for a 2-color space).*

9.1.3 Reddening-free indices

Reddening-free indices $\{c_i\}$ can be constructed in an infinite number of ways. The coefficients a_{ij} appearing in Eqs. (1.30) and (9.13) have to satisfy the relation

$$\mathbf{Z}^{\star} = \{\sum_j a_{ij}\zeta(\lambda_j)\} = 0 \tag{9.18}$$

However, conditions such as (9.18) may come into conflict with the requirement that the a_{ij} must yield color indices which physically are as significant as possible. Relations (1.31) must also be verified. This condition is automatically met when the reddening-free indices are build as linear combinations of preexisting indices.

We now consider two typical examples of reddening-free indices.

Sometimes indices are built specifically to measure the strength of a spectral line. Two filters are used *with the same central wavelength* λ, but with different bandpasses. The color index is defined as

$$c = m(\lambda_1) - m(\lambda_2) \tag{9.19}$$

so that

$$\zeta^{\star} = \zeta(\lambda_1) - \zeta(\lambda_2) = 0 \qquad (a_1 = -a_2 = 1) \tag{9.20}$$

The index so defined is automatically reddening-free. This is the case of the β index of the Strömgren system (see Section 16.4.5), which is used to evaluate the strength of that Balmer line.

The so-called m_1 index of the Strömgren *uvby* system is defined in Section 16.4.5 (Eq. 16.6) as $m_1 = v - 2b + y$ (the symbol m_1 used for this specific color index should not be confused with the usual notation for magnitudes). Observations and computer simulations have shown (Crawford & Mandwewala 1976) that

$$E_{m_1} / E_{b-y} \approx -0.33 \tag{9.21}$$

Consequently the quantity

$$[m_1] = m_1 + 0.33(b-y) \tag{9.22}$$

is reddening-free. Indeed one can write

$$\begin{aligned}
[m_1] &= (m_1)_{\rm obs} + 0.33(b-y)_{\rm obs} \\
&= (m_1)_{\rm int} + E_{m_1} - E_{m_1} / E_{b-y} \left((b-y)_{\rm int} + E_{b-y}\right) \\
&= (m_1)_{\rm int} - E_{m_1} / E_{b-y}(b-y)_{\rm int} \\
&= [m_1]_{\rm int}
\end{aligned} \tag{9.23}$$

From Eqs. (16.6) and (9.21) we deduce

$$[m_1] = v - 1.67b + 0.67y \tag{9.24}$$

so that, from Eq. (9.18)

$$\zeta^{\star}_{[m_1]} = \zeta(\lambda_v) - 1.67\zeta(\lambda_b) + 0.67\zeta(\lambda_y) = 0 \tag{9.25}$$

Eq. (9.24) defines only one of the many reddening-free indices that can be constructed for the *uvby* system. Others are $[c_1]$, $[u-b]$ and c_0. The latter is derived from the observed $c_1 = u - 2v + b$ (see Section 16.4.5, Eq. 16.5) by means of the iterative procedure of Crawford (Crawford & Barnes 1974).

9.2 Wide-band extinction

In a first approximation the effects of interstellar absorption in wide-band photometry are similar to those present in the case of narrow-band photometry. The wavelength to be considered is the mean wavelength as defined in Eq. (1.34). Wide passbands manifest themselves in the same way as the atmospheric extinction: there is a dependency on the shape of the stellar spectrum. Red stars undergo less extinction than blue stars because proportionally more energy is concentrated in the long wavelength part of the band. The non-linear atmospheric extinction, known as Forbes effect, is also present here. As the amount of interstellar extinction increases, the radiation is more reddened and more monochromatic. Hence additional absorption has less and less influence.

The reddening vector is a function of the spectrum and, consequently, of the position of the star in the color diagram (this correspondance is not univocal: different spectra can give the same indices). Moreover, the vector can only be defined for moderate values of the extinction because of the non-linearity effect. A curvature term may be included for larger extinction. A detailed analysis in the *UBV* system was carried out by by Blanco (1953, 1955) for different spectral classes and total amounts of absorption. He derived the relation

$$\frac{E_{U-B}}{E_{B-V}} = 0.76 + 0.03 A_V \tag{9.26}$$

which clearly shows that the reddening vector does not keep a constant direction when the absorption increases.

Hence Eq. (9.9) is only approximately valid for wideband systems. It is commonly written as

$$m_{\rm obs}(\lambda) = m_{\rm int}(\lambda) + \zeta(\lambda) A_V \tag{9.27}$$

where A_V is the interstellar extinction in the V band of Johnson-Morgan's system, at $\lambda \approx 550$ nm. Hence

$$V_{\rm obs} = V_{\rm int} + A_V \tag{9.28}$$

Since A_V depends on the spectrum of the object, the use of Eq. (9.27) to define $\zeta(\lambda)$, i.e., to normalize the extinction law, is not perfectly legitimate. The same can be said of a quantity which is frequently used to characterize the interstellar law

$$\mathcal{R} = A_V / E_{B-V} \tag{9.29}$$

where B is the blue band of Johnson-Morgan's system. $\mathcal{R}$ is commonly called *ratio of total to selective absorption.*

The distance modulus, and hence the distance, of an object then follows from Eqs. (1.27) and (9.29)

$$m_V - M_V = \mathcal{R} E_{B-V} + 5 \log D - 5 \tag{9.30}$$

Let us note that the reddening law is often represented by the function

$$F(\lambda) = (A_\lambda - A_V)/(A_B - A_V) \tag{9.31}$$

which is just another normalization of $\zeta(\lambda)$, giving 0 at λ_V and 1 at λ_B. The disadvantage of this function is that it is based on wide bands. The relation between both definitions—if we treat Johnson-Morgan's bands as monochromatic—is simply

$$F(\lambda) = \mathcal{R}(\zeta(\lambda) - 1) \tag{9.32}$$

Chapter 10 Principles of data analysis

10.1 Introduction

The reduction of photometric data consists in calculating the magnitudes and colors of celestial objects on a standard scale. The influences—either random or systematic—of the Earth's atmosphere and of the instrumentation on the signal have to be removed. The evaluation of the accuracy of the results is the final part of the reduction process.

Unfortunately the literature holds very few specific methods for careful and systematic reduction of photometric data. Many (individual) reduction programmes are based on the review article of Hardie (1962). Unfortunately, few go beyond the matter treated there (for exceptions, see for example Harris et al. 1981, Manfroid & Heck 1983, 1984).

A major source of errors, often underestimated, comes from the standard scale itself. How appropriate is the adopted scale? Is the equipment sufficiently close to the standard (ideal) system? In other words: aren't we trying to calibrate apples with oranges?

10.1.1 Old or new style reductions

There has always been some confusion about the best way to reduce data because of the many potential traps. Although a large diversity of computer reduction-programmes exist, many of them are deceptively simple and do little more than reading digitally stored measurements, performing elementary operations such as sky subtraction, and displaying or printing the final data. Sophisticated algorithms exist, however, and their use is certainly to be recommended since they can handle the data very efficiently. Before going into the details of those algorithms, we shall examine the reasons why many astronomers still prefer classical methods.

- Most astronomers like to have their data reduced on the spot. However, powerful computers are needed to run the new programmes, and not all observatories have powerful computers or work-stations. Moreover, sophisticated photometric reduction packages that can be run quasi-online virtually do not exist.

- Finalized data can be computed only after weeks or months of observations have been completed, so simpler programmes are needed anyway for on-line or for daily reduction. Often those first-order reductions yield "publishable" data, and there is a strong temptation for the observer to feel content with this preliminary work and to switch to new observations.

- Many experienced astronomers consider photometry as a craft, where the human touch is essential in making decisions about some parameters and about some data. So the role of the computer is downplayed: it is seen mainly as an aid in performing tedious calculations and in displaying the results as efficiently as possible. We shall show that the computer can be instructed to examine many relations between the data, and to make use of these data in a rigorous way no human can do. In fact the computer may discover subtle contradictions or anomalies that astronomers would completely overlook, for instance between measurements made several nights apart. But ultimately, dubious and doubtful cases still may have to be dealt with by the astronomer.

- Some of the new published methods are severely flawed. One unfortunate feature was to rely only on standard-star measurements to perform the reductions (see Harris et al. 1981, and discussion of this method by Popper 1982). This approach overlooks the usefulness of all the other constant stars, and increases the already laborious task of the observer who has to measure a lot of standards every night. Moreover, the success of such standard-based methods depends strongly on the validity of the standard scale, not only for the color transformations but also for the extinction calculation. The latter point seems irrelevant to many—extinction calculation should be possible even without standards—and explains the scepticism met by some methods.

- The improvement brought by more powerful reduction schemes is not easily measurable. There is certainly no tenfold improvement in the accuracy. But the final results are more homogeneous and less noisy. Last but not least, fewer problems tend to go unnoticed.

In the next sections we shall describe general reduction methods. We shall not go into every detail of more classical methods, but we shall emphasize the principles that should be of concern to the serious photometrist.

10.1.2 Fitting a model

The model describes mathematically the alterations of the signal between entering the upper atmosphere and its recording. This description involves a number of adjustable parameters characterizing:

- the atmospheric extinction (color/passband effects, time variations);
- the detector sensitivity (dead time, non linearity, temperature dependence, time variations);
- diaphragm effects;
- passband conformity (color transformation).

Usually a *merit function* is built which measures the validity of the agreement between the model and the data. The best set of values for the parameters is found when the merit function reaches its minimum. The various reduction schemes differ essentially in the choice of the mathematical model and, consequently, of the merit function.

The choice of the parameters depends on whether we use a simple or a complex theoretical representation. For instance we know, to some extent, the dependency of the extinction effect on the zenith angle. But we can use first- or higher-order extinction coefficients. We can choose to deal with a homogeneous atmosphere model, or with an atmosphere with multiple components of differing distributions.

Other effects are less tractable. This is for example the case when atmospheric extinction exhibits erratic changes with time. A possible treatment of this kind of effect involves step functions, polynomials, or Fourier series. The corresponding parameters may not necessarily have a real physical meaning, and their number can become very high.

Usually the merit function is built as a sum of squared deviations and the problem falls into the categories of least-squares or chi-square fits. These are well-known techniques subject to several assumptions such as, for example, the normal (Gaussian) distribution of the errors. Most incompatibilities due to the deviation of the error distribution from the assumption of normal distribution arise from odd measurements: when a star goes close to the edge of the diaphragm, when a thin cloud has not been seen, or when a drop or a surge in electric power occurs (e.g. as a dome is rotated). But similar effects are caused by actual variations of the stars: even a "standard" star can be an eclipsing binary of long period or another hitherto unknown variable of small amplitude.

Noise introduced by atmospheric turbulence has peculiar characteristics. Scintillation does not change the average irradiance—over sufficiently long time-intervals it does not add nor remove light from the beam—but it does change the observed distribution of photons. It happens that an initially normal distribution in the arrival rate of photons becomes Gaussian in the *logarithm* of that rate. This leads to several interesting conclusions. The good news is that the magnitudes will, in that case, be normally distributed around their mean value. This is an argument in favor of using least-squares algorithms in the reduction procedures. The bad news is that the average magnitude is not the true magnitude, since the average of logarithms does not give the logarithm of the average. The difference is small when the scintillation noise is small, but it increases rapidly when scintillation becomes important. Let n be the rate of arrival of photons. It can be shown that

$$\ln < n > - < \ln n > = \sigma^2/2$$

where σ is the rms deviation of $\ln n$. For other types of noise, the distribution of $\log n$ is not normal, and higher differences can be expected. This error should be kept in mind when averaging magnitudes of faint stars, or other noisy data. If σ is of the order of .05 mag, a systematic error of about .001 in $\ln n$—and hence in the magnitude (see Eq. 1.13)—will occur. For $\sigma \approx 0.1$ systematic errors of about .005 are to be expected. Those high σ are not uncommon at air masses of the order of 2 where much of the information concerning the extinction is gathered. The average magnitudes will then be overestimated at high air mass and so will be the extinction.

Bad data have to be discarded: the least-squares method is very sensitive to poor data, and including them would yield disastrous effects. The problem arises with moderately discordant data: a measurement off by for example 10 times the standard deviation is certainly wrong, but what about a deviation only one-half or one-third of that?

The ideal solution would be to use better merit functions in what are called *robust* algorithms. An example of such a method involves the absolute deviations instead of the squared ones. Known as the *minimum-sum*, this method was first proposed by Edgeworth (1887) and has practically never been used in astronomy. Branham (1982) discusses such alternatives to least squares and shows the superiority of the minimum-sum in the case of deviant data. Another potential improvement would be the use of the median instead of the mean when computing global values. This eliminates some of the problems created by disagreeing values.

Mathematically those methods are more difficult to set up than least-squares ones (just imagine how absolute values in derivatives must be handled, or how to set up an efficient algorithm based on calculation of medians). They are also more expensive in computer resources. For a thorough discussion of robustness, see also Huber (1981) and Stigler (1977).

In actual conditions one has to compensate for the possibility of wrong data by increasing the number of measurements. Systematic errors could be detected by the asymmetric tails of their distribution, but little can be done to correct the situation.

10.2 Monochromatic photometry

The simplest photometric system—at least with respect to reductions—is a single-filter monochromatic system. Though such a system does not make much sense astrophysically, single-filter photometric monitoring is widely applied in variable star research, and yields highly accurate data at high temporal resolution.

The passband has to be narrow around some wavelength λ_0 so that all color effects can be neglected. The spectral response $S(\lambda)$ acts as the Dirac distribution $\delta(\lambda - \lambda_0)$, and we have for the measured irradiance

$$E_m = \int E(\lambda)S(\lambda)d\lambda = \int E(\lambda)\delta(\lambda - \lambda_0)d\lambda = E(\lambda_0) \tag{10.1}$$

In practice this means a bandwidth of a few nanometers in the visible domain.

Because of the peaked profile of the instrumental response $S(\lambda)$, some problems arise when the spectral distribution $E(\lambda)$ is very irregular, or unstable. A narrow-band system dedicated to photometry of emission objects could be extremely sensitive to color (or spectral) effects if the wavelength or the shape of the emission feature varies (Doppler shifts, broadening...). Equally disturbing color effects will occur if the standard and extinction stars have continuous spectra (which is usually the case) and do not behave as the objects under study. Such instances do occur: comets have emission lines or bands with variable Doppler shifts; emission nebulae and active galactic nuclei display still higher Doppler shifts. To the contrary, photometry of main sequence stars in one of the "visible" bands (around 550 nm) encounters few problems, even with a bandwidth of several tens of nanometers.

In the following we assume that we have the necessary information to convert the measurements into raw magnitudes, i.e., we know quantities such as the dead time of a photon counting system, or the gain of the amplifier, or the integrated and calibrated intensity on a CCD or photographic image (for a discussion of CCD and photographic photometry, see Chapters 13 and 14).

Let us assume that we observe a star of constant brightness at different air masses, that the sensitivity of the instrument is stable, and that the atmospheric extinction is stable and isotropic.

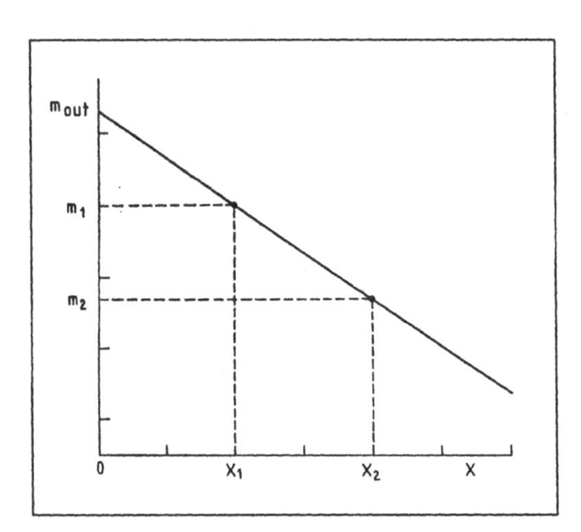

Fig. 10.1 *Bouguer's line. Two observations of the same star (taken several hours apart) provide a determination of the extinction coefficient.*

The raw magnitude $m(X)$ at air mass X obeys the Bouguer law (see Eq. (6.10):

$$m(X) = m_{\text{out}} + kX \tag{10.2}$$

where m_{out} is the—uncalibrated—magnitude outside the atmosphere and k is the monochromatic extinction coefficient. We assume stable atmospheric and instrumental conditions. The extinction coefficient can easily be derived from two measurements at air masses X_1 and X_2 (see Figure 10.1),

$$k = \frac{m_1 - m_2}{X_1 - X_2} \tag{10.3}$$

The magnitude outside the atmosphere is

$$m_{\text{out}} = m_1 - kX_1 \tag{10.4}$$

Note that in order to have a good accuracy one needs $|X_1 - X_2|$ to be as large as possible, with neither X_1 nor X_2 too large. In practice this often means that one observation should be around the meridian and the other at an hour angle of 3 to 5 hours. Remember that the basic assumption is that all observing conditions have remained constant during that time interval.

It would, of course, not be wise to determine two parameters from only two measurements. When more than two observations are available it becomes necessary to use statistical methods: one must fit I measurements $\{m_i, X_i\}$ $(i = 1, \ldots, I)$, to a model with two parameters, k and m_{out}, and minimize the least-squares merit function,

$$\phi(k, m_{\text{out}}) = \sum_{i=1}^{I} [m_i - m(X_i; k, m_{\text{out}})]^2 \tag{10.5}$$

or, more explicitly,

$$\phi(k, m_{\text{out}}) = \sum_{i=1}^{I} (m_i - kX_i - m_{\text{out}})^2 \tag{10.6}$$

This is a straightforward linear-regression problem. Its solution yields the *Bouguer line* (see Fig. 10.2). An approximate fit may be drawn graphically, as was done routinely during the early days of photoelectric photometry. But the analytical solution is to be preferred since the numerical accuracy of the measurements is normally too good to be fully exploited by hand-made diagrams. We must emphasize, though, that a computer plot to visually check the data and the solution is certainly very helpful. Other numerical controls are routinely done, such as checking the largest "residuals" $m_i - m(X_i; k, m_{\text{out}})$ and eventually rejecting the corresponding observations.

The merit function ϕ has a minimum when its derivatives with respect to k and m_{out} vanish.

$$\begin{aligned} \frac{\partial \phi}{\partial k} &= 0 \\ \frac{\partial \phi}{\partial m_{\text{out}}} &= 0 \end{aligned} \tag{10.7}$$

Those equations are rewritten as

$$\begin{aligned} \sum_{i=1}^{I} (m_i - kX_i - m_{\text{out}})X_i &= 0 \\ \sum_{i=1}^{I} (m_i - kX_i - m_{\text{out}}) &= 0 \end{aligned} \tag{10.8}$$

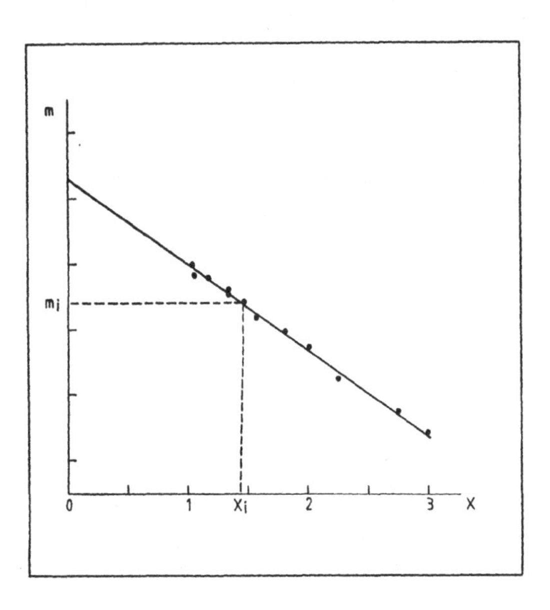

Fig. 10.2 *Bouguer's line. A least-squares fit through many observations of the same star results in a more reliable determination of the extinction line.*

or,

$$\begin{aligned} kS_{XX} + m_{\text{out}} S_X &= S_{mX} \\ kS_X + m_{\text{out}} I &= S_m \end{aligned} \tag{10.9}$$

where

$$\begin{aligned} S_{XX} &= \sum_{i=1}^{I} X_i^2 \qquad & S_{mX} &= \sum_{i=1}^{I} m_i X_i \\ S_X &= \sum_{i=1}^{I} X_i \qquad & S_m &= \sum_{i=1}^{I} m_i \end{aligned} \tag{10.10}$$

Equations 10.8 are easily solved for k and m_{out},

$$\begin{aligned} m_{\text{out}} &= \frac{S_{XX} S_m - S_{mX} S_X}{\Delta} \\ k &= \frac{-S_X S_m + S_{mX} I}{\Delta} \end{aligned} \tag{10.11}$$

with

$$\Delta = I S_{XX} - S_X^2 \tag{10.12}$$

In practice the straightforward application of these equations may lead to the loss of numerical accuracy when differences are calculated (e.g., in the case of a small range of

X-values). In many problems of this type the choice of the particular numerical method is very important.

10.2.1 The multi-night technique: an intuitive view

The principles of photometric reductions presented in this book are markedly different from those exposed in most textbooks, including recent ones. Our opinion is that there is no reason to propagate old and inefficient methods and that astronomers no longer have any excuse to avoid more complete techniques. Digital detectors are now so much more accurate than before, that one simply cannot afford to spoil those data with an inadequate reduction technology.

In order to show very simply the drawbacks of the classical methods, let us assume we made the following measurements of three stars in two successive nights. A and B are known to be constant, A being a standard of magnitude 5.0. Star C is an unknown object.

Night 1	Star	Air mass	Instr. magn.	m_{out}
	A	1.0	6.50	6.40
	B	1.0	6.40	6.30
	B	2.0	6.50	6.30

Night 2	Star	Air mass	Instr. magn.	m_{out}
	A	1.0	6.35	6.20
	B	2.0	6.40	6.10
	C	2.0	6.60	6.30

The extinction evaluation for B during night 1 gives a coefficient $k_1 = 0.1$ and an extra-atmospheric magnitude $m_{1,\mathrm{B}} = 6.3$. Using that same extinction we deduce $m_{1,\mathrm{A}} = 6.4$. Nothing can be said for night 2 with the usual methods. However, it is clear that the necessary information is present in the data, and that it is possible to evaluate the magnitude of C. Stars A and B differ by 0.1 magnitude. Hence we know that if star A had been measured at air mass 2, it would have been of mag 6.50. Consequently the extinction during night 2 was $k_2 = 0.15$. We find thus the extra-atmospheric instrumental magnitudes during night 2, $m_{\mathrm{out},2,\mathrm{A}} = 6.2$, $m_{\mathrm{out},2,\mathrm{B}} = 6.1$ and $m_{\mathrm{out},2,\mathrm{C}} = 6.3$. There has been a shift of -0.2 magnitude in the *zero point* (see Section 10.2.5) between night 1 and 2. In the system of night 1 we would have

$$\begin{cases} m_\mathrm{A} = 6.4 \\ m_\mathrm{B} = 6.3 \\ m_\mathrm{C} = 6.5 \end{cases}$$

Since the standard value of A is 5.0, we have in the standard system

$$\begin{cases} m_\mathrm{A} = 5.0 \\ m_\mathrm{B} = 4.9 \\ m_\mathrm{C} = 5.1 \end{cases}$$

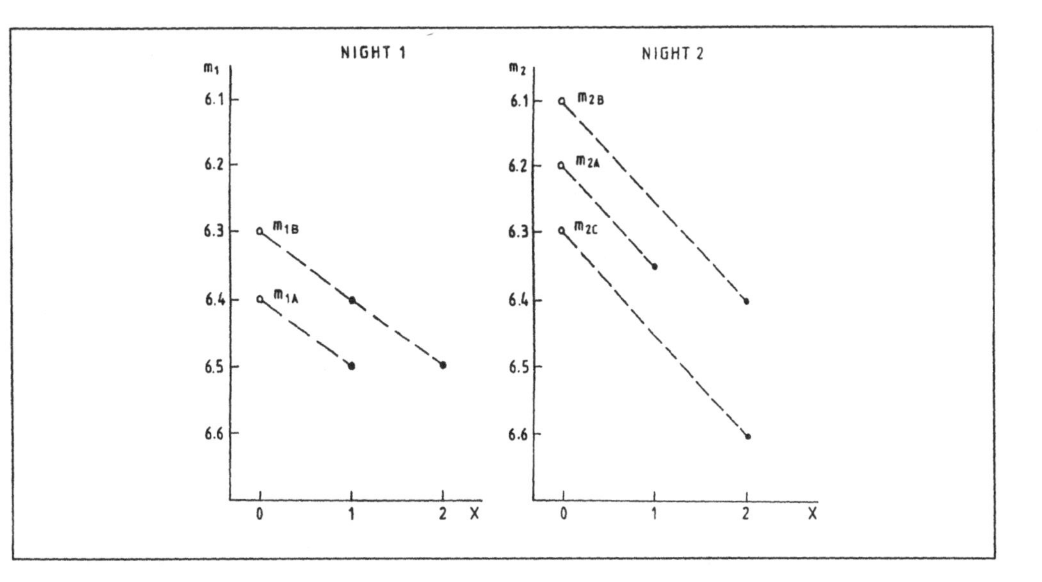

Fig. 10.3 *A multi-night calculation. The extinction can be evaluated on night 2 because night 1 yields the magnitude difference between A and B.*

The procedure is easy to understand on Fig. 10.3. Although the example is not quite realistic, it shows that individual reduction for every night overlooks valuable data. This hidden information is fully exploited by the methods we present here. Note that only one standard value was known, so a method based on standard values only would be totally powerless in even such a simple situation.

10.2.2 The simplest multi-night case

In the previous examples, we assumed stable observing conditions during the night (otherwise we could not use the Bouguer law). The usual practice indeed is to consider that the extinction stays constant during the night and changes during daytime only, unless there is evidence to the contrary. This is a somewhat arbitrary assumption since meteorological phenomena do not follow such regular patterns, but one could argue that serious observers would stop working as soon as the conditions do change or degrade significantly. Indeed the extinction may stay constant for hours or for days. In the latter case we can adopt a more accurate determination of the parameters of the Bouguer equation by observing the star over several nights, and piling up more data. The above equations still hold as long as *the instrumental system is stable* (we shall come to that point later). One has only to extend the summation over i in Eq. (10.5) over several nights.

If k varies from one night to the other, but can be considered as constant within each observing session, we can still benefit from reducing all nights as a whole. This is true

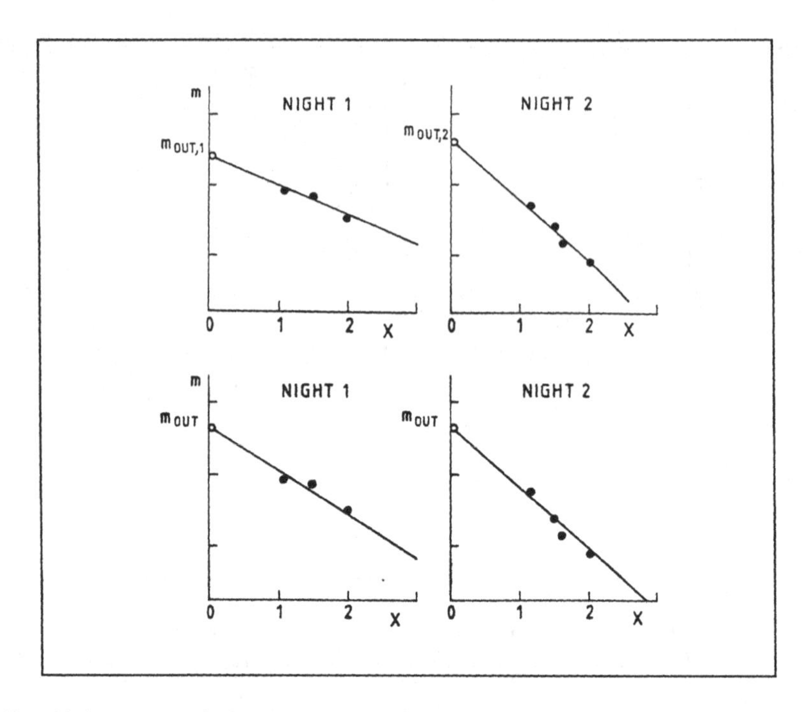

Fig. 10.4 *Bouguer's line. A simple example shows that it is useful to combine information obtained during different nights in order to obtain better reductions. In the top half of the graph, two nights are reduced independently, giving different extra-atmospheric magnitudes. In the bottom half, the same value is imposed to* m_{out}.

even if the instrumental system is not stable, as demonstrated by the simple example presented in the preceding section.

Another interesting example is presented in Fig. 10.4. A star has been observed on two different nights. The system is supposed to be stable (no change of zero point). The extinction has obviously changed and two different m_{out} could be obtained from data of each night. If there are no reasons to suppose that the star varies from night to night, one wishes to have only one value for m_{out}, likely $(m_{\mathrm{out},1} + m_{\mathrm{out},2})/2$. This is certainly better than any of the two individually determined extra-atmospheric magnitudes, and it is useful to plot m_{out} as an additional point at $X = 0$ and to repeat the extinction determination. This procedure is valid for any number of nights. Each night has to be treated twice, once without and once with the extra point at zero air mass. It is interesting to note that this technique was well adapted to graphical techniques: in the early days of photoelectric photometry astronomers used to draw Bouguer lines for different nights on the same graph and tried to force them through the same point ($X = 0, m = m_{\mathrm{out}}$). We now show that a simple algorithm can do the job in a single step.

We denote by k_n the extinction coefficient during night n, ($n = 1, \ldots, N$). The

Bouguer law is rewritten for each night,

$$m_n(X) = m_{\text{out}} + k_n X \qquad (n = 1, \dots, N) \tag{10.13}$$

The merit function is

$$\phi(k_1, \dots, k_N, m_{\text{out}}) = \sum_{n=1}^{N} \sum_{i=1}^{I_n} (m_{n,i} - k_n X_{n,i} - m_{\text{out}})^2 \tag{10.14}$$

I is now a function of n. The parameters corresponding to the minimum are found by solving the system

$$\begin{aligned} \frac{\partial \phi}{\partial k_n} &= 0 \qquad (n = 1, \dots, N) \\ \frac{\partial \phi}{\partial m_{\text{out}}} &= 0 \end{aligned} \tag{10.15}$$

There is a tremendous advantage in adopting this "multi-night" approach. Suppose that we observe the same star five times on every night. In the single-night case we have to determine two parameters, k and m_{out}, from 5 observations. Over five nights we deal with six parameters k_n $(n = 1, \dots, 5)$ and m_{out}, instead of ten. The ratio of the number of observations to the number of unknowns, O/U, is $(\sum_n I_n)/(N + 1)$ in the multi-night case and $(\sum_n I_n)/(2N)$ in the classical case. The advantage of the former method can be measured by a coefficient

$$\eta = \frac{(O/U)_{\text{multi}}}{(O/U)_{\text{class}}} = \frac{2N}{N + 1} \tag{10.16}$$

Here η is almost doubled, which results in a more accurate solution. This is a general remark that holds true for the more global procedures that we shall review later in this chapter. Many computer solutions, still in use in large observatories, do not take advantage of those simple rules.

Let us recall that what we call a "night" is, more generally, an interval of time during which the parametrization holds true. Hence it can be longer or shorter than an actual night. One week of perfectly stable weather can be grouped in a single "night", while a few hours of observing in bad conditions may be split into several "nights".

10.2.3 Bouguer's law for several stars

We have shown that it is possible to improve the efficiency of the single-star algorithm by grouping several nights together. We shall now generalize the method the other way: since all stars obey the same extinction law, observing an additional star will add only one extra parameter, its out-of-the-atmosphere magnitude, instead of two parameters. For the sake of simplicity we start with a single-night period. Figure 10.5 shows data on two stars observed during the same session. With graphical methods, it is certainly not easy to force the slope of the Bouguer lines to be equal. What was done before the computer era—and what is still done in quite a few places—was to find the solution for each star

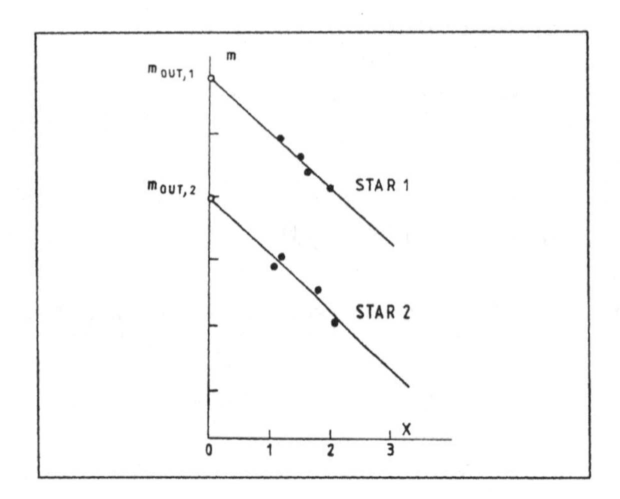

Fig. 10.5 Bouguer's line. During a stable night, two stars should yield the same slope of the Bouguer line.

independently, and then to take the average value. Thereafter, graphically or through least-squares fitting, it becomes possible to obtain refined values of the m_{out}. Here again, it is possible to reduce all that work to a single-step algorithm.

We denote by $m_{\text{out},c}$ the extra-atmospheric magnitude of star c ($c = 1, \ldots, C$), supposed to be constant. The Bouguer law is written for each of them,

$$m_c(X) = m_{\text{out},c} + kX \qquad (c = 1, \ldots, C) \tag{10.17}$$

A merit function is easily built

$$\phi(k, m_{\text{out},1}, \ldots, m_{\text{out},C}) = \sum_{c=1}^{C} \sum_{i=1}^{I_c} (m_{c,i} - kX_{c,i} - m_{\text{out},c})^2 \tag{10.18}$$

where I is now a function of c. Again the effectiveness of the algorithm is vastly improved. Suppose we observe five stars five times on a single night. In the single-star case we have to determine two parameters, k and m_{out}, from 5 observations. With five stars, we jump to six parameters k and $m_{\text{out},c}$ ($c = 1, \ldots, 5$) instead of ten. The ratio of observations to unknowns is again almost doubled, $\eta = 2C/(C+1)$. An important advantage of the method is that every constant star having at least two observations can contribute to the solution, while, graphically, only stars with several well-spaced observations are of practical interest.

Obviously, the multi-night and multi-star analyses can be combined to further increase the efficiency of the parameter adjustment. We leave to the reader to imagine how such an extrapolation would become awkward with classical methods. So we concentrate on the improved modern algorithms.

The following merit function is perfectly adapted to our problem,

$$\phi(k_1, \ldots, k_N, m_{\text{out},1}, \ldots, m_{\text{out},C}) = \sum_{n=1}^{N} \sum_{c=1}^{C} \sum_{i=1}^{I_{n,c}} (m_{n,c,i} - k_n X_{n,c,i} - m_{\text{out},c})^2 \tag{10.19}$$

As a thought experiment, suppose we observe five stars five times on each of five nights. In the single-star case we have to determine 25 Bouguer lines, i.e., 50 parameters from 125 observations. The direct method shows a spectacular improvement since all ten meaningful parameters, $k_n \quad (n = 1, \ldots, 5)$ and $m_{\text{out},c} \quad (c = 1, \ldots, 5)$, are directly evaluated.

Classically, supposing that all C stars were observed every night, one would have $O/U = \sum I_{n,c}/(2CN)$ ($= I/2$ if I constant) while the multi-night, multi-star method gives $\sum I_{n,c}/(C + N)$ or, when I is constant, $CNI/(C + N)$. The coefficient η is $2CN/(C+N)$. One has $\eta = 1$ if and only if $C = N = 1$. The efficiency increases rapidly with N and C. For a large number of nights or a large number of stars one gets

$$\begin{aligned} \lim_{N\to\infty} \eta &= 2C \\ \lim_{C\to\infty} \eta &= 2N \end{aligned} \tag{10.20}$$

If we adopt 5 as the desirable O/U, the classical method needs $I = 10$, while $I = 1$ would be perfectly acceptable in the general method as soon as $\eta \geq 10$ (e.g., ten nights and ten stars). This shows that short nights can be used effectively and that the extinction law can be obtained with only one observation per star and per night, or even less. The information on the extinction is carried by the whole set of stars and can be retrieved even when all measurements are done in a short period of time. Evidently the observations should not be done at about the same air mass: they must include both high and low X. Mathematically, the extinction law being linear, the extreme values are most useful and observations at intermediate air mass are of lesser interest. Practically, though, one should not go beyond $X = 2$, since atmospheric extinction and scintillation deteriorate the signal.

10.2.4 Variable extinction

Extinction is often variable in time and in direction. So it is not sufficient to determine its value for a given night and then assume that it is constant throughout the night in all directions.

The remarks in the previous section lead us to another advantage of the general procedure concerning eventual time variations. Since it is possible to measure the extinction in a short time, without having to wait hours for the rising or the setting of a few stars, it is also possible to analyze rapid variations of extinction. It is sufficient to evaluate the extinction from time to time and to interpolate the data. Or, instead of using discrete k_n, we may define a function of time $k(t)$ which may conveniently be split into N functions

$k_n(t)$, each corresponding to one night. If there are no abrupt variations, the $k_n(t)$ can be approximated by continuous functions such as polynomials. We may write

$$k_n(t) = \sum_{j=0}^{J_n} b_{n,j} t^j \tag{10.21}$$

where J_n is the order of the polynomial representing the extinction during night n. The coefficients of the polynomial expansion will be found by minimizing a least-squares merit function such as

$$\phi(b_{1,0}, \dots, b_{n,j}, \dots, b_{N,J_N}, m_{\text{out},1}, \dots, m_{\text{out},C}) = \sum_{n=1}^{N} \sum_{c=1}^{C} \sum_{i=1}^{I_{n,c}} (m_{n,c,i} - X_{n,c,i} \sum_{j=0}^{J_n} b_{n,j} t^j_{n,c,i} - m_{\text{out},c})^2 \tag{10.22}$$

where $t_{n,c,i}$ is the time of the observation i of star c during night n. One has then to solve the linear system obtained by setting the partial derivatives equal to zero. In order to reach a good accuracy, a wide range of X and t should be available. Hence constant stars have to be observed throughout all nights.

If the extinction varies linearly in time, air mass–magnitude diagrams will show curved Bouguer lines of different slope, for different stars. Diagrams obtained for two stars, one rising and the other setting in a same interval of time, will also show different slopes.

10.2.5 Unstable instrumentation

In the preceding sections (except for the example given in Fig. 10.3) we assumed that the instrumental conditions were stable. However, the sensitivity of the equipment may drift slightly during the observations (see the above-mentioned example) and this has to be taken into account. Instead of the simple law

$$m_{\text{out}} = m - k(t)X \tag{10.23}$$

we have to use

$$m_{\text{out}} = m - k(t)X + z(t) \tag{10.24}$$

where $z(t)$ is the *zero point* characterizing the instantaneous sensitivity of the equipment.

Slow drifts, present in the data because of unstable instrumentation (but also due to improper corrections for time-dependent extinction), may cause spurious night-to-night amplitude variations in the observations of short-period variable stars, such as β Cephei stars (Jerzykiewicz & Sterken 1987). From a Bouguer diagram, however, it is difficult to tell whether the extinction lines with different slopes originate from a linearly variable extinction coefficient, or from instrumental drift.

Again we start with the case of a single star observed during a single night. We assume that the zero point variations can be described by a polynomial expression

$$z(t) = \sum_{l=0}^{L} a_l t^j \tag{10.25}$$

Some instruments are equipped with apparatus allowing the measurement of the instantaneous sensitivity. For instance, an internal calibrated light source can give a regular check of the zero point variations. Or a temperature gauge can be useful when one knows the temperature dependency of the sensitivity. When such corrections are included in the equations, the time-dependent term may well vanish. However, this is rarely the case because none of those methods are perfectly adequate and it is wise to keep the variable zero-point.

We formulate our criterion as

$$\phi(a_1,\ldots,a_L,b_0,\ldots,b_J,m_{\rm out}) = \sum_{i=1}^{I}(m_i - X_i\sum_{j=0}^{J} b_j t_i^j + \sum_{l=1}^{L} a_l t_l^j - m_{\rm out})^2 \qquad (10.26)$$

Why did we not include the independent term, a_0, of the instrumental drift? Because so far we still have to normalize the magnitude scale, and any arbitrary constant may be added to our results. Mathematically a_0 would merge with $m_{\rm out}$ and it would be impossible to evaluate these parameters separately. Hence, we consider only the time-dependent part of the polynomial development. As soon as we introduce a standard scale, i.e., a reference value for $m_{\rm out}$, the ambiguity about the coefficient a_0 will disappear.

In peculiar conditions, the minimization of the above merit function is likely to give poor results because of the similarity of the time dependencies for the extinction and for the zero point. A variation in the zero point could mimic an extinction variation if the air masses are not well distributed. If we can approximate X by a polynomial in t, both series expansions for the zero point and the extinction will merge and become undistinguishable. Unfortunately, when only a single star is involved, it is always possible to represent X as a function of time, and develop it in Taylor's series or fit it by a polynomial. This simply demonstrates the obvious fact that, when only one star is observed, one can attribute its—non-intrinsic—variations either to the Earth's atmosphere, or to the observing equipment, or to both. Moreover, one may rightly object to the high number of observations needed to determine $U = L+J+2$ unknowns. With $O/U = 5$, and with second-order polynomials, one needs at least 30 observations. The solution will only reflect the particular model, not the physical phenomena, and it will be totally unsatisfactory for any other star observed during the same night, unless it is very close to the first one.

We come to the conclusion that the single-star problem is ill-conditioned and that two or more stars are needed to remove the under-determination of the general reduction problem. This is no new result. Indeed, it is the basic idea of the successful M&D method of the Geneva photometry team. Basically this method requires observing two extinction stars, one of them rising ("Montante") and the other one setting ("Descendante"), simultaneously.

Rufener (1964) developed the M&D method (see also Mandwewala 1976) to take into account the effects of variable extinction. He assumes that if the extinction is variable it varies slowly and isotropically in the whole solid angle considered, which is a hypothesis much less stringent than are the necessary conditions for using the Bouguer method. Figure 10.6 illustrates the situation with simulated observations for an M and a D star (open circles) for the case of a slowly decreasing extinction coefficient $k(t)$.

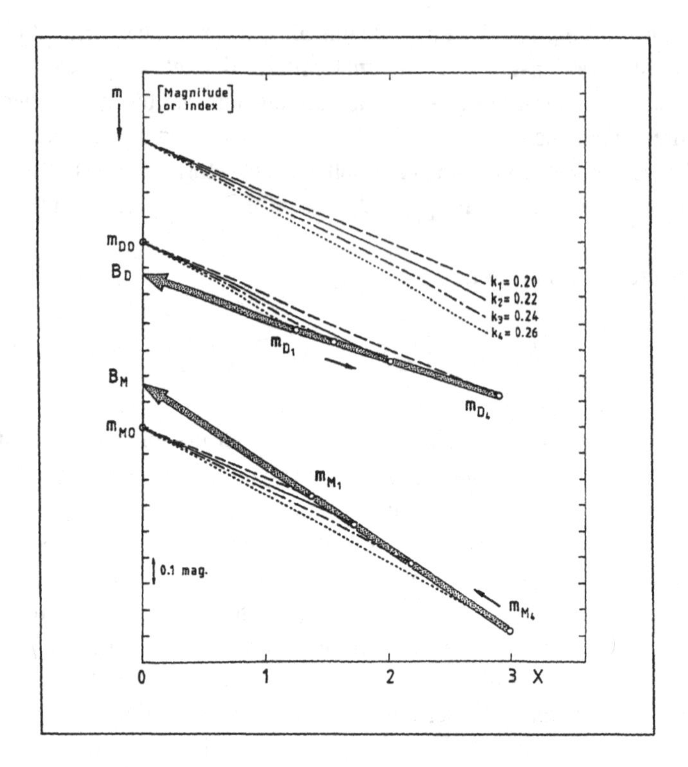

Fig. 10.6 *The "Rufener M&D" method. Variability of extinction coefficient $k(\lambda, t)$ is visualized by four k-values at moments of time t_i (based on Fig. 2 of Rufener 1986).*

m_{D_0} and m_{M_0} are the magnitudes outside the atmosphere for the stars M and D, and B_D and B_M are the out-of-atmosphere magnitudes that would be obtained if the classicial Bouguer method were applied on the datapoints obtained at times $\{t_1, \ldots, t_n\}$. Following the hypothesis that atmospheric extinction is the same for the measurements of the stars M and D at the time points t_i, the extinction coefficient can be eliminated from (Eq. 10.2) applied to each set of M, D measurements. This yields a system of linear equations, allowing the determination of the out-of-atmosphere parameters m_{D_0} and m_{M_0}. The M&D method thus rests on the assumption that any variation in extinction occurs isotropically so that the whole solid angle in which the observations are made is affected by the same variation of transparency, and on the assumption that equipment response is stable, and that the stars M and D are non-variable. Clearly, with such a scheme, k cannot be a single-valued function of time and no ambiguity remains.

Moreover, the first problems we have treated in the previous sections show that multi-night and multi-star algorithms are able to drastically increase the η coefficients. Hence we shall generalize the method to include time drifts, both of zero point and of extinction. The merit function is written

$$\begin{aligned}\phi(a_{1,0},\ldots,a_{n,l},\ldots,a_{N,L_N},b_{1,0},\ldots,b_{n,j},\ldots,b_{N,J_N},m_{\text{out},1},\ldots,m_{\text{out},C}) \\ = \sum_{n=1}^{N}\sum_{c=1}^{C}\sum_{i=1}^{I_{n,c}}(m_{n,c,i} - X_{n,c,i}\sum_{l=0}^{L_n} a_{n,l}t^l_{n,c,i} \\ - \sum_{j=0}^{J_n} b_{n,j}t^j_{n,c,i} - m_{\text{out},c})^2\end{aligned} \tag{10.27}$$

Note that coefficients $a_{n,0}$ do appear in the expression. They represent the zero-point shifts from night to night. Over N nights there can be only $N-1$ degrees of freedom for the $a_{n,0}$ since the $m_{\text{out},c}$ fix the overall zero point. We can choose to fix any one of the $a_{n,0}$, or any linear combination of them (e.g., the average zero point). In the following we shall arbitrarily assume that

$$a_{1,0} = 0 \tag{10.28}$$

Consequently, the other $a_{n,0}$ represent the zero points relative to the first night. With this convention, the corresponding linear system to be solved is

$$\begin{aligned}&\frac{\partial\phi}{\partial a_{n,l}} = 0 \quad \begin{cases} n = 1, N \\ l = 1, L_1 & \text{if } n = 1 \\ l = 0, L_n & \text{if } n \neq 1 \end{cases} \\ &\frac{\partial\phi}{\partial b_{n,j}} = 0 \quad \begin{cases} n = 1, N \\ j = 0, J_n \end{cases} \\ &\frac{\partial\phi}{\partial m_{\text{out},c}} = 0 \quad c = 1, C\end{aligned} \tag{10.29}$$

10.2.6 Normalization to a standard scale

Usually the equipment used by the observer is designed to reproduce some standard system as closely as possible. Standard values are published for several constant stars all over the sky so that they may be measured over a wide range of intrinsic colors, allowing anyone to have all observations converted to that system. Measurements made at different observatories can then be directly compared. This means that the wavelength of the filter should be identical to the wavelength of the standard filter and that the shapes of the transmission profiles should be similar.

A simple example will show how failing to meet that condition can be disastrous. Suppose our system is centered in a wing of the Hα line. A small shift of λ_0 would create confusion when we measure both emission- and absorption-line objects. Of course, not all photometric systems are located in such critical areas. But the actual spectral characteristics (slope of the continuum, presence of emission features or molecular bands) of all objects will certainly differ, and this will introduce errors. Even in the rather "quiet" green region we can find perturbing features (e.g., the 520 nm depression in the spectra of Ap stars).

Knowing the standard magnitudes of some stars is a great advantage for the reduction procedure. Besides locking the natural system to the standard one—which is all we can do when we measure a single star—we benefit from the constraints existing between the $m_{\text{out},c}$.

We consider a general case where S standard stars ($s = 1, \ldots, S$) are observed during N nights. We call y_s the standard magnitude of star s. The system transformation writes

$$y_s = m_{\text{out},s} + K \tag{10.30}$$

where K is a constant (*zero point*). We have to minimize the differences between the y_s and the computed $m_{\text{out},s} + K$. Hence an additional term appears in the merit function

$$\begin{aligned}\phi(a_{1,1}, \ldots, a_{n,l}, \ldots, a_{N,L_N}, b_{1,0}, \ldots, b_{n,j}, \ldots, b_{N,J_N}, m_{\text{out},1}, \ldots, m_{\text{out},C}, K) \\ = \sum_{n=1}^{N} \sum_{c=1}^{C} \sum_{i=1}^{I_{n,c}} (m_{n,c,i} - \sum_{\substack{l=0 \\ (n,l)\neq(1,0)}}^{L_n} a_{n,l} t_{n,c,i}^{l} \\ - X_{n,c,i} \sum_{j=0}^{J_n} b_{n,j} t_{n,c,i}^{j} - m_{\text{out},c})^2 \\ + \sum_{s=1}^{S} (m_{0,s} + K - y_s)^2\end{aligned} \tag{10.31}$$

(again we imposed the condition $a_{1,0} = 0$). Since we may not assume that all constant stars have standard values, the summations over s and c have different limits ($s \leq c$). From now on we shall simplify the notations. We write

$$\begin{aligned}\mathbf{a} &= \{a_{n,l}\} \quad \begin{cases} n = 1, N \\ l = 1, L_1 & \text{if } n = 1 \\ l = 0, L_n & \text{if } n \neq 1 \end{cases} \\ \mathbf{b} &= \{b_{n,j}\} \quad \begin{cases} n = 1, N \\ j = 0, J_n \end{cases} \\ \mathbf{m}_{\text{out}} &= \{m_{\text{out},c}\} \quad c = 1, C\end{aligned} \tag{10.32}$$

and, for the computed extra-atmospheric magnitude,

$$U_{n,c,i} = m_{n,c,i} - \sum_{\substack{l=0 \\ (n,l)\neq(1,0)}}^{L_n} a_{n,l} t_{n,c,i}^{l} - X_{n,c,i} \sum_{j=0}^{J_n} b_{n,j} t_{n,c,i}^{j} - m_{\text{out},c} \tag{10.33}$$

With those notations the merit function is

$$\phi(\mathbf{a}, \mathbf{b}, \mathbf{m}_{\text{out}}, K) = \sum_{n=1}^{N} \sum_{c=1}^{C} \sum_{i=1}^{I_{n,c}} (U_{n,c,i} - m_{\text{out},c})^2 + \sum_{s=1}^{S} (m_{0,s} + K - y_s)^2 \tag{10.34}$$

The linear system of equations is written in a simplified form

$$\begin{array}{ll} \dfrac{\partial \phi}{\partial \mathbf{a}} = 0 & \dfrac{\partial \phi}{\partial \mathbf{m}_{\text{out}}} = 0 \\ \dfrac{\partial \phi}{\partial \mathbf{b}} = 0 & \dfrac{\partial \phi}{\partial K} = 0 \end{array} \tag{10.35}$$

10.2.7 On the meaning of the function ϕ

By now the merit function ϕ has become pretty complicated. Using the notations just introduced, we shall write it in a more explicit way. Consider the simplest case of Eq. (10.6). Equation $\partial\phi/\partial m_{\text{out}} = 0$ in (10.8) gives

$$m_{\text{out}} = \frac{1}{I} \sum_{i=1}^{I} (m_i - kX_i) \tag{10.36}$$

Using our new notation we rewrite this as

$$m_{\text{out}} = \frac{1}{I} \sum_{i=1}^{I} U_i = < U > \tag{10.37}$$

the brackets denoting the average value. Then Eq. (10.6) becomes

$$\phi(k) = \sum_{i=1}^{I} (U_i - < U >)^2 \tag{10.38}$$

and represents the squares of the deviations of the U_i to their mean value. One parameter, m_{out} has disappeared. This procedure can be repeated for more complex cases. For instance, using

$$\frac{\partial \phi}{\partial \mathbf{m}_{\text{out}}} = 0 \qquad \frac{\partial \phi}{\partial K} = 0$$

from system (10.35) one transforms function (10.34) into

$$\phi(\mathbf{a}, \mathbf{b}, K) = \sum_{n=1}^{N} \sum_{c=1}^{C} \sum_{i=1}^{I_{n,c}} (U_{n,c,i} - < U >_c)^2 + \sum_{s=1}^{S} (< V >_s - y_s)^2 \tag{10.39}$$

where

$$< U >_c = \frac{\sum_{n=1}^{N} \sum_{i=1}^{I_{n,c}} U_{n,c,i}}{\sum_{n=1}^{N} I_{n,c}}$$

is the computed mean magnitude out of the atmosphere. Similarly we denoted by

$$V_s = < U >_s + K \tag{10.40}$$

this mean magnitude transformed to the standard system.

Here too, the parameters $\mathbf{m}_{\text{out}}$ disappeared. The linear system to be solved is formally similar to (10.35) except for equation $\partial\phi/\partial\mathbf{m}_{\text{out}} = 0$ which merged into the other ones. The price to pay for the reduction in the number of parameters is a much higher complexity of the equations.

The meaning of the merit function (10.39) is obvious. It adds the squared deviations of all measurements of each constant star to their averages and the squared deviations of the transformed averages to the standard values—when the latters exist. Hence the best use is made of every measurement, and every constraint is taken into account.

10.3 Multifilter photometry

The logical step after the study of narrow-band single-filter photometry would be to go for broad-band single-filter photometry. However, such photometry makes no sense (except for applications in differential photometry of variable stars). If the bandwidth is so large as to produce color effects—and this is what one means by broad-band photometry—these effects will drastically decrease the accuracy of the measurements. These inaccuracies are inherently uncorrectable since single-band photometry carries no information on the color or on the spectral types. Hence we proceed directly to multifilter photometry.

10.3.1 Narrow-band multifilter photometry

The single-filter reduction scheme can be adopted as it is, where each color is analyzed separately, and where color indices are computed by combining the results afterwards. This solution is not recommended because a great deal of information is lost that could be gained from the inter-color relationships. We mention below three important factors:

- The extinction coefficients in each wavelength are not independent. They are related by a law

  $$k = k(\lambda, \mathbf{h}) \tag{10.41}$$

  which depends on a series of parameters $\mathbf{h}$. Normally the use of such a law involves fewer parameters than the independent determination of each extinction coefficient would, and it increases the efficiency η. For instance, the extinction variations in the visible domain are often perfectly gray (see e.g., Olsen, 1983, for the *uvby* system)—which means that these variations are equal at all wavelengths. For a system with F filters one may write

  $$k_{n,f}(t) = k_{n,1}(t) + h_f \quad f = 2, \ldots F \tag{10.42}$$

If the h_f are not known, each color introduces only one additional parameter instead of J_n. On the other hand, if those coefficients are accurately known— when based on previous observations— they eventually cease to be considered as free parameters. Hence, in almost every case it is possible to decrease the number of b's by introducing instead a few h's.

- The zero points may behave similarly in each color if a single detector is used for all colors or if external effects (for example, because of temperature effects) affect each band in a similar way. Consequently, one reduces the total number of parameters by writing, e.g.,

$$z_{n,f}(t) = z_{n,1}(t) \tag{10.43}$$

or, in a polynomial development such as (10.25),

$$a_{n,l,f} = a_{n,l,1} \tag{10.44}$$

Those simplifications are not allowed when the spectral response of the equipment changes (variations of the filter profile, or of the photomultiplier response curve). Such variations, of course, make any serious photometry impossible.

- Multi-filter photometry carries some information on the spectrum of each object. This allows a certain amount of color correction which would be impossible if the K_f were treated separately. Instead of rewriting Eq. (10.40) for every filter

$$V_{s,f} = < U >_{s,f} + K_f \tag{10.45}$$

or, in vector notation,

$$\mathbf{V}_s = < \mathbf{U} >_s + \mathbf{K}_f \tag{10.46}$$

one introduces a more precise color transformation,

$$\mathbf{V}_s = \mathbf{Q}(\mathbf{K}) < \mathbf{U} >_s \tag{10.47}$$

Q stands for an operator (with free parameters **K**). Thus, additional parameters have been introduced, but they will contribute to obtaining a more accurate solution.

10.3.2 Color transformations

Combining all colors in the reduction procedure, as suggested in the previous section, increases the complexity of the merit function and of the minimization procedure. Normally this will no longer involve a system of linear equations. We shall now review some of the simplifying assumptions which are usually accepted.

In almost every photometric system the color transformation operator **Q** is *assumed* to be linear, which allows the writing of Eq. (10.47) in matrix form.

$$\mathbf{V}_s = \mathbf{M} < \mathbf{U} >_s + \mathbf{K} \tag{10.48}$$

Instead of the magnitudes $< \mathbf{U} >$ one generally uses color indices or a set of magnitudes and indices (e.g., b–y, m_1, c_1 or y, b–y, v–b, u–v instead of y, b, v and u in the Strömgren photometric system). The transformation to this set of magnitudes and indices is described by a matrix $\mathbf{H}$

$$\mathbf{U'}_s = \mathbf{H} < \mathbf{U} >_s \tag{10.49}$$

so that the color transformation becomes

$$\begin{aligned} \mathbf{V}_s &= \mathbf{M} < \mathbf{U'} >_s + \mathbf{K} \\ &= \mathbf{MH} < \mathbf{U} >_s + \mathbf{K} \end{aligned} \tag{10.50}$$

$\mathbf{H}$ is not necessarily square, the number of independent indices being $F' = F - 1$. In the above mentioned *uvby* examples, one has

$$\mathbf{U} = \begin{pmatrix} y \\ b \\ v \\ u \end{pmatrix} \tag{10.51}$$

$$\mathbf{H} = \begin{pmatrix} 1 & 0 & 0 & 0 \\ -1 & 1 & 0 & 0 \\ 0 & -1 & 1 & 0 \\ 0 & 0 & -1 & 1 \end{pmatrix} \quad \text{for} \quad \mathbf{U'} = \begin{pmatrix} y \\ b\text{–}y \\ v\text{–}b \\ u\text{–}v \end{pmatrix} \tag{10.52}$$

and

$$\mathbf{H} = \begin{pmatrix} -1 & 1 & 0 & 0 \\ 1 & -2 & 1 & 0 \\ 0 & 1 & -2 & 1 \end{pmatrix} \quad \text{for} \quad \mathbf{U'} = \begin{pmatrix} b\text{–}y \\ m_1 \\ c_1 \end{pmatrix} \tag{10.53}$$

The elements of matrix $\mathbf{M}$ change with the equipment used, but, for a given instrumental configuration they are quite constant in the short to medium term. The stability of photometric equipment has been investigated by several authors (see e.g., Rufener 1967; Olsen 1977); the latter author showed that the four-channel *uvby* photometer at ESO remained stable over at least a five year period. Fluctuations in $\mathbf{K}$ are absorbed by the zero-point parameters $\mathbf{a}$ so that $\mathbf{K}$ is also constant.

From the above considerations we deduce a general merit function

$$\begin{aligned} &\phi(\mathbf{a}, \mathbf{b}, \mathbf{M}, K) \\ &= \sum_{f=1}^{F} \sum_{n=1}^{N} \sum_{c=1}^{C} \sum_{i=1}^{I_{n,c}} (U_{f,n,c,i} - < U >_{f,c})^2 + \sum_{f=1}^{F'} \sum_{s=1}^{S} (< V >_{f,s} - y_{f,s})^2 \end{aligned} \tag{10.54}$$

where $y_{f,s}$ is the standard value of star s in index f. The associated system of equations is

$$\begin{aligned} \frac{\partial \phi}{\partial \mathbf{a}} = 0 \qquad \frac{\partial \phi}{\partial \mathbf{M}} = 0 \\ \frac{\partial \phi}{\partial \mathbf{b}} = 0 \qquad \frac{\partial \phi}{\partial K} = 0 \end{aligned} \tag{10.55}$$

Together **M** and **K** introduce but a few free parameters and do not appreciably degrade the efficiency η. Hence *the determination of the color transformation is a procedure which does not require much additional work from the part of the observer.*

Let us insist on this statement, which contradicts much of what is said on photometric observing. The color transformation involves a small number of parameters (about 10 for the Strömgren system). This is negligible compared to the hundreds of parameters describing the extinction and the zero point during a few weeks of observing. About a dozen four-color observations of well-chosen standard stars would suffice to satisfy the rule of thumb of a *ratio of five measurements per free parameter* (in *uvby*). Those measurements could be done on a single night, or they could be distributed over the whole observing run. Of course, the choice of the standard stars should be made in such a way as to allow safe and accurate reductions, i.e. they should have widely different indices. This is not always easy, since the lists of standards are often biased toward some spectral types. Moreover, the standard values are not perfectly accurate and, to be safe, it is recommended to enlarge somewhat the available sample of standards (see Section 11.4).

10.3.3 Wide-band multifilter photometry

Wide passbands introduce color effects into the extinction model. Equation (10.24) is no longer valid and should be replaced by

$$m_{\mathrm{out},f} = E_f(\mathbf{m}, \mathbf{a}, \mathbf{b}, X, t) \qquad (10.56)$$

The main feature is the dependency on the other magnitudes, i.e., on the color of the measured objects. This additional link between the different colors does not show up in the formal writing of (10.54) and (10.55). However, the explicit developments will be still more complex and yield a large system of (generally) non-linear equations.

Some of the parameters in Eq. (10.56) may be known after some time (just as in the case of the monochromatic extinction). This reduction in the number of unknowns increases η and allows improved and more robust reductions. However, this complicates the programme code which has to account for all possibilities, such as freeing or freezing certain parameters, or properly handling "missing values".

10.3.4 Fractioning the standard set

We assumed that a single color transformation was valid for all stars, but, in practice, it is possible to define such a transformation only in limited domains of spectral types and classes. Hence we divide the standard stars into several groups ($r = 1, \ldots, R$) and we assume that within each such group there exists a transformation

$$\mathbf{V}_s = \mathbf{M}_r \mathbf{H} < \mathbf{U} >_s + \mathbf{K}_r \qquad (s = S_{r-1}, \ldots, S_r) \quad (r = 1, \ldots, R) \qquad (10.57)$$

We generalize ϕ as

$$\begin{aligned}\phi(\mathbf{a},\mathbf{b},\mathbf{M},K) = &\sum_{f=1}^{F}\sum_{n=1}^{N}\sum_{c=1}^{C}\sum_{i=1}^{I_{n,c}}(U_{f,n,c,i} - <U>_{f,c})^2 \\ &+ \sum_{r=1}^{R}\sum_{f=1}^{F'}\sum_{s=S_{r-1}}^{S_r}(<V>_{f,s} - y_{f,s})^2\end{aligned} \quad (10.58)$$

On the left-hand side of (10.58), $\mathbf{M}$ and $\mathbf{K}$ stand for all matrices $\mathbf{M_r}$ and vectors $\mathbf{K_r}$. The number of unknowns has been increased, but in most cases R does not exceed 3. Here again we see the tremendous benefit of global solutions. If the reductions corresponding to each subset r were made separately, a large portion of the information concerning the extinction and the equipment would be lost.

10.3.5 Diaphragm effects

Diaphragms of different sizes may be used in a single observing run according to the specific programmes or depending on the atmospheric conditions. Instead of making separate reductions for each diaphragm size, one calibrates the effect as accurately as possible. Ideally, one assumes that the change of the observed magnitude is a discrete function of the diaphragm size and that there are no erratic variations due to seeing. One supposes also that the spectral dispersion at non-zero zenith angles has a negligible effect. We denote by $q_{d,f}$ the differences between the raw magnitudes which would be measured with diaphragms 1 and d ($d = 1, \ldots, D$) in filter f.

$$m_{d,f,n,c,i} = m_{1,f,n,c,i} + q_{d,f} \quad (10.59)$$

$m_{d,f,n,c,i}$ is the magnitude which would be obtained if the observation were carried out with diaphragm d. The merit function (10.54) may be generalized as

$$\begin{aligned}\phi(\mathbf{a},\mathbf{b},\mathbf{q},\mathbf{M},K) = &\sum_{d=1}^{D}\sum_{f=1}^{F}\sum_{n=1}^{N}\sum_{c=1}^{C}\sum_{i=1}^{I_{n,c}}(U_{d,f,n,c,i} - <U>_{f,c})^2 \\ &+ \sum_{f=1}^{F'}\sum_{s=1}^{S}(<V>_{f,s} - y_{f,s})^2\end{aligned} \quad (10.60)$$

The $q_{d,f}$ behave as additional zero points. The average extra-atmospheric magnitude is now defined as

$$<U>_{f,c} = \frac{\sum_{d=1}^{D}\sum_{n=1}^{N}\sum_{i=1}^{I_{d,f,n,c}} U_{d,f,n,c,i}}{\sum_{d=1}^{D}\sum_{n=1}^{N} I_{d,f,n,c}} \quad (10.61)$$

The associated system of equations is

$$\frac{\partial\phi}{\partial\mathbf{a}} = 0 \quad \frac{\partial\phi}{\partial\mathbf{b}} = 0 \quad \frac{\partial\phi}{\partial\mathbf{q}} = 0 \quad \frac{\partial\phi}{\partial\mathbf{M}} = 0 \quad \frac{\partial\phi}{\partial K} = 0 \quad (10.62)$$

It is quite easy to justify the introduction of the diaphragm calibration in the general reduction procedure. The efficiency is increased since regular obervations of constant (standard or comparison) stars are all that is needed to calibrate the diaphragm effects. Of course this calibration can be done relatively quickly by spending a few observing hours—in perfect conditions—on a single star around the meridian. And once this is done, it is valid for many nights.

10.3.6 Bright and faint stars

A problem often encountered by astronomers working at large telescopes is that bright stars are out of reach. But most original standard stars are very bright and often saturate the equipment. To overcome the problem, careful attempts have been made to establish sequences of faint standard stars in several photometric systems. This means using small telescopes to observe faint stars with a high accuracy, which is not very easy. Moreover, problems of compatibility between the original standard system and the faint standard system arise when the natural system of the small telescope is not sufficiently similar to the original system. Hence a more direct solution is recommended.

Whenever possible, it is always wise to go back to the original standards. And this can be done by using neutral-density filters (see Section 5.7; note that front-diaphragms are inadequate for this purpose, see Section 15.6.5). Such filters can be calibrated in the laboratory, but for a variety of reasons it is not always possible to reproduce exactly the actual observing conditions. Hence the exact calibration has to be done on the stars. When the filters are perfectly gray, the situation is exactly analogous—from the data-analysis point of view—to that of the variable diaphragms discussed in the previous section.

The calibration can be done with a small series of dedicated observations or it can be built into the global reduction scheme. Equations similar to (10.60) and (10.62) are easily written by adding a vector **P** of free parameters describing the zero point shifts due to the filters.

Imperfect neutral-density filters introduce color effects. Hence the effective response of the equipment is different with and without filters. This is usually much less dramatic than the differences existing between different sets of filters supposed to represent a same photometric system. A linear color transformation may be applied between the data obtained with and without filters. This transformation has to be applied to the raw magnitudes $\mathbf{m}'$ obtained with the gray attenuator g

$$\mathbf{m}_{g,i} = \mathbf{G}\mathbf{m}'_{g,i} + \mathbf{P} \tag{10.63}$$

The merit function will now include **G** and **P** as free parameters. The accuracy of the reduction will depend on the number of constant (not necessarily standard) stars observed with and without attenuator. Complete freedom is allowed in the choice of those stars. However, when a color effect is suspected, a wide range of spectral types should be represented in the sample.

We emphasize that this procedure allows the direct use of the original standard stars, instead of a second- or third-generation set. The homogenization problem of many photometric systems is sufficiently acute to make this feature invaluable (see Chapter 11). Moreover, it leaves the possibility of establishing a catalogue of secondary standards directly with a large telescope.

10.4 General remarks

We have seen that building a merit function is not always a straightforward matter. One has to consider several factors and decide whether implementing some particular feature is worth the extra work in programming and computing. As shown in the previous section, one can include the calibration of diaphragms and of gray attenuators in the general reduction procedure, and this is certainly worthwhile. But would it be useful to introduce the calibration of a non-linear effect such as the dead-time correction of a pulse-counting system? The answer depends largely on the kind of work performed by the observers. If it involves many bright objects requiring this correction, it is certainly useful.

No constraint is imposed on the filter sequence: the measurements of the various colors need not necessarily be done simultaneously. In principle, it is even possible to spend long periods (even nights) observing in a single filter and still obtain correct reduction parameters from this limited information. This is an asset for observational techniques which are basically sequential, such as CCD photometry or photographic photometry.

Very faint stars at the limit of detection are often observed which have noise larger than about 0.1 mag. According to the discussion in Section 10.1.2 systematic errors due to averaging over magnitudes are introduced. Those stars are not included in the reduction procedure, but their observations are reduced by the values of the parameters derived from the general reduction. A proper averaging procedure should involve the count rates themselves. An alternative method, which has the advantage of minimizing the effect of deviant data, is to calculate the median value (with virtually similar results either in magnitude or count rate, unless there are very few points).

In all the functions we have omitted weighting factors. We considered the usual least-squares fit which supposes the measurement errors to be independent and to follow a normal distribution with the same variance. When the standard deviation is not constant one can weigh each residual with the inverse of its standard deviation. The merit function becomes what is commonly known a "chi-square". But to what extent can we estimate the standard deviation of the measurements? In most cases it is quite impossible, but a few rules can be adopted. The errors introduced by the atmosphere increase with air mass X, and this dependency as well as the color dependency ought to be taken into account.

It is well established that the error due to scintillation grows like X^2 in the wind azimuth, and $X^{3/2}$ at right angles to it (see Section 7.1). If this were the only cause of error, observations should then receive weights proportional to X^{-3} or X^{-4}; hence, close to air mass 2 weights of only 0.06 would be attributed to some measurements! Actually other errors exist which show less variation with zenith distance. In fact, the actual error law can be approximated by combining the scintillation error and a constant term which has to be empirically determined but is generally somewhere between 0.003 and 0.010. If faint stars have to be included in the reduction algorithm, higher errors may appear and lower weights should be adopted.

Instead of using a complex error law, a common practice is to establish a discrete scale of weighting factors (e.g., 1, 1/2, 1/4, 1/8) and to attribute them in a rather empirical way (e.g., all faint stars would receive the lowest rating, measurements above air mass 1.5 and 2 would receive weights of, respectively, 1/2 and 1/4...). Such an approach is often used to qualify the precision of measurements of stars that do not contribute to the reduction process (e.g., variable stars, see Sterken et al. 1986).

10.5 Examples of reduction algorithms

The reduction method explained above is very general and covers a wide variety of reduction schemes. One such scheme specifically developed for *uvby* photometry (the PHOT2 programme) has been described by Manfroid (1985a) and is mainly used in reducing data from the *Long-Term Photometry of Variables* project (LTPV) at ESO (Sterken 1983, 1986; Manfroid et al. 1991).

PHOT2 uses a single merit function which basically is the sum of the squared deviations of average standard star observations from standard values, and of the squared deviations of all constant stars (including the standards) from their average values. The extinction for each passband is handled separately, i.e., it makes no use of any relation between the four channels. Fluctuations in the zero points as well as in the extinction coefficients are allowed, and they are represented by second-order polynomials in a time-dependent variable. This variable can be time itself, or a continuous function of time, or—to allow for sharp variations—a discontinuous function of it. The color tranformation is represented by a 4×4 matrix with either 6 or 12 non-zero coefficients. For instance, in the simplest, and most common case, matrix **M** of Eq. (10.50) is written as

$$\mathbf{M} = \begin{pmatrix} m_{11} & 0 & 0 & 0 \\ m_{21} & 1 & 0 & 0 \\ m_{31} & 0 & m_{33} & 0 \\ m_{41} & 0 & 0 & m_{44} \end{pmatrix} \tag{10.64}$$

The optimization of the merit function is treated by an iterative second-degree method, which converges in only a few steps. The algorithm treats a series of up to a few months of observations in a single run.

As a second example we shall describe in more detail another programme directed at multi-filter narrow-band photometry. It skips entirely the color transformation problem, i.e., only zero points are computed, instead of complete color transformations. Such a method is followed, for example, when one sets up a new photometric system, or when one works with insufficiently accurate standards. Moreover, our experience with a variety of narrow-band systems shows that, whenever possible, it is much better to set up one's own standard in the natural system, and to use these data without color transformations, rather than using universal standards. This obviously supposes that one has a rather large amount of data to work on. A color transformation can be done afterwards, if necessary. A special case, where this method is applied, is the well-known case of one-filter (differential) photometry of variable stars, an approach that is frequently put to use when monitoring variable stars with periods of minutes to hours, and where high temporal resolution is imperative (see also Section 15.3.2).

Let m be the observed raw magnitudes, m_0 the estimated actual magnitudes or the standard magnitude in the case of standard stars (at least one standard is necessary to fix the zero point), X the air mass, k the extinction coefficient, f the filter, and z the zero point. We designate the night by subscript n, the constant stars (including standard stars) by subscript c, the standard stars by s. The extinction coefficient as well as the zero point vary from night to night. The following merit function describes the model:

$$\phi(k_{n,f}, m_{0,c,f}, z_{n,f}) = \sum_{n}\sum_{c}\sum_{f}(m_{c,f} - k_{n,f}X + z_{n,f} - m_{0,c,f})^2 \tag{10.65}$$

In this equation the summation is to be carried over all observations. The m_0 are free parameters for non-standard stars only. In order to improve the model we may impose some constraints. First, the extinction coefficients do not vary independently: additional extinction is often gray (see Section 10.3.1), at least over some wavelength ranges r. This means that for filters f inside domain r we can write (see Eqs. (10.41) and (10.42))

$$k_{n,f,r} = k_{1,f,r} + k'_{n,r} \tag{10.66}$$

Next, the zero points are not quite independent from one night to the other (see also Section 10.3 and Eqs. (10.43) and (10.44)). They represent the spectral response function of the instrumentation, and they often vary simultaneously in each band. As a first approximation we write

$$z_{n,f} = z_{1,f} + z'_n \tag{10.67}$$

In many cases, however, the overall response of the instrumentation is modified during the observing run: the detector has been changed, or its operating conditions are modified, or the instrumentation is transferred to another telescope, a fresh aluminum coating is deposited on the mirrors, etc... Hence distinct Eqs. (10.67) may have to be applied during groups of nights, that we designate by g.

$$z_{g,n,f} = z_{g,1,f} + z'_{g,n} \tag{10.68}$$

Equation (10.65) is rewritten as

$$\phi(k_{1,f,r}, k'_{n,r}, m_{0,c,f}, z_{g,n,f}, z'_{g,n}) = \sum_{g}\sum_{n}\sum_{c}\sum_{f}(m_{c,f} - k_{1,f,r}X - k'_{n,r}X + z_{g,1,f} + z'_{g,n} - m_{o,c,f})^2 \tag{10.69}$$

The estimated magnitude of each observation is

$$U_{c,g,n,f,r} = m_{c,f} - k_{1,f,r}X - k'_{n,r}X + z_{g,1,f} + z'_{g,n} \tag{10.70}$$

and, with simpler notations, Eq. 10.65 becomes

$$\phi(k_{1,f,r}, k'_{n,r}, m_{0,c,f}, z_{g,n,f}, z'_{g,n}) = \sum_g \sum_n \sum_c \sum_f (U_{c,n,g,f,r} - m_{0,c,f})^2 \tag{10.71}$$

As explained earlier an alternative and equivalent form for ϕ is obtained by replacing the m_0 relative to non-standard stars with their average estimated value

$$< U >_{c,f,r} = < m_{c,f} - k_{1,f,r}X - k'_{n,r}X + z_{g,1,f} + z'_{g,n} > \tag{10.72}$$

(see below). The terms relative to the standard stars can then be singled out and one would write

$$\begin{aligned} &\phi(k_{1,f,r}, k'_{n,r}, z_{g,n,f}, z'_{g,n}) \\ &= \sum_g \sum_n \sum_{c \neq s} \sum_f (U_{c,n,g,f,r} - < U >_{c,f,r})^2 + \sum_c \sum_f (< U >_{s,f,r} - m_{0,s,f})^2 \end{aligned} \tag{10.73}$$

After determination of the parameters by minimizing this function, the m_0 can be obtained directly through Eqs. (10.72).

Contrary to the PHOT2 method, both functions (10.71) and (10.73) can be directly minimized since the system of partial differential equations is linear. We have

$$\begin{aligned} &\frac{\partial \phi}{\partial k_{1,f,r}} = 0 \quad \begin{cases} f = 1, F \\ r = 1, R \end{cases} \\ &\frac{\partial \phi}{\partial k'_{n,r}} = 0 \quad \begin{cases} n = 2, N \\ r = 1, R \end{cases} \\ &\frac{\partial \phi}{\partial z_{g,n,f}} = 0 \quad \begin{cases} g = 1, G \\ n \text{ not first in group} \\ f = 1, F \end{cases} \\ &\frac{\partial \phi}{\partial z'_{g,n}} = 0 \quad \begin{cases} g = 1, G \\ n \text{ not first in group} \end{cases} \end{aligned} \tag{10.74}$$

with or without the additional equations

$$\frac{\partial \phi}{\partial m_{0,c,f}} = 0 \quad \begin{cases} c = 1, C \quad c \neq s \\ f = 1, F \end{cases} \tag{10.75}$$

according to which form of ϕ is used (F is the number of filters, R the number of groups of filters having similar extinction variations, N the number of nights, G the number of groups of nights, C the number of constant stars; the partition in ranges r is implicit).

Equations (10.75) lead directly to $m_{0,c,f} = < U >_{c,f}$ which explains why both formulations are equivalent.

We can estimate the number of parameters as $F + N - 1$ for the extinction terms (neglecting the subdivision into ranges r), $GF + N - G$ for the zero points and eventually $(C - S)F$ for the magnitudes. For relatively large numbers of nights and few groups, this gives a total of about $2N$ or $F(C - S) + 2N$ parameters. This is very small when compared to the large quantity of parameters classical methods would have to determine (F extinction coefficients and zero points for each night).

The choice between formulations (10.71) and (10.73) is not critical. For the ease of programming, Eq. (10.71) is to be preferred, although it results in rather large, sparse, matrices in the optimization procedure. If confronted with a very large number of constant stars and many filters, and with access to a small computer, it certainly is wiser to use Eq. (10.73). Let us also point out that numerical accuracy can be degraded when solving big linear systems. Fortunately, specific techniques exist for such sparse systems and they ought to be adopted.

One could argue that because of the larger number of parameters that are involved in Eq. (10.71) (as compared to Eq. 10.73) these parameters would be determined with a lower accuracy. This is not the case. The extinction and zero-point coefficients obtained in both calculations are exactly the same. The additional undetermination due to the increased number of unknowns is entirely redistributed on the m_0. As soon as Eq. (10.72) is used to complement Eq. (10.73), the m_0 are found to have exactly the same values as with Eq. (10.71). Both methods are entirely equivalent.

A programme based on this method has been used with great effectiveness in several problems, e.g., the set-up of a new photometric system appropriate to Wolf-Rayet stars. It shows all the advantages of a global multi-night, multi-star algorithm. Since it does not include the color transformation, it easily deals with incomplete data, i.e., missing filters. This means that an extremely flexible observing schedule can be adopted, with, for instance, entire periods dedicated to only one or a few particularly interesting filters. The constraints between extinction coefficients and between zero points proved to be very useful, and considerably reduces the noise on those parameters.

As the number of filters increases, it is seen that the algorithm is actually suited to the reduction of spectrophotometric data. The method is also easily adapted to multi-telescope observations (on a same site) by allowing separate variations of zero point Eq. (10.67) for each telescope. Extinction information is then shared between the telescopes. This extension is also valid for photometers where different channels have different zero-point variations.

Chapter 11 Homogenization

11.1 Introduction

Photometry consists of quantitative observations. Besides measurement errors, which are discussed in another section, these observations are affected by the quality of the instrumentation. In most experiments, the measuring apparatus has to be calibrated in order to ensure conformity to the established standards. For instance, land-surveyors and geometers need to know whether they use an accurate unit of length. It is quite easy to apply a relative correction of e.g., 10^{-4} to all distances, if it appears that one uses a tape-measure which is off by that amount. Such differences would be more difficult to correct if they were found to be temperature-dependent: one could write down the temperature during each measurement and take that into account. More complex cases arise, e.g., when the temperature varies along the length of the tape. This would occur when surveying a field with a long tape-measure. The ground temperature varies according to the nature of the surface (rock, sand, grass...), to the wind speed and direction, to insolation, etc. We are then confronted with a situation where the measurement depends in an intricate way on the measured object itself. And unfortunately the measure gives almost no clue on the nature of the subject of the analysis, so an iterative correction procedure cannot be applied.

The photometrist is in a similar position when he aims a telescope towards a star. The measurement is expressed by an integral equation such as Eq. (1.14). This equation involves the shape of the stellar spectrum, which is unknown, and the instrumental and atmospheric functions. Consequently, different instruments with different spectral transmissions perform in different ways and there is no way to compare their measurements, unless the stellar spectral irradiance and the instrumental functions are perfectly known—but who would need photometry in this case? Theoretically if the wavelength sampling by the photometric system is sufficiently dense, i.e., if the passbands overlap generously and have quite smooth transmission curves, the color transformation and the extinction correction can be accurately performed. However, this is the case for none of the existing systems—though some interesting proposals for such instrumentations have been made.

The symmetry of Eq. (1.14) in the stellar and instrumental functions is obvious. The result is taken as a measure of the energy distribution of the stars, but by measuring a given star with different photometric systems—which is actually done in different observing runs—one could describe the result as a "measurement" of the photometric systems. Those measurements differ when different stars are used in the same way as the magnitude of a star varies when different systems are compared. The same could be said of the atmospheric transmission. One could even establish a photometric system with a single passband instrument for measurements obtained at air masses 1, 2 and 3. On the other hand, measuring a single star at those same three air masses, in the same band, amounts to a measure of the atmospheric transmission. It is actually a direct method for evaluating the extinction.

Summarizing the situation, we have an integral equation wherein the kernel is made of a product of functions, all of which can be variable, and we try to estimate the properties of one of those functions from the result of the integration. This requires that the other functions be kept as stable as possible. All of them, except the atmospheric extinction, can be controlled. To some extent the atmospheric extinction can be kept constant if all observations are made at constant air mass. Observational constraints are obvious, but the quality of the result is well worth it.

As to the instrumentation itself, let us simply state that:

> *photometric measurements cannot be compared unless they are obtained with rigorously similar instrumental configurations.*

This statement should be slightly amended to make room for zero-point variations. Any instrumental variation introducing a simple proportionality coefficient in the detected signal can be tolerated. For instance, a larger telescope or a higher-transmission filter, but with the same profile, can be substituted in any system. All other variations are to be strictly avoided. Those include transformations that are sometimes judged beneficial, such as:

- Going to a higher observing site: this modifies the atmospheric transmission and creates difficulties when the extinction varies appreciably over the passbands. This is particularly true in the Johnson-Morgan *UBV* system: originally the blueward flank of the *U* filter was defined by the glass envelope of the 1P21 tube, but, in practice, tubes with quartz envelopes are often used and the atmospheric window sets the blue cutoff. Most broad-band photometric systems suffer from this problem in the ultraviolet, but also in the infrared, since the atmospheric spectrum becomes quite irregular there. Therefore, compatible photometric sites must necessarily be at similar altitudes (generally 2000 to 2500 m).

- Replacing the old-fashioned bell-shaped passbands with transmission functions characterized by steep flanks and rippled tops, as produced by high-transmission multiple-cavity interference filters. In extreme cases, filters are even discarded and replaced by mechanical slots defining sharp passbands at the output of a dispersing element, in order to get the highest signal.

- Replacing the original detector with one having extended blue or red response. A good example of such a detrimental move is the use of EMI9789QB cathodes (giving extended blue response) for *uvby* Strömgren photometry. All *u* data obtained with such a combination are not in the Strömgren system.

11.2 Conformity

The above discussion was intended to show that, above all other considerations, astronomers should commit themselves to adhere strictly to rigorous technical specifications regarding the implementation of a photometric system. When setting up photometric equipment, great care should be given to select filters and detectors which perfectly reproduce those of the standard equipment, i.e., the equipment which was used in defining the photometric system. Transformation equations between various systems are valid only for sufficiently close systems, and for selected groups of objects. They should not be trusted in general cases; hence, perfect compatibility between equipments is always advisable.

Needless to say, this conformity of systems is seldom met in practice. Among all important existing photometric systems only the one set up by the Geneva group has kept strict specifications over the years. Its technical parameters are well documented and each realization of the system has been carried out with utmost care. We believe that, today, it is the only existing photometric system which is not accompanied by "clones". This is not only the consequence of the great care exerted by the originators of that photometric system, but also because the system is a "closed" one, i.e., it is available only at one or two sites and it is used almost exclusively by scientists belonging to the team (for a discussion on this topic, see Rufener 1985).

Most other systems are "open" and used by many different people in many places. Proper attention is seldom given to the problem of conformity. In the most common situation, a multi-purpose photometer serves all photometric systems used at a given telescope or observatory. Before an observing run, apparently adequate filters are chosen from the observatory stock, depending upon the current availability. A choice of a few photomultipliers may be proposed, or one may be forced to be content with a single wide-band one for all observations. Damaged or lost filters are replaced according to very loose rules, often with more emphasis on maximum transmission than on passband shape. In such conditions a visiting observer may well make each of his observing runs with different instrumentations, sometimes without even being aware of this fact. Needless to say, such a situation has created numerous clone systems which are only barely compatible with each other (and with the prime standard system).

The effect of inhomogeneous passbands has been investigated through numerical simulations and direct observations. Problems with broad-band systems were discussed by Bessell (1986) and Taylor et al. (1989). Results concerning the *uvby* system have been published by Manfroid (1984, 1985a, 1985b), Manfroid & Sterken (1987) and Sterken & Manfroid (1987); these authors carried out numerical simulations using the Gunn &

Stryker (1983) spectrophotometric atlas and various filter- or slot-defined systems. In addition, observations were done with five photometric systems at two telescopes of the European Southern Observatory, La Silla; the passbands of the five systems are shown in Fig. 11.1 together with the standard passbands. Standard and program stars were observed in all systems, and the data were reduced to the standard system via linear transformations. Examples of the resulting differences between data obtained with these various systems are shown in Figs. 11.2 and 11.3. Similar results were also reported by Kurilienė & Straižys (1987), who compared the standard *uvby* system with a system having slot-defined *uvby* passbands.

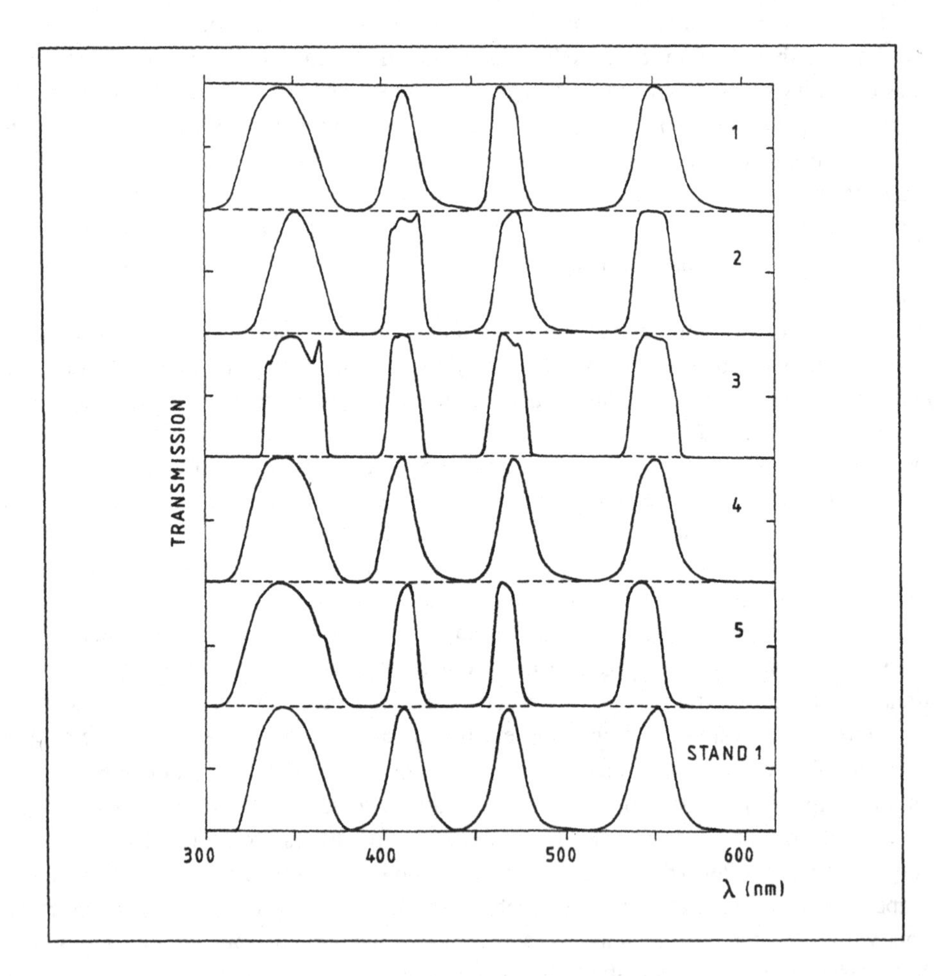

Fig. 11.1 *Transmission curves of five* uvby *filter sets. The bottom curves are standard passbands (Manfroid & Sterken 1987b).*

Those analyses simply represent an empirical investigation of Eq. (1.14) for the family of functions $E(\lambda)$ describing actual spectra, when reasonable variations of the

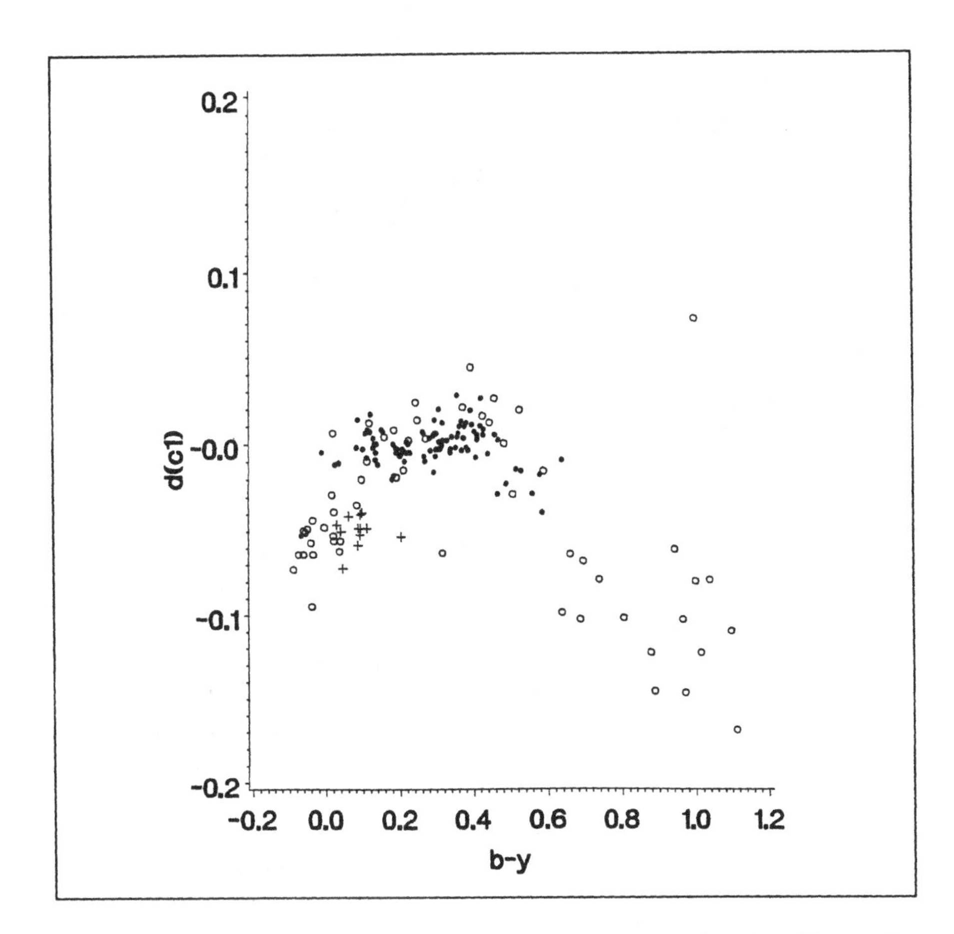

Fig. 11.2 *Inter-system deviations in* c_1 *between filter-sets 1 and 2. Observed differences in the index* c_1 *are shown as a function of* b–y *when the various filters of Fig. 11.1 are used (Sterken & Manfroid 1987). The symbol • represents standard stars, while o represents program stars. Only unreddened main-sequence stars are shown. Crosses are stars belonging to the young open cluster NGC3293.*

function $s(\lambda)$ are explored. The main conclusions for the *uvby* system are the following:

- Color transformations are only valid for restricted domains of stellar types and classes and for small ranges of interstellar reddening. For instance unreddened stars with $b–y < 0.0$, $0.0 < b–y < 0.2$, $0.2 < b–y < 0.4$ and $0.4 < b–y$ need to be treated separately. Those boundaries depend slightly on the instrumental system and mostly reflect spectral properties. Due to the underrepresentation of very hot stars, the lower limit was first found by numerical calculations, and not by observations.

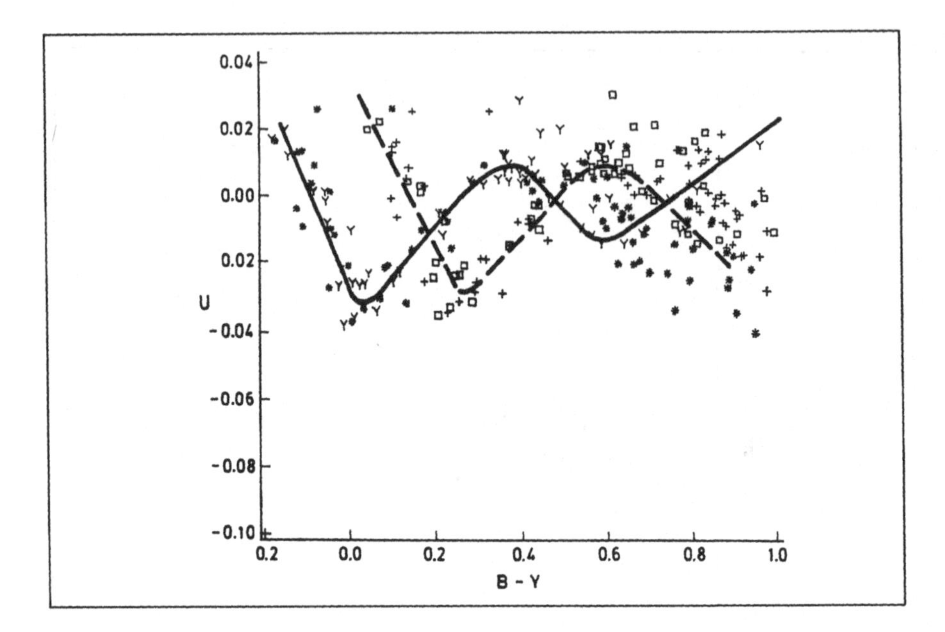

Fig. 11.3 *Inter-system deviations in* u. *Differences in the* u *magnitude as a function of the* b–y *index for filters similar to the standard ones, but shifted 2 nm towards the UV compared to the standard ones (numerical simulations, see Manfroid 1985a and Sterken & Manfroid 1987). Symbols + and ⋆ respectively represent unreddened main-sequence stars and unreddened other stars.* Y *and* ▫ *respectively represent reddened (A_V = 1) main-sequence stars and other reddened stars. The lines are hand-drawn mean curves.*

- Relatively small instrumental differences can result in large deviations in the indices of some stars, particularly for B stars. Very few standard stars exist in that domain, so evaluating the accuracy of photometry of hot stars is very difficult.

- Different transformations are needed for main sequence stars and for more luminous stars, especially for stellar types G and later.

- Reddened stars present special problems: *most often, objects affected by interstellar extinction do not follow the same color transformations as unreddened stars.* Moreover, the separation points on the *b–y* axis are shifted according to the value of interstellar extinction. The observer may be led to believe that since indices of the program stars fall inside the range of indices of standard stars, exact transformations will come out. This is unfortunately not the case.

- Color indices of unresolved binary stars with components belonging to different subsets but with comparable luminosities cannot be transformed into a standard system. The problem is particularly serious for eclipsing binaries, where the relative contribution of the components to the total light varies throughout the eclipses. This is even true when the components refer to the same subset of transformation equations.

Dividing the stars into four groups according to the b–y index is a severe constraint since a sufficiently large number of standard stars have to be found and observed within each group. Taking into account the fact that luminous stars also require a special analysis and that reddening classes should be separated, one may appreciate the very high complexity of the color transformation procedure. Fortunately the multi-night reduction method based on Eq. (10.57) relaxes considerably the observational constraints by distributing the information on atmospheric extinction over all constant stars. A large number of standard stars is required but they need not be observed every night.

Most photometric work has been done without caring about those difficulties. There are two main reasons. First, many astronomers were not aware of the problem, let alone of its complexity. Second, there were no adequate tools to tackle the reduction procedure. Fortunately, in the quest for higher accuracy, the compatibility issue is now better acknowledged. More sophisticated photometric reduction algorithms are also appearing. This is especially important in the present time, with the increasing occurrence of coordinated multi-site and multi-wavelength campaigns, where there is increasing need to compare, combine and merge data coming from one observer's site with results provided at another scientist's telescope. The ultimate problem of observational networks is not communication, nor lack of financial support, nor the handling of these huge amounts of measurements. *The fundamental problem of observational networks is bringing the resulting data to a common standard:* it is a problem of homogenization.

There is also a tendency among astronomers to believe that future photometry will be done uniquely with CCDs and that the old problems will evaporate. This is an optimistic view. On the contrary, CCD photometry has led to the appearance of new filters, with all types of passbands (for an extensive discussion, see Sterken 1992). The variations of sensitivity on the blue part of the spectra of many chips is also detrimental.

Moreover, an old disease of photometry did develop, viz the "secondary standard syndrome": standard stars are bright, because they were originally observed with modest equipment; they cannot be measured with large telescopes or/and very sensitive detectors. This problem was almost eradicated in classical photometry because very few large telescopes were available for doing such photometry. The big instruments were reserved for spectroscopy or imaging, where the largest light-gathering power or the highest spatial resolution are needed. Hence photometrists almost always ended up working at small and medium telescopes because there is still so much to do on bright stars—mainly for studying stellar variability. CCD cameras, on the other hand, are often available only on large telescopes (although they have begun to appear at telescope sizes down to the 1 meter level). Together with the intrinsically high sensitivity of those detectors, this leads to an impossible situation since the original standards are too bright and are observable only with photoelectric photometers.

The usual solution is to define secondary standards. Those have to be bright enough to be observed with the classical photometers, and faint enough not to saturate the

CCD cameras. Apart from the fact that one could rather say that those stars are too faint to be observed with the classical photometers and too bright for the CCDs, the homogeneity problem enters with full impact. The secondary standards are often defined using an instrumental system rather different from the standard one. The errors are then aggravated by the use of still another system for subsequent observations. In practice not only secondary standards but tertiary and higher-order standards are used. The resulting accuracy of the final data is certainly awful. To add to this sad situation, there is no way to check those errors. To the contrary, the simple tests done on the standard deviations of the residuals may yield very small values, and large systematic errors go unnoticed. But they may show up as exotic astrophysical properties of very distant objects.

In the same vein comes the "improved standard syndrome". Here a new list of standard values, based on numerous and accurate data, is presented as a promising alternative to the original one. Again, even if the various fits look very nice, the data include the properties of a different instrumental system, and the accuracy that one believes to be brought to the third digit is illusory. And there are many classes of objects or reddening values for which the deviations could become excessively large. Observations with the same instrumental system could benefit from those new standards—in terms of *internal* accuracy only—but in other cases one has to be aware of their limitations.

An additional feature of those improved standards is that in general they have been subjected to clever color transformations, in order to fit best the original standards. Stellar subgroups may have been defined for which different empirical corrections have been applied. This makes the overall fit very good but, here again, the last digits are not very significant. But systematic errors are still lurking, and a new trap is created: strictly spoken the standards so-defined do not represent any existing or any physically possible instrumental system. The systematic errors introduced are such that no function $s(\lambda)$ could ever comply with the integral equation. Even if one observes in the same instrumental system as the one used to define the secondary standards, the data are not easily transformed to those standards: one has to follow strictly the same procedure as for the definition in order to get compatible data. Garrison (1985), discussing standards for spectral classification, pointed out:

> "When using a system of standard stars, it should be obvious that the standards should be taken under the same conditions as the unknowns, but some astronomers still violate this basic rule. The use of large telescopes has actually increased the temptation, since telescope time is at such a premium and since the new detectors are too sensitive to be used on stars brighter than 10th or 12th magnitude. Thus there is a danger that the use of poor secondary standards, and fewer observations of them, will increase in the future."

11.3 Homogenization

In practice, how can we compare data obtained during different observing runs? The strict answer to this homogenization problem has already been given: we cannot, unless all observations were made with exactly the same instrumentation. This means that the Geneva observing team can produce catalogues with consistent values for all their observations. But averaged or "homogenized" *UBV* or *uvby* data can only be much less accurate.

Of course, for some studies one needs only very crude values, and in these cases a general compilation of photometric indices can be useful. But one has to be aware of the severe limitations of such works. First, data from various sources are merged and, although weighting factors corresponding to the apparent accuracy of each source are generally introduced, the systematic errors due to the system incompatibilities never cancel. Next, the input catalogues have differing scopes. Some stars are listed in one source only, and the full systematic deviation of that catalogue will show up in the average. Some stars are present in several catalogues and different blends of systematic errors appear in the "homogeneous" catalogue. Clearly, such lists are anything but homogenous. Let us forget the improper label, but remember that using those catalogues requires the utmost care and that they cannot yield accurate astrophysical conclusions. Even for statistical work, one must be aware that any sample may be biased toward some of the input catalogues and reveal systematic trends.

This constitutes what Garrison (1985) calls the "tyranny of the mean", or the unfortunately common method of taking data for stars or for groups of stars randomly from the literature and then to use averaged values as "standards". Too often the term "standard" is used too loosely, and everyone should know that, no matter how many such measurements support that mean value, it can only be used as a standard when it is part of a well-defined system using standard stars and standard techniques. This point, especially, holds for comparison stars selected for differential photometry of variable stars. Such comparison stars are chosen only to eliminate (or at least to minimize) differential effects, but such stars are virtually never genuine standards of the system and should never be used as such: *constancy is a necessary condition for a star to be a standard star, but constancy of light is far from being a sufficient condition.* In practice, the following remarks should be considered if actual homogenization is needed.

- Data obtained with the same instrumental system can be merged, but in that case the same color transformation procedure should be applied to the whole data set. This procedure will be treated in detail in Section 11.4.

- Data obtained with slightly differing instrumentations can be compared if each set has been carefully transformed to a genuine—or at least a well-defined—standard system. This means that robust transformations have been obtained for well-defined stellar subgroups.

- In the determination of the corrections for reducing one catalogue to the system of another, attention should be paid to differences that are due to the fact that the distribution of the number of stars, common to the two catalogues, as a rule varies with magnitude. The form of this frequency function will influence the value of the statistical differences.

- Magnitudes and color indices obtained with strongly different intrumentations, even inside one photometric system, cannot be compared (with the exception of some well-defined cases). For instance the "visual" band (V in many systems), is rather well reproduced for most stars, even when the shape of the filter considerably varies. The slope of the spectrum in the blue-green region ($B–V$ of the Johnson-Morgan system) is also fairly well reproduced in other systems (e.g., $b–y$ in the Strömgren system), when one considers specific stellar groups, for instance A or F unreddened main-sequence stars. In such circumstances, useful comparisons can be made and average values can be computed.

- In general, data from different systems should not be merged. Even in a single photometric system, separate catalogues relative to the various intrumental systems should be kept apart. The data may—or may not—be transformed to the standard system, but in the latter case all parameters of the transformation should be given explicitly. In such a way other data appearing in the same systems can be directly compared.

11.4 Merging data from a single system

When several runs have been made with the same instrumental system, it is clear that the general multi-night method explained in Chapter 10 is most welcome. This means that all the original data should be analyzed together in order to extract better values for all parameters. When the size of the cumulated data sets increases, the problem becomes rapidly intractable, even for powerful computers, the main constraints being the amount of central memory and the total CPU needed. Both increase more or less as the square of the number of nights. The space occupied by the data (original and reduced), on the other hand, increases only linearly with the number of nights.

It is thus necessary to split large data sets into smaller units (e.g., of 20 to 50 nights each) and to reduce each of them separately. In order to show the errors introduced by this procedure, we shall discuss some actual examples concerning the *uvby* system. These examples come from observing runs carried out with 5 different implementations of the *uvby* system at the La Silla Observatory, in the framework of the LTPV project (Sterken 1983, see also Section 10.5). They represent an average of about six months per year (over a time span of almost 10 years) at one of the small photometric telescopes at La Silla (the ESO 50 cm, the University of Bochum's 61 cm, and the University of Copenhagen's 50 cm telescopes).

The color-transformation matrix **M** is written in the form of Eq. (10.64). A statistical analysis of this matrix obtained for each period shows that the rms values of the m_{ij} (given in 0.0001 in Table 11.1) for any of the five instrumental configurations stayed constant within a few percent: there is no indication of any significant regular variation of the equipments. The small variations are random fluctuations mainly caused by the particular distribution of the standard stars, by inaccuracies in the standard values, and by measurement errors. Figures 11.4 to 11.6 show how the uncertainties on the color coefficients translate into errors on the color indices themselves. The space of the dependent variables b–y, m_1, c_1 was scanned and the corresponding standard indices were calculated for each observing run. The graphs show the standard deviation of the inter-run variations that can be expected within a given particular system (the plots concern measurements obtained in system 7 and have been chosen because a larger amount of data is available; graphs made for other systems are very similar).

Table 11.1 rms values of the m_{ij} ($\times 10^4$) for each of the five instrumental configurations. System 7 corresponds to the Danish 50cm SAT telescope (for a more complete identification of the systems in column 1, see Manfroid et al. 1991). System #8 data come from the ESO 50 cm telescope equipped with a sequential photometer, an RCA9789RA photomultiplier and new, high-transmission, interference filters. n is the number of periods.

System #	$\sigma(m_{11})$	$\sigma(m_{21})$	$\sigma(m_{31})$	$\sigma(m_{41})$	$\sigma(m_{33})$	$\sigma(m_{44})$	n
4	36	80	135	266	205	43	4
5	134	101	396	232	287	61	8
6	70	49	186	168	243	89	7
7	61	53	132	275	226	64	13
8	144	172	275	293	278	79	5

The y magnitude and the zero-point terms were not considered: zero points show variations (due to instrumental settings, status of mirror coating, temperature effects, etc.) which are not associated with any color effect and are not related to problems discussed here. Moreover, zero-point variations affect every star by the same amount and they are easily corrected. The inter-run variations presented here consist only of the color dependent terms. They apply to data corrected for the zero-point shifts, or to differential data (differences between a star and a comparison star, which are free from any zero-point shift).

Table 11.1 shows that some parameters are better defined than others. For instance, the m_{11} coefficient representing the b–y transformation can be easily determined. This is also the case for m_{44}, corresponding to c_1. However, the c_1 transformation is not so well determined because of the relatively large uncertainty affecting the non-diagonal coefficient m_{41}. Both other non-diagonal coefficients show a similar instability affecting y and m_1. But the m_1 index is the most sensitive because the diagonal term m_{33} is rather large. Let us recall that by definition $m_{22} = 1$ and thus shows no variance.

Since measurement errors are comparable in each band we conclude that errors on the m_1 index come mainly from the distribution of the values of the standard star indices. The range of m_1 is smaller than for other indices. There are few standard stars with extreme values, they change according to the epoch of the observations, and they have a strong weight in the calculation of the solution. All this contributes to increased inter-period variations.

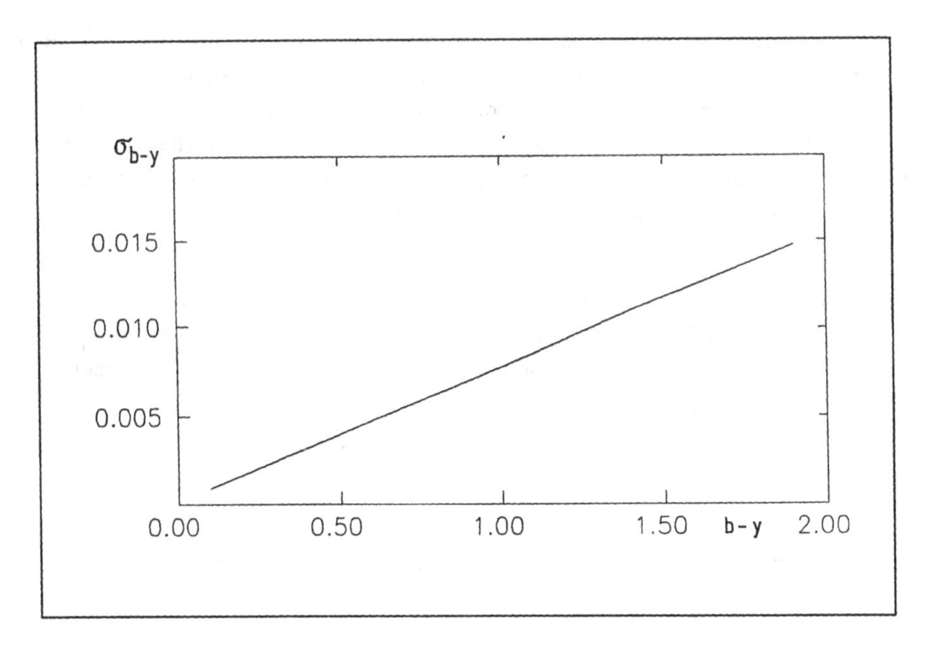

Fig. 11.4 Inter-run variations. Standard deviation of the inter-run variations σ_{b-y} *of observations obtained at the Danish 50cm telescope, as a function of* b–y *(after Manfroid & Sterken 1992).*

The role of the particular distribution of the standard star indices is obvious when the positions of those stars are plotted on the contour diagrams of Figs. 11.5 and 11.6. The orientation of the ovals is correlated with the shape of the standard sequence.

The average inter-run systematic variations easily amount to a few percent for small values of the color indices, e.g., for differential measurements of a star whose indices differ only slightly from those of the comparisons. The effects can reach .10 mag or more when large color indices are involved. This is always the case for many types of exotic objects (emission-line objects, cool stars...) for which it is impossible to find suitable comparison stars with similar colors since they would be equally exotic and also variable. The positions of the programme stars measured in the LTPV project are superimposed on the plots of Figs. 11.5 and 11.6. They are spread over a much larger area than the standard stars (some of them even lie outside the frame). On the other hand, the distribution of the bona fide (non-variable) comparison stars (which are not plotted for sake of clarity) observed in the LTPV closely mimics the standard star distribution.

The preceding discussion applies to the average inter-run variations (rms value). Statistically, about one third of all runs exhibit still larger effects, and huge variations appear in a few unfortunate cases.

The individual data sets typically contain 50 to 100 different standard stars, each of them with about 5 to 10 measurements. Those stars constitute a representative sample of the existing *uvby* standards. Reducing the numerical errors by increasing the frequency

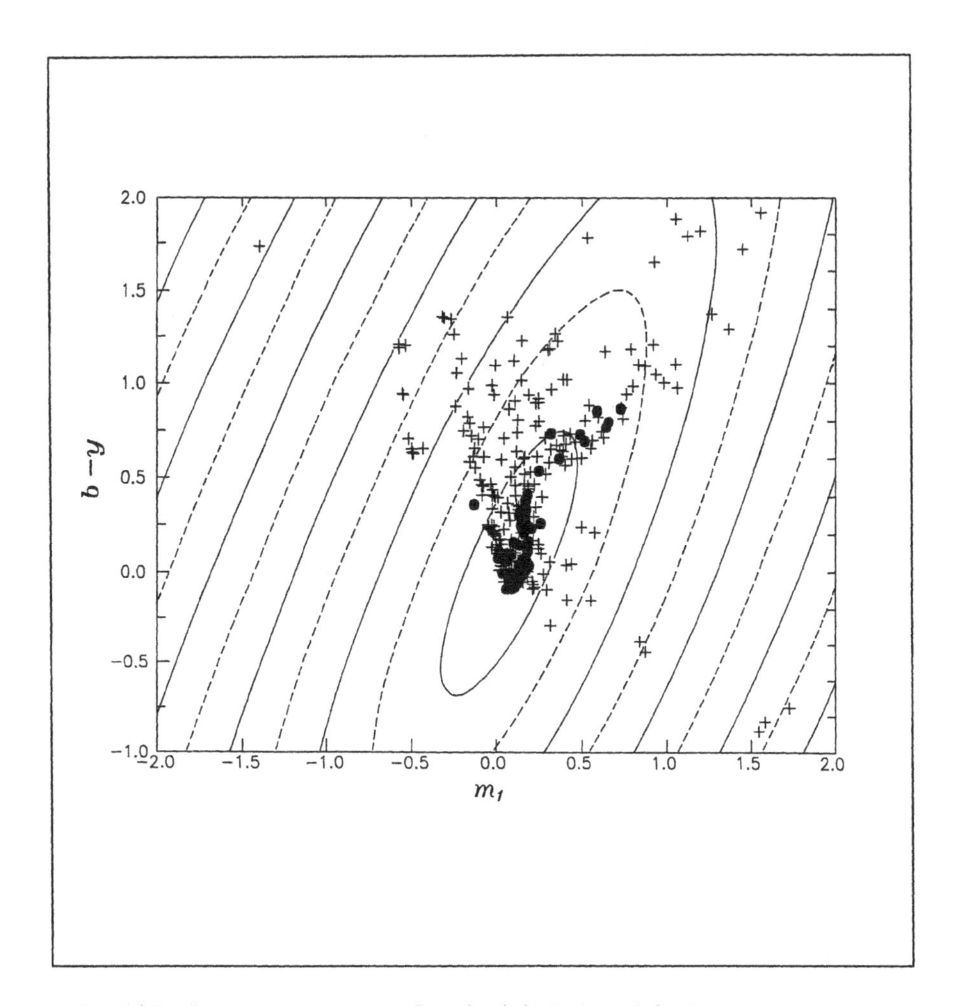

Fig. 11.5 *Inter-run variations. Standard deviation of the inter-run variations* σm_1 *of observations obtained at the Danish 50cm telescope, as a function of* b–y *and* m_1. *Adjacent contours are separated by 0.005 mag. The average positions of variable stars (+) and of standard stars (•) observed in the LTPV project are superimposed. The orientation of the ovals shows that the dependence on* m_1 *is dominant. This reflects the narrow range of standard* m_1*-values (after Manfroid & Sterken 1992).*

of measurements (by a substantial factor of 5 or 10) would impose too heavy a burden on the observing schedule, and is totally unrealistic.

A significant improvement could be achieved when a wider set of standard stars, including many more constant stars with peculiar or extreme indices, becomes available. Figures 11.5 and 11.6 show that mainly the ranges in $b-y$ and m_1 need to be extended. That is not an easy task: it would involve late-type stars, deeply reddened stars or

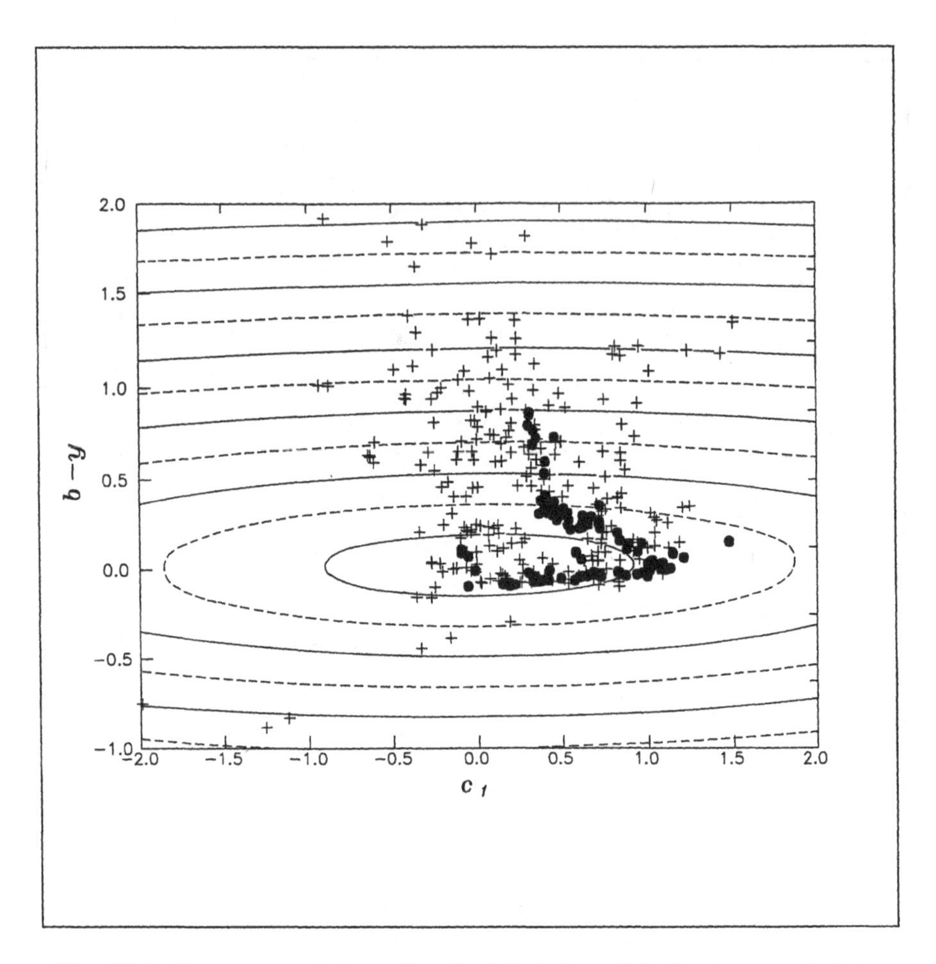

Fig. 11.6 Inter-run variations. Standard deviation of the inter-run variations σc_1 of observations obtained at the Danish 50cm telescope, as a function of b–y *and* c_1. *Adjacent contours are separated by 0.005 mag. The average positions of variable stars (+) and of standard stars (•) observed in the LTPV project are superimposed. The orientation of the ovals shows that the dependence on* b–y *is dominant in this case. The range of standard* c_1 *values is very wide but the corresponding* b–y *do not show such a large dispersion and do not allow a good determination of* m_{41} *(after Manfroid & Sterken 1992).*

emission-line objects, many of which could prove to be variable. In fact, a large number of slightly variable stars with extreme indices could be used to derive more accurate color coefficients than a dense pack of constant normal stars could provide.

Fortunately, it is not difficult to implement a method which merges the various data sets of the same photometric system in such a way that most photometric parameters are improved. The method which may come to mind is to average the color transformation parameters (with proper consideration for the individual zero points) and to recalculate the final data with those coefficients. Doing so for large, already well-defined, data sets can only give good results. However, one sees immediately that the same reasoning that we applied to go from the single night to the multi-night procedures can be used. The asumptions are that a single transformation scheme is valid for the global data set, and that constant and standard stars are available.

The data of each smaller period are reduced to the instrumental system by applying the inverse color transformation, which is supposed to be known. This implies, by the way, that the original transformation is single-valued, which is not an obvious issue when stellar subgroups are treated independently. A global transformation is then obtained by minimizing a merit function defined as the sum of the squared deviations of standard star observations to standard values, and of the squared deviations of the constant stars to their average values. These constraints are the same as those imposed by the multi-night procedure, but they are applied to data already corrected for atmospheric extinction. The simpler algorithm allows one to handle a quasi unlimited number of nights. The advantage over a simple averaging of the transformation relations is that new information about the constant stars, i.e., their constancy over the whole period, is included. Long-term variations of some stars, and particularly of standard stars, can be detected so that the final reductions are more accurate.

The adopted procedure allows a continuous updating of the data sets. Every time additional measurements are obtained in the same instrumental systems, the complete set corresponding to this system may be reprocessed.

11.5 Conclusions

The transformation of photometric data between two different systems (often between an instrumental and a standard system) introduces errors of two kinds that one could designate respectively as *conformity errors* and *reduction errors*.

Conformity errors arise from the very fact that the passbands of both systems are different and that there is no rigorous way to estimate to how much those differences can amount to for any object that could be observed.

The reduction errors, on the other hand, are of a purely methodological nature, and are due to the limited range of stellar types used in the color-transformation procedure. Uncertainties in the determination of the reduction parameters yield systematic errors which are largest for stars with extreme indices. Hence, measurement errors of a few millimagnitudes, and inaccuracies of the same order in the standard values, cause much larger uncertainties in the indices of many such exotic stars. Those systematic effects exist even between sets of observations that have been obtained using identical standard stars. They are enhanced when the observations which are compared include different subsets of a general standard star list. Note that we are speaking of a well-defined standard system, where the standard values were obtained with a single equipment, in an unbiased way. If the standard star data are not homogeneous, still larger effects can be expected, since they include both reduction and conformity errors.

Homogenization techniques can address the reduction errors, but they are powerless in confronting conformity errors.

Chapter 12 Infrared photometry

12.1 Introduction

Herschel (1800) was the first to describe the infrared portion of the solar spectrum. The infrared wavelength region is a less well-defined part of the spectrum than is the visible. While the visible region covers all wavelengths to which our eyes are sensitive, the infrared deals with radiation beyond the red cut-off wavelength of the visible domain, say from about 800 nm up to about 1 mm. Beyond 800 nm the human sensory system switches from the eye to other receptors which are sensitive to heat radiated by warm and cool objects. In the 800 nm–1.1 μm wavelength region, red-sensitive photographic emulsions, silicon CCDs and also red-sensitive photocathodes can still be used. However, longward of 1.1 μm Si becomes completely transparent, and these detectors become ineffective and special detectors must be used instead (see Section 12.3).

The infrared region is subdivided into near-infrared (1.5-5 μm), infrared (5-30 μm), and far-infrared (30 μm to 1mm). Beyond 1 mm we enter the realm of radio astronomy. This subdivision originates from the different detectors and observational techniques used for each of these spectral ranges. At wavelengths from about 30 to 300 μm, measurements from the ground are impossible to achieve. In addition, at wavelengths beyond 30 μm, interference filters are difficult to make and are not widely available commercially.

In what follows we confine ourselves to wavelengths below the 30 μm limit, where most infrared measurements are still made by photometry, which—in principle—goes along exactly the same principles as those we have described in the preceding chapters, but transposed to longer wavelengths. That shift brings very specific differences and complications (but also advantages) at various levels, viz.

- The atmosphere in the infrared region has a complex and strong absorption spectrum, mostly due to H_2O (all over the spectrum), CO_2, N_2O and O_3. The concentration of these constituents (particularly H_2O) varies from site to site and from time to time at a given site, which results in a spectrum of broad and deep absorption regions which, in some cases, totally block all radiation in that specific wavelength band (see Fig. 12.1). Hence, the extinction effect in the infrared is much more difficult to correct than it is in the visible domain. There fortunately exist a number of narrow regions (called atmospheric windows) between these molecular bands, where the radiation can reach the ground. In these windows narrow sections have been selected to define standard photometric systems (*JHKLM*, see Chapter 16). A large part of the disturbance by the atmosphere can be eliminated by doing infrared observations by means of aircraft-, balloon- or rocket-borne instruments. Obviously, space is the best site for infrared work.

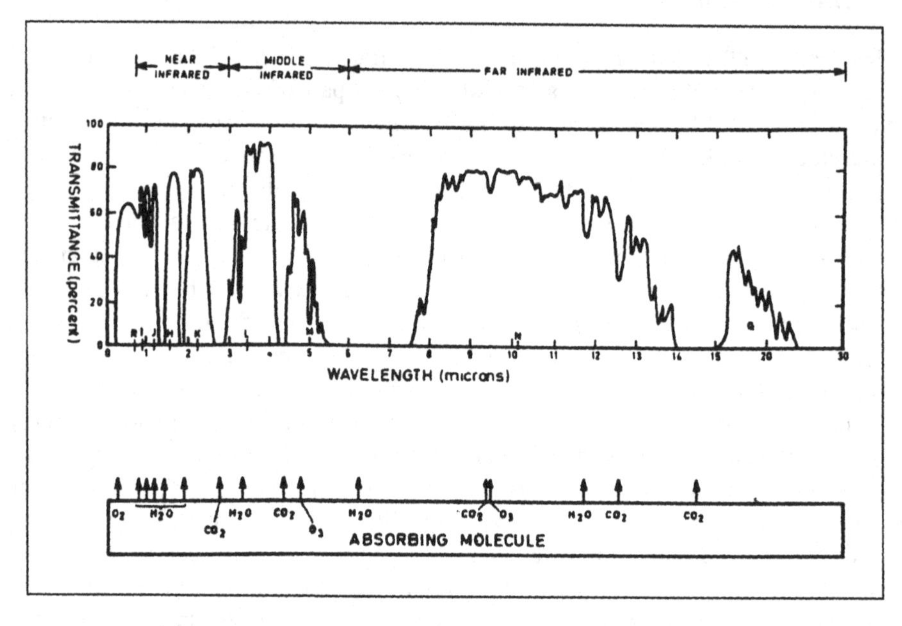

Fig. 12.1 *Atmospheric transmission in the infrared. Infrared atmospheric transmission at La Silla, corresponding to 1mm of precipitable H_2O above La Silla (taken from The ESO Users Manual, 1989). The central wavelengths and designations of the principal infrared photometric bands are indicated.*

- Not only absorption is a problem, the sky too emits strongly in the infrared as a non-black body at a temperature of about 260 K; this emission peaks at about 15 μm, close to a very good atmospheric window. This atmospheric emission is not only strong, but also variable (this is even so at balloon altitudes, where absorption becomes very low) and sets the minimum background level which is received by the detector. The major problem arising from the sky emission is that it is rather nonuniform and affected by large-scale spatial gradients, which translate into temporal variations. In other words, the noise associated with sky emission is not restricted to photon noise only. Since the sky emission is much less than that from the telescope itself, such gradients can be detected only through a *beam-switching* technique (see Section 12.2). Nonuniform emission combined with scintillation results in excess noise, commonly called *sky noise*. This is the dominant source of noise in the 5–30 μm region.

- Local radiation at ambient temperature may also disturb the measurements. At 10 μm, objects at room temperature radiate prominently, and so do telescope mirrors and telescope support structures. Therefore the telescope must be of specific design so that no "warm" surface is seen by the detector. All components which cannot be hidden (filters for example) have to be cooled to very low temperatures. Still, the mirrors of the telescope, and the telescope support structure, contribute to background radiation.

- Another factor of importance for infrared observations is the seeing. As the diameter of the seeing disk decreases slightly with longer wavelengths, inexperienced observers tend to think that, in case of bad seeing, it would be better to observe at longer wavelength bands than at shorter wavelengths. Unfortunately, this is usually not true since bad seeing is frequently associated with high sky noise: sky noise increases rapidly with increasing field of view; therefore, for large telescopes (for which the minimum field of view is determined by the seeing and not by the diffraction limit, see section 12.4) the seeing directly affects the background incident on the detector.

Because of its very specific requirements, infrared photometry has been developed and applied only at those very few sites where excellent atmospheric conditions combined with high technological capabilities are available.

12.2 Chopping and nodding

At 10 μm, objects at room temperature radiate prominently, and so do telescope mirrors, telescope support structures and domes. Another aspect of infrared photometry is that the photon energies are so low that enormous numbers of photons are required to give detectable power levels, so signal shot noise becomes secondary to the background shot noise.

As a matter of fact, the background power (coming from within the telescope and from the atmosphere) received by the detector is, for a typical telescope, of the order of 10^{-7} W, whereas the signals to be measured may be as small as 10^{-15} W. The main problem is that the background is highly variable, not only because of the variation of the sky background, but also because of the variation of the telescope's own radiation (originating from constantly changing thermal gradients).

One must apply a technique called *discrimination* to differentiate between the signal coming from the source and the signal coming from the background. The procedure consists in moving the telescope from (sky+star) to sky. Taking the simple difference of the two signals, however, would not cancel all the changes in the telescope and sky backgrounds, and would even keep all the low-frequency noise. To suppress the excess noise at low-frequencies the signal has to be modulated. A technique called *phase sensitive (synchronous) detection* allows to distinguish between the modulated signal coming from the source and the one coming from the background. This is usually done through an apparatus called *"Lock-in Amplifier"*, but can also be achieved with adequate software. The modulation of the signal is obtained by a procedure known as *chopping*: two neighboring areas of sky (A, containing the object, sky and telescope, and the reference beam B, containing sky and telescope) are alternately "seen" by the detector, and this switching happens at a frequency that is compatible with the time constant of the detector (typically 8–13 Hz). As such, the background (which in the ideal case is identical in the two beams) cancels out by subtraction, and one is left with the pure signal from the source beam.

However, identical background signals are difficult to achieve, and when there exists a difference between the beams due to imperfect cancellation of the background, the phase-detected signal of the sky is not zero (in infrared jargon this is called the *offset* of the sky). Clearly, the (variable) offset will be included in the phase-detected signal from the source. To overcome this problem, the telescope is switched to the symmetric position, where A becomes the reference beam, and B holds the object. This beam-switching, or *nodding* is repeated every 5 to 10 seconds. As a result, both sky emission and any difference in telescope background between the two fields are eliminated (but not the noise on the background radiation!). The closer on the sky are the source and the reference beams, and the faster is the beam-switching made, the more efficient is that procedure. The chopping produces two images of the source which differ in phase by 180°. Therefore, the phase-detected signal from one of these images contains the sky emission, the gradient and the source), while the phase-detected signal from the other image contains only the sky emission and the gradient. As the two beams pass through very nearly the same air path, both the sky emission and any spatial gradients in the sky emission are cancelled by subtracting the two signals. This difference, then, will give twice the signal from the source.

Both processes of chopping and nodding are performed automatically by using a focal-plane chopper (rotating or vibrating mechanical chopper), or a wobbling secondary mirror, and by offsetting around the right ascension or the declination axis.

The requirements of beam-switching, but also the need for a "cool" telescope, make only a few existing telescopes useful for infrared observation. Most telescopes, for example, have sky baffles to reduce the influence of stray light, but at infrared wavelengths these baffles (which are at ambient temperature) introduce background radiation instead of suppressing it. Large telescopes nowadays are by design better adapted for infrared work, especially if they are to be installed at high-altitude sites in dry climates.

12.3 The detector

Until the late 1980s, virtually all infrared observations had to be made with single-pixel detectors (for a long time, imaging-type observations in the infrared had to be made by raster-scanning of the brightest objects). Since then there has been a tremendous development in the field of infrared array detectors. Starting with (32 × 32) arrays of elements, modern technology now provides (256 × 256) and even larger infrared array devices. However, no detector exists that completely covers the 1–5 μm spectral region, available detectors having cut-off wavelengths of about 2.5 μm and 4 μm.

Single-pixel detectors are mostly indium-antimonide (InSb) *photovoltaic* detectors (up to 5 μm) and *bolometers* (beyond 5 μm). In a photovoltaic detector the incident radiation produces a photovoltaic potential which sustains a current. This current is proportional to the number of incident photons per second. InSb are the most widely used detectors for observations in the 1.1–5 μm range. The bolometer is a *thermal* detector which senses the heating effect of infrared radiation by a slight change in its temperature-dependent electrical resistance. The predominant noise in such a detector comes from the fluctuations in the sensing element. All uncooled thermal detectors are limited by the ambient thermal fluctuations and by their large heat capacity, and for these reasons do no longer play a significant role in astronomy. The low-temperature Germanium bolometer is most widely used for observations in the 3–30 μm range (it is also sensitive at shorter wavelengths, but has been superseded in the near-infrared by the photodetectors). All bolometers are background limited (i.e. the noise is dominated by photon noise).

An infrared array stores the electrical charges at each pixel, until the array is read out. This happens by a transfer of the charge on each pixel to an individual outlet, and these charges are removed sequentially as a voltage which is digitized and fed into a computer. It is important to realize the fundamental difference in the way an infrared array transfers charge, and the way this is done in a CCD (see Chapter 13). For a more detailed description of the construction of infrared array sensors, and for a review of recent developments, we refer to McLean (1989) and Wade & McLean (1989). Infrared arrays are very complex devices, both from the point of view of construction as from the point of view of application. New arrays, offering both larger formats and improved performance, will certainly become available in the near future, and we therefore limit our discussion to classical infrared photometers only.

12.4 The photometer

Infrared photometers in principle do not differ from a classical visual sequential photometer, as described in Section 3.2: they contain a detector, filter and diaphragm wheels, Fabry optics (lens or mirror), and viewing eyepieces. However, in order to reduce the background radiation, a general optical design consists in avoiding that the central obscuration of the telescope be seen by the detector. Infrared photometers are bulky because of the specific requirements of infrared cryogenics, and eventually, because of the need for a focal-plane chopper and/or a dichroic filter to allow for guiding on the visual light of the source while the infrared light is being measured. In that respect, one has to note that dichroic filters (beamsplitters) usually reflect all of the infrared part of the light onto the detector, and let the visible light pass through. However, part of the visible light is also reflected and a faint object which could be seen in the absence of a dichroic filter could be invisible when the beamsplitter is used. Furthermore, due to the thickness of the filter, the centering of an object onto the detector is not the same with and without the filter in use. However, many objects observed in the infrared are not visible, so one must have a photometer designed in such a way that offset guiding on a nearby visible star is possible. Therefore, the viewing eyepiece is mounted on a movable stage which has micrometer controls for exact positioning off the optical axis.

It is essential not only to cool the detector, but also its associated optical parts. Hence, a prominent feature of an infrared photometer is the dewar, which prevents the cryogenic liquids from evaporating away. Inside the dewar sits the detector, which is in good thermal contact with the coolant liquid. Other components, such as filters and diaphragm wheels, which are seen by the detector, are mounted in the dewar too. In such a way, the detector sees only the uncooled optical train of the telescope. A schematic outline of a bolometer dewar is given in Fig. 12.2. The rigid temperature control offers an additional bonus because it enhances the stability of filter transmission characteristics (see Chapter 5). Infrared telescopes aboard rockets and spacecraft are being completely cooled to cryogenic temperatures (this cannot be achieved for ground-based instruments because of the unavoidable consequence of deposition of dew on the cooled surfaces).

Infrared detectors are cooled either with liquid nitrogen or liquid helium. Nitrogen is colorless, odourless and tasteless, and cannot be detected by the normal human senses. Nitrogen is non-flammable, and boils at 77 K. Helium, an inert gas, has the lowest boiling point of all cryogenic liquids (4 K). Helium is present in the air in the ratio of one part per 200,000. Commercial helium is obtained from natural gas where it exists in concentrations of several per cent by volume. The use and manipulation of those liquid gases puts some specific hazards to the observer: besides the risks of burn and frostbite, there is the fact that prolonged inhalation of cold vapour gas can damage the lungs. The manipulation of these substances can create oxygen-deficient atmospheres, which in extreme cases (oxygen reduced to about 10 per cent by volume) may abruptly hinder breathing; small amounts of these cryogenic liquids can be converted into considerable amounts of gas (about a factor of 700; in the case of helium, vaporization very easily happens). Both gases, however, are non-toxic. They are brought into the dome in portable dewar flasks. In the case of helium, utmost care must be taken to prevent air from entering down the dewar neck, and solidifying. Such a situation would cause slow but continuous increase in pressure inside, which would unavoidably lead to explosion.

12.5 Observation and reduction of infrared data

Observing sequences and data reduction are the same as those discussed in Chapters 6 to 11. Standard stars must be observed for determination of the extinction, and transformations matrices must be set up.

Infrared photometry has more error sources than has photometry in the visual spectral region. Besides the problems encountered in the former, extra errors occur due to the difficulty of centering optically faint sources, and due to the particularities of the atmosphere in the infrared.

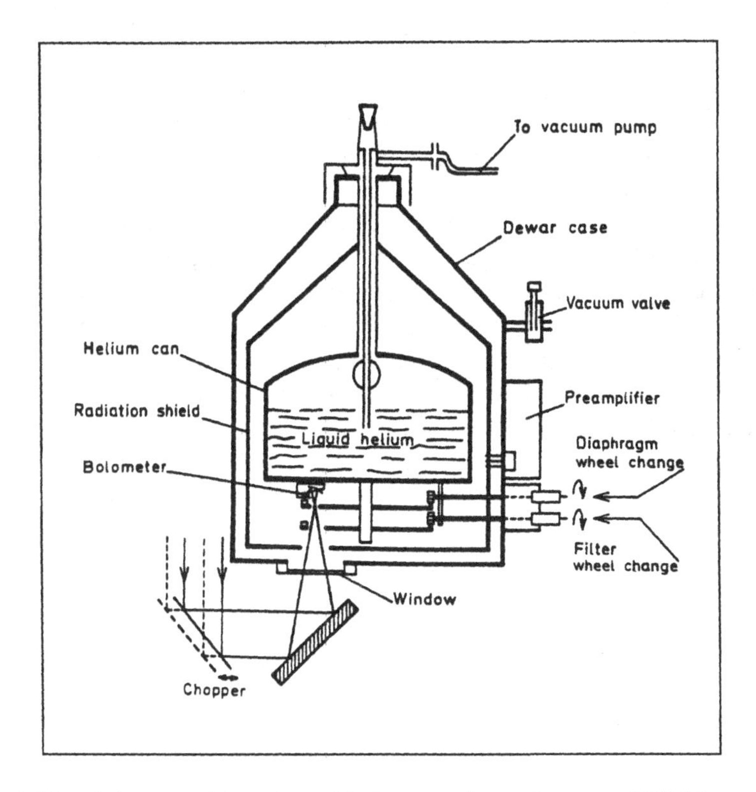

Fig. 12.2 *Bolometer. Side view of bolometer dewar (source: ESO Users Manual, 1989).*

One particular problem is the determination of infrared extinction. In Chapter 6 we adopted the assumption that atmospheric extinction is proportional to air mass, so that the observed magnitudes could be extrapolated to zero air mass. In the infrared, however, extinction is due to the presence of a large number of often saturated and overlapping molecular lines. Conventional corrections for extinction by linear extrapolation to zero air mass thus will cause an extinction error (see Manduca & Bell 1979) yielding stellar magnitudes which are too faint. Moreover, the *JHKLM* bands are wide and introduce color effects (see Chapter 8). Atmospheric fluctuations also cause large time-variations of these effects. Consequently, a good method of observing is to bracket program star measurements between standard star observations. An additional complexity comes from the uneven distribution of the various absorbers with altitude. Hence the absorption coefficient κ is no longer constant and the definition of air mass is inaccurate.

12.6 Day-time observing

An interesting aspect in infrared photometry is that, due to the fact that light scattering at infrared wavelengths is negligible, the infrared sky is as "black" in daytime as it is at night. This does not mean, however, that day-time observing is as good as night-time observing.

Observers have to be aware that image quality degrades in day-time, which results in enhanced noise. As the seeing decreases with longer wavelengths, day-time observations should thus be performed in the infrared region rather than in the near-infrared domain.

Also, the sky brightens and becomes more variable in the near infrared, both spatially (depending of the proximity of the Sun) and in time. As the level of background sets the sensitivity of the detectors operating in that spectral region, it is necessary to use, whenever it is possible, a low pre-amplifier gain. This recommendation applies also for observations taken during sunset and sunrise. Inexperienced infrared observers tend to believe that they can start observing before, and finish after, the observers in the visible. That is true only if special care is taken. Note also that sunlight on any part of the telescope should by all means be avoided.

Chapter 13 Charge-coupled devices

13.1 Introduction

In classical photoelectric photometry all the light falling through a diaphragm is integrated. When the image of a double star or part of a stellar cluster lies inside the diaphragm all contributions are automatically added. At the same time the sky background is also included in every measurement, making necessary separate observations of the background. Consequently the objects have to be measured sequentially. But there exist two-dimensional detectors which allow one to obtain images of several objects in a single exposure. If those images can be accurately calibrated in terms of irradiance—relatively or absolutely—then a much better use of telescope time will result. The first attempts with photographic emulsions gave poor results in that respect. The photographic plate can store a tremendous amount of spatial information but with a very poor photometric accuracy (see Chapter 14). In addition, the photographic emulsion cannot reliably record large magnitude differences and has such a low quantum efficiency that long exposures are generally required.

Fortunately, the development of electronic detectors has changed this situation and among these detectors the *charge-coupled device* (CCD) has emerged as an undisputed leader.* As an example, the CCD has a dynamic range (ratio of maximum detectable light intensity to minimum detectable light intensity) which covers more than 10 magnitudes. The CCD features high quantum efficiency and excellent linearity together with the ability of recording simultaneously the light intensity distribution of an extended field. It provides good performance in the red spectral region. Most CCD photometric work is made on stellar clusters and on galaxies. A few frames provide data on a very large number of stars up to very faint magnitudes, thus allowing detailed studies of stellar populations. Photometry of nebulae can be done without resorting to the slow and

* For an excellent description of CCDs and their applications, see McLean 1989.

intricate procedure of raster-scanning the area with the telescope. CCDs are much closer to the concept of perfect detector than is any other detector; they combine the accuracy of the photomultiplier with the extended field view of a photographic plate. Internal signal amplification, however, is not present.

13.2 The detector

In a *photoconductor* such as silicon, the absorption of a photon releases an electron from the valence band and promotes it to a conduction band (where it can be moved by an electric field). The release of such electrons happens with a very high quantum efficiency.

> *A CCD is a (metal-oxide semi-conductor, MOS) solid-state integrated circuit that stores the electrons produced by incident photons as discrete packets of charge in potential wells maintained by an electric field.*

A CCD consists of a 2-D array of discrete tiny photodiodes (often silicon). The detector is covered with a silicon dioxide insulating layer. On its surface an array of conducting electrodes is deposited. The CCD imager is subdivided into closely spaced columns, which are separated from one another by barriers a few micrometers wide. These columns, called “channels”, are at right angles to the electrodes. These channels prevent any charge movement sideways. The grid of channels and electrodes (also called “gates”) forms a two-dimensional array of pixels. A pixel is thus defined by a channel, and one set of parallel electrodes.

Usually the semi-conductor is of type p, which means that the charge carriers are holes. A positive voltage is applied to each cell and so creates a depletion zone. Any electrons which happen to be produced there—thermal electrons or photoelectrons—are stored in each cell, while the corresponding holes vanish through the substrate.

The amount of electric charge in each element represents information (a quantity which is proportional to the infalling radiant flux). *Charge coupling,* a relatively new concept in semiconductor electronics, manipulates this information. Charge coupling is the collective transfer of all mobile electric charges stored in a semiconductor storage element to a nearby element by the external manipulation of voltages.

Those successive voltages changes are known as *clocking*. When high-speed imaging is not needed, this process is generally done only once, at the end of every measurement, in order to minimize the electronic readout noise. High clock rates are quite possible (e.g., in TV cameras) and can be adopted in sufficiently high light-level conditions.

The principle of functioning of a CCD and the concept of charge coupling is easily visualized by a mechanical analogy (Kristian and Blouke 1982). One may compare the recording of an image with a CCD with the concept of fully automatic measurement of the distribution of rainfall over a remote, rugged field. To do this, one could construct a sophisticated pluviometer array (see Fig. 13.1).

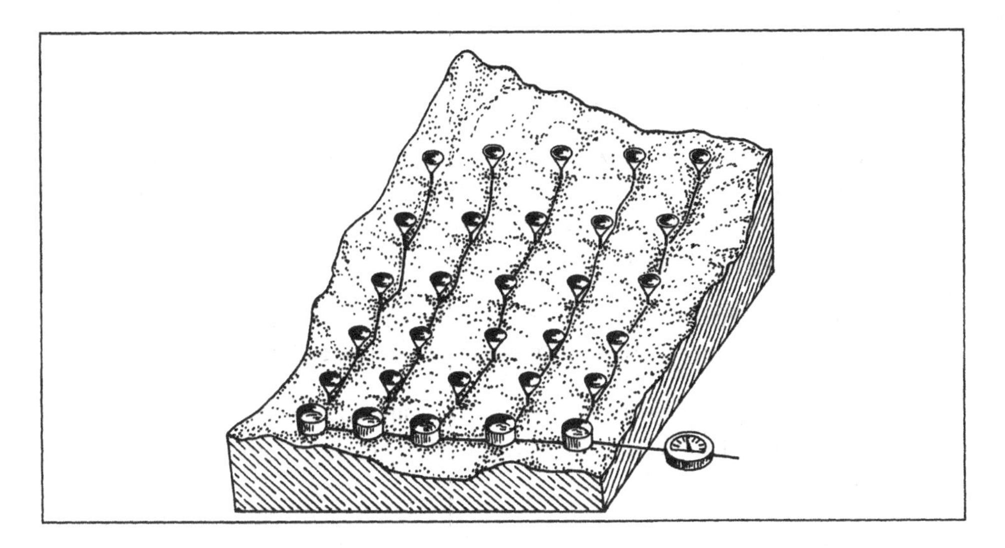

Fig. 13.1 *Principle of functioning of a CCD. In this diagram the CCD is explained by a mechanical analogy.*

In order to measure the amount of precipitation at each individual spot, one might consider installing a remotely controlled valve at each receiving element, and connecting each valve with an individual pipeline to a collecting water meter. Each pluviometer element would then be "read out" by sequentially letting its collected water drain towards the meter. This procedure would pose an extreme hardware problem in the sense that an enormous amount of water pipe would have to be fitted, insulated and maintained. For a square field of $n \times n$ evenly spaced collectors, the total pipe length would be approximately $n^2(n+1)/2$ units of separation. A more economic and feasible solution would be to connect all receiving elements columnwise by a pipe ("channel") , and to construct a row of output buckets with one such bucket at the end of each column. These individual buckets would then be connected by a row, forming an "output register", which in turn is coupled with a water metering device. The proper measurement would be a procedure which starts by emptying the nearest bucket to the output line in the output collector, and then, by opening the valves at each bucket, transferring all individual charges of water from every bucket to the bucket downhill. As soon as the first charge arrives in the leftmost collecting bucket, that bucket is emptied with an analogue cascade procedure, until the amount of water passes the meter. The time moment associated with every measured amount of water unambiguously determines the location the water comes from. With proper timing (clocking), a complete readout of the whole collected charge of precipitation could be done. Note that this solution only requires about n^2 lengths of pipe, so that the relative gain is of the order of a factor $n/2$. For large fields (for example 1000 elements square) the first solution needs a factor of 500 more hardware piping than the second solution does. This gain is extremely important, especially for extended fields.

This mechanical model also illustrates some of the associated problems, viz.

- individual sensitivity properties of each receiving element (due to manufacturing imperfections) may be present;

- if the time interval between two successive reading-out actions (exposure time) is too long, receiving elements will completely fill and even overflow. This saturation effect will cause registration of maximum capacity, whatever the real amount of precipitation is;

- whenever a valve problem between the output register and a column occurs, the registration of that whole column will be in trouble, and one has a *bad column*;

- collected amounts of water will not always completely flow to the adjacent bucket, and some may be left behind (but not lost); in the case of a CCD this is called a *charge trap*;

Figure 13.2 gives the structure of a so-called "three-phase" CCD; the scheme represents a 4 row by 5 column detector with 20 pixels.

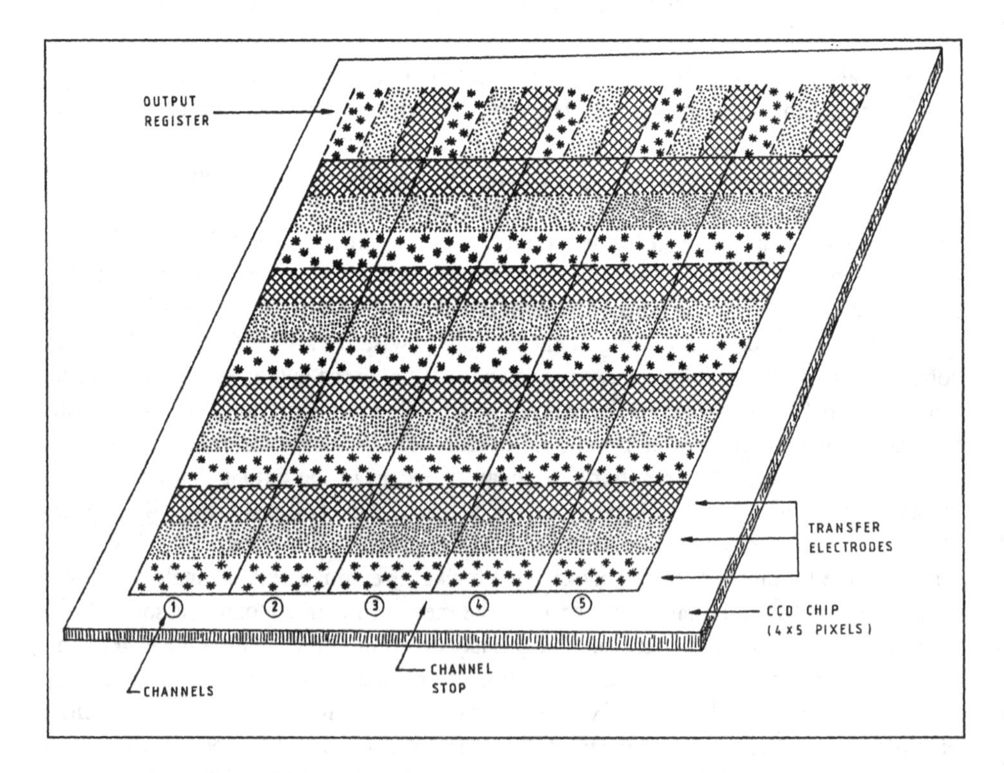

Fig. 13.2 Three-phase structure of a 20-pixel CCD.

During exposure, one set of electrodes (each member of the triplet in Fig. 13.2 is identified by a same type of shadow) is held at a positive voltage; the other electrodes are held at an almost zero potential. This creates a potential well under each of the positively charged electrodes, and this region can serve as a collector for electrons. Any electron in that collector well will move to an adjacent collector if electrode voltages are changed appropriately. When an electron is moved to the conduction band by an impinging photon, it is attracted by a nearby potential well, where it is kept. At the same time a hole is created by the removal of that electron; this hole is forced away from the well. During prolonged illumination of the chip, electron charges build up in the potential well associated with each picture element. The collected charge is directly proportional to the number of incident photons at each pixel.

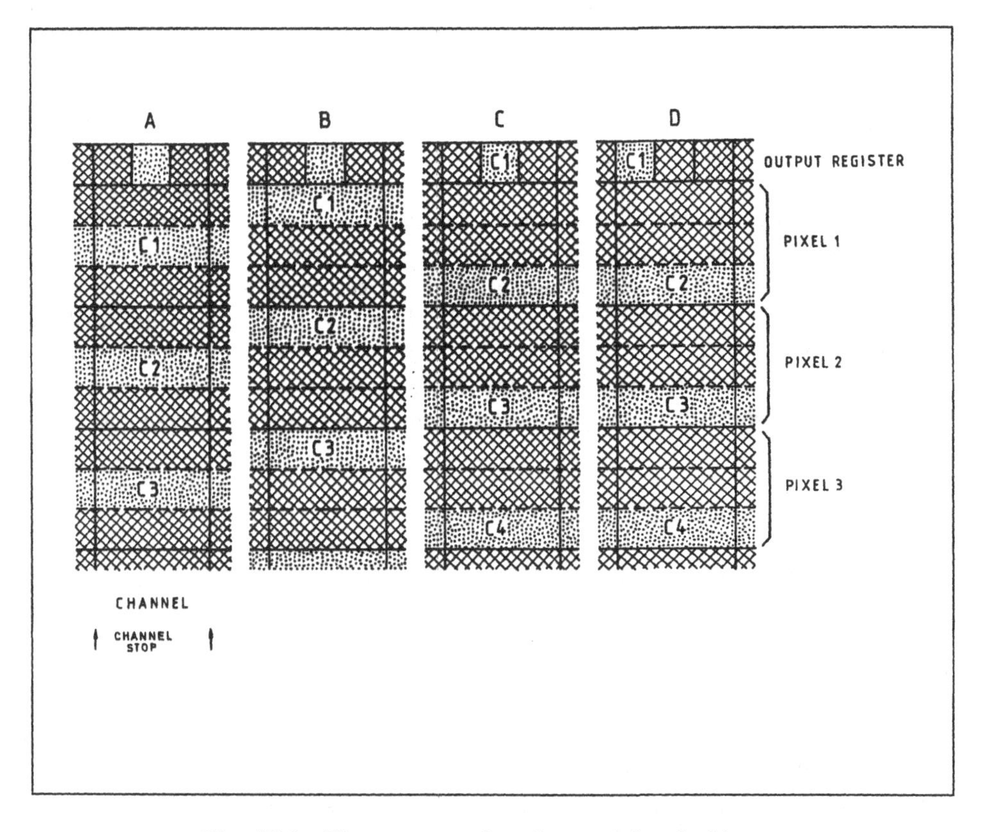

Fig. 13.3 Three consecutive phases of the clocking.

At the end of exposure the camera shutter is closed, and the 2-D charge distribution is read out. This reading out (clocking) is illustrated in Fig. 13.3, which shows the content of one and the same channel during three consecutive clock phases (A, B and C), and one readout phase (D). The potential wells with their trapped charges (C1, C2, C3,...) are moved systematically in the direction of the output register by lowering the level of the next collector toward the output register to the same level as the original well, after which the level of the original well is raised, building a barrier. This creates situation B out of situation A: electron charges are moved 1/3 of a pixel upward. The same procedure is then repeated again and again, until the original charge arrives in the output register. Each full clock cycle (three phases) will move the charge of each cell into an adjacent cell. The charges in the output register are transferred one at a time to the amplifier in the same way as they were transferred across the detector. Now, all original charges from one row (like C1) move towards the left to an amplifier. The whole process is repeated with each row, until the whole distribution of charges has been read. The amplifier allows the measurement of the charge present in each pixel.

The reading out of the charge pack closest to the amplifier invokes two transfers; for reading out the charge farthest away from the amplifier, $3n + 3m$ transfers are needed for a $n \times m$ array (this means 4500 transfers for a 500×1000 chip).

13.3 The camera

CCDs are not mounted in a photometer, but in a camera. Many CCD-cameras are in use now; small versions are even available off-the-shelf. In principle, at least, a CCD-camera is of simpler design than is a good photometer. Long time-exposures required for dim objects suffer from thermal dark current. This can be efficiently reduced by cooling the detector, which is often achieved using liquid-nitrogen vapor which is blown across a metal support on which the CCD is mounted. The temperature of the chip should be regulated to within 1°C.

The data acquisition system should allow simultaneous acquisition of the observation in process, and also the displaying and elementary handling of images taken earlier.

13.4 Characteristics of CCDs

The CCD has superior *quantum efficiency* compared to other detectors such as the photographic plate and the photomultiplier. This is clearly illustrated in Fig. 13.4 which compares the quantum efficiency (in percent) in function of wavelength for a typical CCD, a photomultiplier, a photographic plate, and the human eye. Peak efficiencies can exceed 80%.

The redder photons pass farther into the silicon because Si is more transparent to light of longer wavelength. CCDs are less sensitive for blue photons, which are easily absorbed in the overlying electrodes. One therefore uses "thinned" CCDs, very thin detectors used upside down (photons arrive from the back side). Blue response can be improved by coating the device with a thin layer of a chemical that absorbs blue photons and reexcites redder light. This procedure, however, remains not very effective.

Several sources of noise limit the CCD performance. The number of electrons stored in every pixel and attributed to photons coming from the object (plus the sky background) follows a Poisson statistics and the noise is equal to the square root of the electron number, whatever their origin (see Section 1.1). The dark current provides noise of its own.

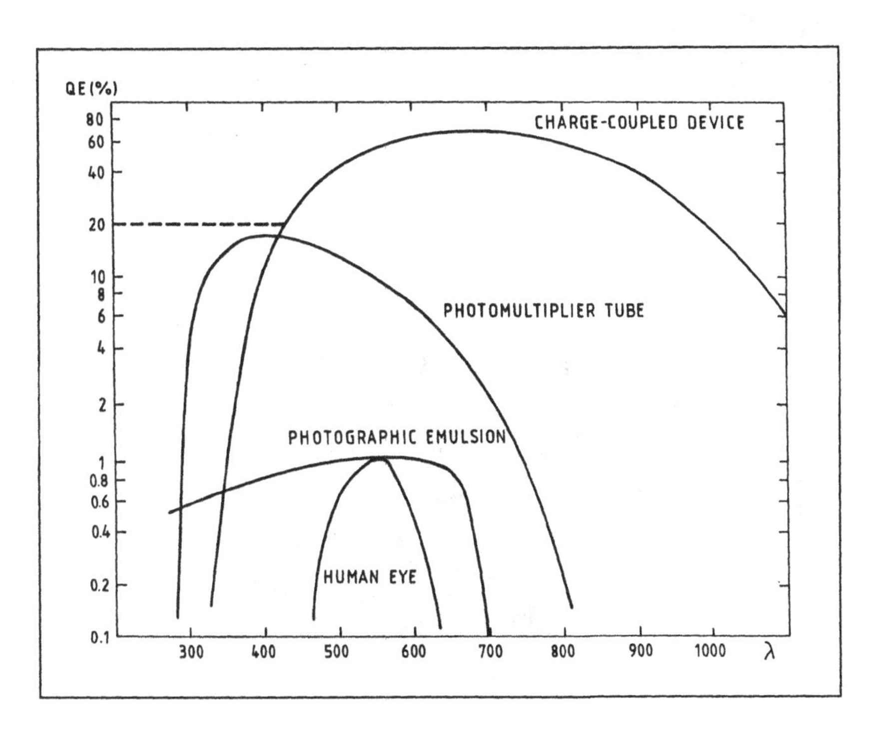

Fig. 13.4 *Quantum efficiency of CCD compared to other detectors (after Kristian & Blouke 1982).*

The number of counts actually recorded involves an additional and irreducible noise, which is present in all frames, the *readout noise* (RON) from the CCD output registers. Manufacturers give a numerical value for the single-pixel RON. Since stellar images easily may contain a dozen pixels, the RON per star will be, in that case, much larger.

Mackay (1986) describes the three different noise regimes as the signal level varies: at the lowest level the readout noise dominates; at higher signals the shot noise (square root of the signal in electrons) is added to the readout noise; near saturation the S/N finally levels off due to non-linearity. A *variance diagram* can be constructed from ratios of identical images (by taking at least a pair of calibration exposures at several different signal levels, but preferably several exposures at many different levels). The variance diagram is a straight line whose parameters yield the readout noise and the ratio between *data-numbers** and actual numbers of electrons.

Charge Transfer Efficiency (CTE) refers to the fact that not all electrons advance with the packet on each transfer, the residual charge appearing in a trailing package (generally in columns, but poor performance in the serial register manifests itself in rows). The charge-transfer efficiency is a function of the design. Reasons for low CTE

* Data numbers are expressed in DN, the unit adopted for convenience in a particular acquisition system for the analogue-to-digital conversion. For example a 16 bit system may use a maximum of 65536 DN. The acronym ADU, for *analog-to-digital unit* is also used.

are multiple: the electrons are inhibited from moving because of lock regions of lower potential energy, or the frequency of operation is too high so that there is not enough time for all the electrons to follow the moving potential wells. If a fraction ϵ of charge is left behind at each transfer, the result after n transfers is a loss of $1-(1-\epsilon)^n$ of total charge.

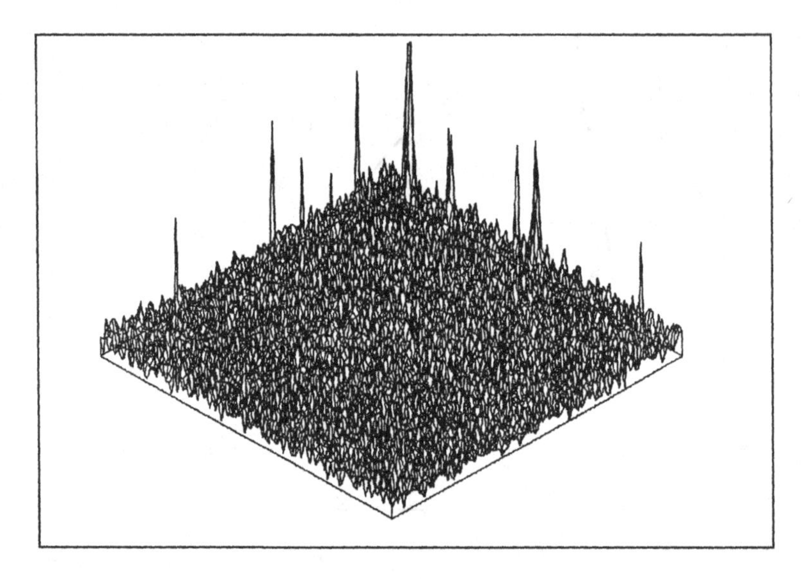

Fig. 13.5 *Raw output (partial) of a dark exposure. The vertical axis represents the signal present in each pixel. The vertical spikes (shown along the rim only) are spurious signals due to cosmic ray events.*

Because of its random fluctuations, the readout noise would generate spurious negative values in the signal. To prevent this situation, an electrical offset is applied to the device before readout by adding a constant amount of charge to each pixel. This level is called *bias*, and can be determined by a very short (ideally 0 sec) exposure; the bias level must be subtracted from every exposure in the reduction procedure.

The minimum detectable level is determined by noise, typically a few electrons. The maximum level is fixed by the saturation at high exposure. Each potential well can only hold a certain number of electrons before the well is full; this sets the saturation level. This generally occurs at a value of a few 10^5 electrons, giving a dynamic range nearing 10^5. Before saturation occurs, non-linearity effects show themselves as a drop in quantum efficiency. Hence realistic values for the dynamic range of a CCD are of a few 10^4. At saturation, the electrons begin to spill to adjacent pixels in the same column, and in extreme cases, the whole column may be affected. This is known as *charge bleeding* or *blooming*. Adjacent rows are not affected.

Charge traps are impurities or defects in device structure which prevent charges in certain columns from being transferred to the next pixel; the charge may be released

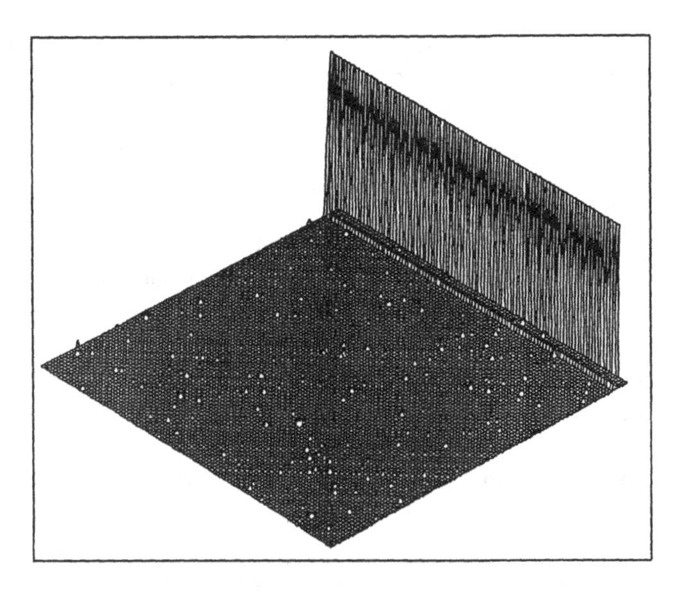

Fig. 13.6 *Occurrence of a bad column is readily apparent upon first inspection of output. These columns must be removed before any further signal processing and reduction takes place.*

later, and this phenomenon is known as *deferred charge*; one way to avoid deferred charge—at the cost of increased noise—is to *preflash* the CCD so that the traps are filled with electrons;

Bad columns are columns of the CCD array which do not read out charge. Such defects are readily visible at first inspection of the raw output of the frame (see Fig. 13.5).

Some pixels are more sensitive than others, but each pixel is approximately linear in its response. This spatially non-uniform response is due to manufacturing imperfections, geometrical variations between pixels (this includes dust grains or any object projecting a shadow on the photosensitive surface), inhomogenities of the material (typically seen as a tree-ring pattern due to the doping process of the semiconductor), and mechanical stresses (possibly induced by the thinning process). Two kinds of non-uniformities can be distinguished: pixel-to-pixel variations (high-frequency, with values of a few per cent) and overall gradients across the chip (low-frequency, amounting to tens of percents).

In order to eliminate those spatial variations the image is divided arithmetically, one element at a time, by a "flat-field" that is made by exposing the CCD to uniform irradiance. In such a way one gets an image with consistent response (this effect can also be visualized in the mechanical model of Fig. 13.1).

One can use a white diffusing screen, or take a diffusely illuminated spot on the inside of the telescope dome and view it with the CCD in its normal observing configuration, using exactly the same optical path as is used for the science exposure. Particularly, since flat-fielding is done with the use of a cool incandescent lamp, a blue flat-field may be severely biased by the presence of a red leak (unless appropriate color-balance filters are used to modify the spectral characteristics of the flat-field lamp). An alternative is to use a CCD image of the clear twilight sky—just after sunset or just before sunrise—for flat-fielding. But clear sky has a different color temperature than the night sky (there are also differences due to emission lines and polarization), and this introduces another systematic effect. The color of the night sky can be approximated by using color correction filters (see Stetson 1989, Djorgowski & Dickinson 1989) so that spectral mismatch between the data and the flat-field is avoided.

Correct flat-fielding is very important: residual flat-flielding errors are a limiting factor in CCD photometry since incorrect flat-fielding will cause the magnitude of a star to depend on the place of the detector the star image fell on. These non-uniformities are strongly color dependent and one should in fact perform several flat-fields using the same filter as for the science exposures, and with spectral characteristics that match the color of the objects. This generally includes a flat-field with a color that matches the sky.

The response of the various pixels may also show different non-linear effects: flat-fields obtained at different light-levels are not consistent with each other. Consequently flat-fields should be done at irradiance levels comparable to the science exposures.

Interference effects (*fringes*) are wavy patterns due to the layered structure of the detector which causes multipath interference effects in thinned backside-illuminated CCDs, and this modulates the sensitivity. For photometry with broadband filters these effects would be almost negligible, unless emission-line objects are observed. For instance, the night sky produces fringes related to emission lines beyond 700 nm which are absent in the laboratory illumination of the flat-field screen. With narrow-band filters, fringe effects may become a serious problem.

Dark current. Photons are not the only source of electrons registered in a CCD; thermal agitation in the silicon layer produces electrons as well. These thermal electrons are indistinguishable from photoelectrons. In absence of any light thermal effects give rise to a dark current I_d which follows the law

$$I_d = A \exp(-B/kT) \tag{13.1}$$

(Mackay 1986), where A and B are constants, k is the Boltzmann constant, and T is the device temperature. A 10°C change in working temperature may decrease the direct current by a factor of 3. But remember that cooling also reduces response to red light, so that for applications in the red spectral region, it may be necessary to work at a higher temperature. Dark exposures, of different lengths, will (at constant operating temperature) yield accurate determination of dark current. Dark current can sometimes be represented by a mean value for the whole detector surface, but very often CCDs are non-uniform in dark current properties.

Standard stars are bright, and bright stars have the best determined magnitudes. Since such exposures (even at small telescopes) will be short compared to the exposures of faint stars or even of extended objects, one must be sure that the exposure-time absolute error is only a couple of milliseconds. If this condition is not met, non-uniformity effects between edge and center of the field will be introduced. The shutter-timing error becomes significant if high-precision ultra-short (order of seconds) exposures of bright-star fields are taken. Correction factors for shutter timing defects can be determined by taking dome or sky flat-fields with sequences of different integration times (assuming that the dome flatfielding-lamp has constant irradiance).

Short exposures on standard stars raise another problem when compared to long exposures of other objects: during long exposures the air mass may change by a substantial amount, and this effect is difficult to correct for. One solution is to observe several faint "secondary" standard stars, both during short and longer exposures (with air mass range comparable to that of the scientific exposures). This should allow one to determine a correction factor. In order to cancel out residual pixel-to-pixel variations, standard stars should be observed several times at different locations on the detector, and also at different air masses. Stetson (1989) approximates the mean air mass of the exposure by

$$X = (X_0 + 4X_{1/2} + X_1)/6 \tag{13.2}$$

where X_0, $X_{1/2}$ and X_1 are the air masses at beginning, the middle and at the end of the observation, respectively. X is independent of the extinction coefficient.

Cosmic ray and X-ray events are produced by cosmic rays (mainly muons produced by the interaction of high energy protons in the upper atmosphere) that hit the CCD and cause a signal of several thousand electrons which are concentrated in only a couple of pixels. This occurs at a rate of the order of 1–2 cm^{-2} min^{-1}; the intensity increases with increasing height of the observing site, and is lowest at geographic latitudes below about 20° (Compton 1933). X-rays emitted by the material of some UV-transmitting glasses can have a higher rate of events, and particles emitted by radioactive nuclei in the metals used in the cryostat may also yield unwanted signals.

Because of the linear response and the large dynamic range of CCD detectors, standardization poses fewer problems than is the case when using photographic emulsion: whereas the centers of the photographic images of galaxies and of bright stars are saturated, CCD frames (if properly exposed) yield unsaturated pixels. This not only increases the information content of the data frame, it also allows the determination of the stellar magnitude of the inherently bright standard stars.

The output of a photomultiplier photometer consists of a few numbers every minute, and can be handled by the most simple equipment (a small printer, or even by taking notes). On the other hand, after each integration a CCD gives a very large output (hundreds or thousands of kilobytes, and the figures are increasing rapidly as larger chips are being manufactured). Since sequences of several frames are made every hour, powerful computers are an absolute necessity to digest that flow of information.

Cheap CCD cameras are now available for amateurs, but a complete high-quality CCD photometric setup (camera and electronics) is still out of reach of the smallest observatories. Consider also that the reduction procedures involve extensive computing facilities.

13.5 Extraction of data from a frame

A CCD is often used to compare objects located on the same frame. This minimizes the need for "all-sky" photometric techniques. In narrow-band photometry it is even possible to calculate differential magnitudes from a single frame, without any need for standard or extinction measurements.

The extraction procedures used to obtain raw instrumental magnitudes from a frame can become quite complicated, especially for crowded fields. Therefore, the term "photometric reduction" has taken a different meaning in CCD work as compared with classical work. Here it designates in the first place a complex image-processing job which does the equivalent of the straightforward sky-subtraction at diaphragm photometers.

Once basic precautions have been taken into account, one can proceed to place each frame on a common internal system, of which the zero point can be determined by digital aperture-photometry on several stars in each field. This instrumental CCD-system can then be transformed to a fundamental standard system. This standardization can be repeated for different apertures, so that a valid transformation to the standard system is constructed for each aperture diameter.

Two methods are used to extract data from a CCD frame.

- *Aperture photometry.* This means performing diaphragm photometry on the frame instead of doing it on the sky with the telescope. Because it is a computer job, complete flexibility is allowed for the choice of apertures. Just as is the case for conventional photometry, this technique is reserved to uncrowded fields.

- *Profile fitting.* For crowded fields—and such frames are common at the faint magnitude levels reached by CCDs—it is necessary to switch to the procedure of profile fitting: A model of the stellar images is fitted to the data via a least-squares algorithm. Many stars can be handled at the same time. The integration of the individual two-dimensional profiles yields the contribution of each star.

13.6 Aperture photometry

Having integrated the signal in a given area around a star, it is necessary to get an estimate of the background in that area. The simplest way is to measure an equivalent area in a star-free region close to the star, but not too close so as to avoid any contamination from the wings of its profile. This is exactly what is done on the sky in conventional photometry. We speak of "background" instead of "sky" because what we are interested in is not the sky value but the total background that would be present in that area if the star were not there. This background includes faint stars, diffuse atmospheric or sky sources, and defects.

It is wise to observe the background at symmetrical locations around the star in order to cancel out any linear trend. In conventional photoelectric photometry, one would make one measurement on two sides with a circular aperture. Since a computer allows simulation of any shape, an interesting option is to use an annulus around the star. In order to achieve high precision, the area of the background aperture should be made larger than the stellar aperture. Just as in conventional photometry it is wise to stay sufficiently far from the star in order to minimize its contribution to the signal. If a ring is used, its inner radius should be large enough.

The contribution to the measured background of a star's own light is harmless as long as it constitutes a stable fraction. This means that all background measurements are to be done in exactly the same conditions, and at a fixed offset from the stars. This is often not the case since the background may have to be measured at some particular place due to the crowdedness of the neighboring field, or to the presence of a nebula. The atmospheric conditions (seeing) may also considerably modify the image profile between successive frames and the fraction of stellar light falling into the aperture will vary accordingly.

Where should one place the aperture on the frame? The natural approach would be to look for a sky area particularly devoid of even the faintest stars, and of any blemish. This is a mistake: one should measure a background that is as similar as possible to the background in the vicinity of the star of interest. If the star measurement happens to include faint background stars, or nebulosities, then the aperture measurement should be done at a similarly "polluted" location. Finding an exact replica of the star field is of course impossible. The selection is mainly a matter of appreciation. The best method is to observe a few such places, in some random way, and to take the average value. The errors will increase with the brightness and with the inhomogenity of the field. A very bright and irregular background cannot be accurately handled this way, and one would need to switch to the profile fitting method.

An automatic procedure may calculate the modal value of the distribution of the signal for the pixels within the comparison area. It represents the most probable value of the signal of a randomly chosen pixel in that area (Stetson 1987). This technique minimizes the probability of a large error in the predicted value of the background and should be used whenever individual decisions are not made manually.

Aperture photometry needs a lot of care and appears to be quite time-consuming. One of the advantages of CCD photometry is the possibility of observing a large number of stars. The amount of reduction work becomes tremendous and automatization of the process is an absolute requirement.

13.7 Profile fitting

13.7.1 The point-spread function

Atmosphere, telescope, filters and detectors each modify the image profile. The real energy distribution has several components and becomes a complicated (time-dependent, position-dependent and signal-dependent) unknown function, the *point-spread function* (PSF). The PSF represents the distribution of the irradiance due to a point source as a function of the position relative to the centroid of its image. Let (x_0, y_0) be the position of the centroid. The PSF at some position (x, y) will be written as $I = I(x - x_0, y - y_0)$, rather like a bivariate Gaussian distribution, but displaying power-law wings. The

PSF may depend on the coordinates of the centroid itself, (x_0, y_0). In the latter case, which may occur because of optical aberrations, a more general expression is necessary: $I = I(x, y, x_0, y_0)$. Four independent variables are involved instead of the two relative coordinates $x - x_0$ and $y - y_0$. It may still be possible to adopt a relatively simple expression in many instances. For example, in the case of a symmetrical bi-dimensional Gaussian distribution, I is characterized by a parameter describing the width of the profile. A slight variation of this parameter with the distance to the center of the frame could easily be included. A description of the wings of the PSF, or of non-symmetrical distributions, requires more parameters. For instance, imperfections in the guiding result in irregular images. This problem may be less severe in observations from space where quasi-perfect tracking can be achieved.

In ground-based observations, the atmosphere introduces a seeing profile which is approximately Gaussian in its central part (King 1971; see also Fig. 3.3).* The PSF has to include this seeing-broadening as well as all other optical effects.

The PSF is color dependent, so stars of different types exhibit different profiles when observed through wide-band filters. Part of the color dependency arises from chromatic effects in the instrumental optics. Another kind of color effect is due to the atmosphere: because of the chromatic dispersion at moderate to large air masses, stellar images are transformed into short spectra. Solving this requires the determination of the color from multifilter observations. This leads to an iterative procedure: an average profile yields a first coarse estimate of the color indices; those indices allow one to pick the right profile for each star, which in turn will yield better colors.

As a result of those imperfections, the calculation of the PSF has to be done for every frame. Sometimes a single frame may require a variety of PSF.

13.7.2 Analytic or empirical PSF

An analytic expression for the PSF may be very difficult to obtain in some situations, for instance when the profile is distorted. It would seem that turning to an empirical determination of the PSF would solve all the problems. The profile is not approximated by an analytical expression, but by an array of values determined by examination of several images. This grid can be moved and scaled to match any stellar image.

What are the advantages of each method?

- Finding appropriate analytic functions to represent non-symmetrical images can be a very difficult task and may involve an inconveniently large number of parameters, though not as large as the number of pixels in an empirical PSF.

- Extended objects can be measured with the aperture technique, but not with PSF fitting. Simple images such as elliptical galaxies can be modelled to some extent and they could be incorporated into a profile-fitting procedure.

---

* We do not consider very short exposures of the kind required for speckle observations, and which show very irregular images.

- The PSF width is usually of a few pixels. This means that the pixel-to-pixel variations of the irradiance are quite large. An analytical expression can be very accurately integrated over each pixel, so comparison with the observed profile is easy. The empirical approximation uses interpolation methods which may not adequately represent the true profile, especially in the vicinity of the center.

- Building the empirical PSF by combining several image profiles needs accurate superpositions of the grids and involves a first series of interpolations. If not accurately done, this process leads to a broader global profile, reflecting the imperfections in the alignment of the centroids. The interpolation process introduces errors, but it may be argued that they may partially compensate for similar errors made on the program stars.

- The numerical integration of analytic distributions is generally more time-consuming than the interpolation in tables.

Since neither method is ideal it was logical to combine them. Stetson (1987), in the widely distributed DAOPHOT computer programme, adopts a PSF profile with two components:

- a bivariate Gaussian distribution* fitting the central regions of the stellar profiles,

- a two-dimensional array representing the observed residuals of the actual data from the normal profile; this table is calculated at half-pixel intervals.

In this hybrid method the analytical function represents the large pixel-to-pixel variations while the empirical residuals generally keep small amplitudes and do not introduce large interpolation errors.

13.7.3 Stellar photometry using the PSF

Aperture photometry on electronic frames is very similar to direct photoelectric photometric work at a telescope. Stars are observed one at a time, and a single quantity, the global signal within the diaphragm area, is recorded for every setting. These quantities are then subjected to the usual steps of background subtraction and photometric reduction, as explained in Chapter 10. Apart from a few particularities explained in the next sections, the basic analysis of photoelectric photometry remains the same.

Profile fitting, on the other hand, constitutes a complex intermediate procedure essentially involving image processing applications. Image processing packages are used everywhere and they range from the basic to the highly sophisticated. But the goals of astronomical photometry are so unconventional that none of the commercially available programmes are of such use for that purpose.

---

* A Gaussian is usually rather slow to compute, unless the machine has a hardware exponential operation. Other analytical forms might be both better and faster.

Once again astronomers have been obliged to devise their own algorithms and to write entirely new codes. A wide variety of algorithms exists. Among the best known programmes we list the above-mentioned DAOPHOT and also ROMAPHOT (Buonnano et al. 1983). Those software packages are constantly evolving.

What are the main features of a photometric image processing algorithm?

- Establish the PSF.

- Given an initial guess for the position of a star, the algorithm centers and scales the PSF on the actual profile. Hence the first step is to compile a list of stars to be measured. This is done automatically by a separate routine which is able to sort genuine stars from the overall noise and from non-stellar objects (nebulae, galaxies).

- The underlying background is evaluated simultaneously, eventually as a function of (x, y). Alternatively, the "sky" derived from aperture photometry may be used. Stetson (1987) gives several arguments in favor of this option, a major advantage being that the background is evaluated from the "best" pixels.

- The advantages of the profile fitting method are only apparent in crowded field photometry, i.e., when stellar profiles overlap so much that aperture photometry is impossible. Consequently, the profile fitting algorithm should be able to handle many stars at once. Ideally the whole frame should be treated at once in order to make the best use of common information (such as the overall background when it is not derived independently) and to keep the total number of unknowns to a minimum. This is somewhat similar to the gain obtained in multi-night photometry versus single-night photometry. For practical reasons (computer power) it is convenient—or necessary—to divide the frame into subgroups. Some clustering routine has to find which stars may be treated simultaneously, according to some selection criterion, which basically is the separation between stars. Nevertheless, the maximum size allowed by the computer program may be exceeded in the case of a very crowded area. The criterion has to be relaxed and the accuracy will be degraded. Thereafter each group is treated separately, with a conveniently small number of free parameters. The global number of parameters used for the reduction of the frame may be increased over a whole-frame solution.

- The fit condition is generally expressed as the minimization of a least-squares type merit function. Solving the problem is done by an iterative procedure which, of course, should converge, and should do so as rapidly as possible. The nature of the problem makes this a very difficult task, as anybody who has ever had to deconvolve a severe spectral blend knows well. Numerical methods should be chosen carefully to avoid the solution oscillating or getting trapped in a secondary minimum. Due allowance should be made for the rejection of bad data or bad pixels. Correction to the initial list of probable stars should be possible.

- When several frames of a same field are combined, information on the position of stars can be shared in the reduction process.

A general algorithm ideally incorporates both aperture and profile fitting photometric techniques. It may also include the final steps of extinction correction and color transformation.

13.8 Extinction

All objects on the same CCD frame have been recorded in similar atmospheric conditions. They are spatially very close, so they have the same zenith distance and the same air mass. Those are ideal conditions for "differential photometry" where variable stars are referred to nearby constant objects (comparison stars) so as to minimize all perturbations of atmospheric nature (see Section 15.3.2).

Nevertheless, the extinction should be calculated when differences of colors among the stars make second-order extinction terms important; this is an inconvenience of wide-band photometry.

The procedures used to calculate the extinction are quite similar to those used in conventional photometry and were treated in Chapter 6. One can make use of the advantages of CCDs by carefully selecting the targets. It can then be possible to collect information on many stars—of different colors—on a single frame. This frame would normally be taken through all filters. Best efficiency is reached when such frames are taken at extreme air mass, i.e., closest to the zenith ($X \approx 1$) and at, say, $X \approx 2$.

13.9 Color transformation

Fundamentally there are no reasons why the magnitudes extracted from CCD data would not accept the same color transformation procedures as the photomultiplier based magnitudes, with the same caveats (see Chapter 8). There are, however, several technical constraints which give rise to additional complications. One has good reasons to believe that owing to the fast technological advances in the area, most of the difficulties will disappear in the near future.

- A major problem arises from the *spatial non-uniformity* of the chips, and eventually of the filters. Dramatizing the situation, one could say that each pixel has its own response curve and needs a specific color transformation. Seen in that perspective the case would be hopeless. Variations between pixels consist mainly of a scale factor and an offset, but there is some wavelength dependency. Part of these effects is dealt with by performing flat-fields of various colors, corresponding to those of the observed objects. Again narrow-band photometry is at an advantage if flat-field exposures are made through these filters.

- *Inadequate spectral response.* We have mentioned earlier that correct transformations need an adequate sampling of the spectrum by the photometric bands. We noted also that all existing systems are undersampled by a factor of 5 to 10. Hence in order to keep the transformation errors as low as possible it is essential to reproduce the standard system as closely as possible. The sensitivity curve of a CCD is quite different from that of most photomulipliers. The combination filter-CCD gives inadequate response curves, especially for the blue and red filters. For instance, Johnson-Morgan's U, Strömgren u and the like are useless, since most of the passbands lie outside the sensitivity curve of the CCD. The application of specific coatings on thick CCDs or the use of thinned CCDs improved the ultraviolet response significantly (QE of the order of 0.25 at 350 nm). However, the resulting response remains rather irregular (see curves 1–3 on Fig. 13.7).

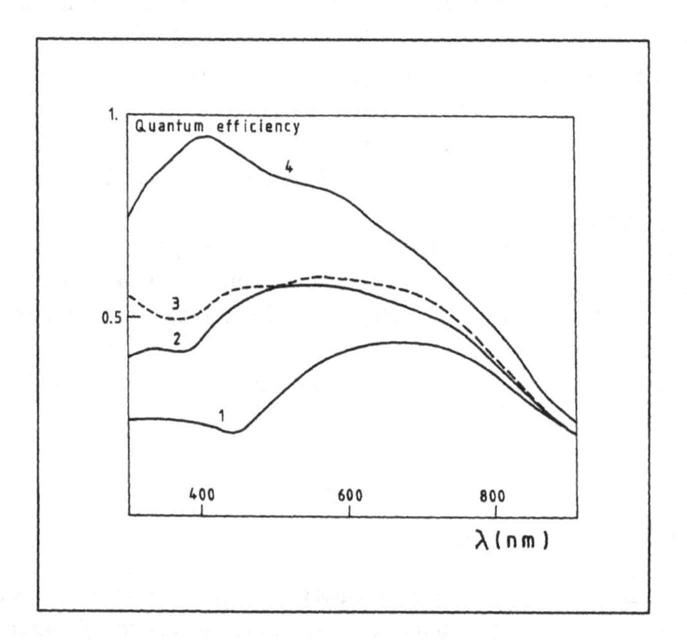

Fig. 13.7 Sensitivity curve of various CCDs (from Fort 1988).
1 UV coated thick CCD
2 & 3 Thinned uncoated CCDs
4 HFO_2 antireflection coated thinned CCD

Since many efforts are being put into increasing the QE toward 1, the response may become smoother, as indicated by curve 4 on Fig. 13.7, corresponding to a thinned backside-illuminated chip, with an antireflection coating.

- *Secondary standards.* CCD photometry is plagued by a more fundamental ailment, the improved, or n-th generation, standards (see Section 11.2). The sensitive chips are saturated in a fraction of a second by the original standard stars of most photometric systems. In addition, there exists no accurately calibrated shutter for such short exposures. The two-dimensional property of the CCD makes it convenient to have many standards on a single frame, in order to save telescope time and computer time. Both arguments have led to the use of conventional photometers to produce catalogues of secondary standards (of second or higher generation) concentrated in small fields.

We have already made clear our opinion about such a procedure, even when it is carried out with care. Let us recall that original standards are generally very bright and can only be observed directly with small telescopes. Hence the secondary standards are observed with the same low light-gathering power instruments. Those have to be pushed to their limits to get reasonable accuracy on the faint stars. This is accentuated if the faint would-be standards are concentrated in a small field. Those conditions are totally inadequate for conventional diaphragm-photometry. Moreover, the instrumental system is never the original standard one, and the standards produced will be used with yet another instrumental system, that of the CCD equipment.

There are two ways out of this situation.

1) It is possible to define completely new photometric systems, tailored to the CCDs, and with CCDs. Those systems can be better suited to extragalactic work than the old *UBV* or *uvby* systems. They can avoid the sky emission lines. In order to be reproducible, the transmission curve and the response curve of the CCD should be well calibrated and preferably be smooth.

2) If an existing system is to be used, it is essential that it be reproduced as closely as possible by an appropriate combination of CCD and filters. In order to avoid intermediate steps, it is advisable to observe the prime bright standards instead of secondary ones. This is quite possible through the use of calibrated, uniform, gray filters. If secondary standards have to be defined, this should be done using that same technique.

Chapter 14 Photographic photometry

14.1 Introduction

A photographic emulsion is a thin light-sensitive layer that covers a film or a glass plate. The emulsion contains grains of silver halide that undergo chemical changes when struck by light. Only about 10% of the incident light is actually absorbed by the grains in the emulsion, and several to several hundred photons must be absorbed by a grain for it to turn black during development. The grain-to-photon ratio is only a couple of percent at most, and even the best modern plates have a quantum efficiency which is not better than that of the eye. The photographic emulsion is the only detector that—after processing—becomes the storage medium itself.

The very low efficiency with which light is converted into blackened grains, and the fact that the dynamic range of an emulsion is so small that brightness differences of more than a factor of a few hundred cannot be faithfully recorded, severely limit the use of this detector. Apart from this, there is the well-known *reciprocity failure*: a given exposure will give less image density than an exposure half as long at double aperture area. That means also that an exposure of 30 minutes on a field is not compatible with a 1-minute calibration exposure.

In addition, photographic emulsion is also nonlinear in the sense that there is not a linear relation between photographic density and amount of incident light: for short exposures and for very long exposures, the response to the total amount of light is less than it is for exposures of optimal duration. So, for every exposure one must elaborately determine the relation between density and intensity, and this calibration is a function of the wavelength of incident light.

Since the routine use of photomultipliers and recently of CCD detectors, these drawbacks have pushed the users of the photographic plate or film as carriers of photometric information onto the defensive. Today, with the advent of high-performance Schmidt telescopes, fast computer-controlled microdensitometers and the increasing power of automatic and semi-automatic image processing, we are witnessing a come-back in the appreciation of photographic photometry. The fast development of this kind of laboratory equipment opens totally new frontiers for research on archival plates, and the high-speed extraction of photographic data in numerical form allows, for the first time, full exploitation of a tremendous amount of hitherto inaccessible information. Also, with the advent of more powerful telescopes being built that will push the limiting magnitudes further and further away from the limiting magnitudes of existing sky surveys, it is likely that a future sky survey will still be carried out with the photographic emulsion as basic medium (methods of plate processing still improve, and, though we have not seen any major advances since the introduction of the Kodak TP-2415 emulsion in the mid-seventies, still more sensitive emulsions can be produced). Finally, the photographic method is still the only method of doing photometry of really large extended objects at high resolution (gigapixel CCDs are still far away). For an in-depth comparison of CCD versus photographic detectors, we refer to West (1991a,b).

In this chapter we do not go into detail about the properties of the photographic medium, nor about observing techniques, since these aspects are well-described in the specialized literature. We do approach the concept of photographic photometry from the viewpoint of the occasional user who studies celestial photographs taken recently or in the far past, or of the scientist undertaking analysis of available Schmidt atlases or digitized sky surveys. So only post-observational manipulation of the photographic medium is treated.

To date, three groups of homogeneous photographic sky surveys exist, or are in progress:

- The Palomar Surveys:
  - POSS-I, the well-known National Geographic Society-Palomar Observatory Sky Survey, completed in 1955 (blue and red regions)
  - POSS-II, the new Palomar Observatory-European Southern Observatory Photographic Atlas of the Northern Sky, to be completed by the end of this decade (blue, red and infrared regions)
- The ESO/SRC Atlas of the Southern Sky (blue and red)
- The UKST-I and -II surveys

14.2 The photographic emulsion as a storage medium

Any single large-scale Schmidt plate contains about 1 Gigapixel of information at 10 μm resolution, and can cover ten thousand stellar images and galaxies. Systematic sky surveys encompass a couple of thousand Schmidt plates in each photographic passband. From this viewpoint, the photographic medium is still far from obsolete, even when compared to CCD images, especially now when a new generation of super-microdensitometer machines is being designed and built.

Blackening of the emulsion is measured by its relative transmission when exposed to incident light from the microdensitometer. Transmittance (see Eq. 5.1) is a function of position, so we define it in terms of the local transmitted irradiance $E(x, y)$

$$T(x, y) = E(x, y)/E_0 \tag{14.1}$$

$T(x, y)$ is the local transmittance of the emulsion when illuminated by a uniform irradiance E_0. The optical density (5.2) is then

$$D(x, y) = \log(1/T(x, y)) \tag{14.2}$$

The only reason for using optical density, instead of other functions of transmission, is historical because of the use of logarithm tables.

The sensitivity of the photographic emulsion is expressed as the necessary exposure time to obtain a given photometric precision. *Graininess* is a subjective visual impression of nonuniformity in the image. It is a basic limitation in photographic photometry: the images are made of developed silver grains, and the graininess is the noise. Graininess is objectively described by *granularity*, the standard deviation of density produced by the granular structure of the material when uniformly exposed. With a fast emulsion, the grains are large, and the number of photons needed per grain is small. Faint star images are then barely discernible from the plate background grain. Fine-grain emulsions furthermore have a stronger contrast, but are inherently slow. Fine grain is more important than speed, and therefore more and more fine-grain emulsions are preferred to the older emulsions. Kodak, for example, gives its astronomical emulsions an indication according to granularity: increasing roman numerals denote decreasing granularity. For specific details about granularity and spectral response, we refer to Kodak technical information publications.

Resolution of emulsions are up to 200 line pairs/mm or 40,000 pixels/mm^2 (CCD has fewer than 50 lines/mm). This does not mean that the resolution actually achieved in astronomical applications equals the resolving power of the base material, since resolution is often degraded by the optics of the telescope, by the optics of the microdensitometer, and by mechanical vibrations occurring during exposure as well as during scanning.

Photographic materials show several astonishing effects which have some importance for any photometric applications. We mention the effect of *solarization*, a decrease of irradiance which appears at greatly increased exposure after the maximum density has been reached (past the shoulder of the characteristic curve, see Section 14.3). This effect for example appears when exposing calibration plates with an ill-designed calibration device. Solarization is sometimes being confused with the *border effect* or *Eberhard effect*, which arises at the border of a dark image and a light background (as is the case with calibration exposures) and results in a depression of the density inside the image compared with the higher density at the extreme edges. Such edge effects are processing effects (one should reduce them by using vigorous agitation during development), and they must be taken into account, especially when establishing the characteristic curve. We finally mention the *Kostinsky effect* that appears when images of double stars (or two close stars) are formed during development. At the place where the images nearly touch, development is hindered because the developer is locally exhausted, and this causes a distortion of the two inner parts of the image with an exaggerated separation of the two images.

14.3 The characteristic curve

The characteristic curve gives the relation between density and amount of radiant energy or intensity of exposure (plotted on a logarithmic scale) that exposed the calibration emulsion.

The characteristic curve is often referred to as the H and D curve, after Hurter & Driffield (1890), who assumed that optical density was proportional to the amount of silver deposited per unit area. The use of opacitance (see Section 5.1) instead of the density strongly reduces the nonlinearity of the H and D curve (as noted by de Vaucouleurs 1968) in the low-density region.

In order to establish the characteristic curve, one must accurately know the intensity ratios of the exposed calibration areas on the plate, and the measurements of the spots must give uniform results (no gradients or asymmetries). Figure 14.1 illustrates a "typical" characteristic curve. The lower left end of the curve is parallel to the horizontal axis; the density is at a pre-exposure background which is independent of exposure. This part is called the *fog level*: a combination of the *chemical fog*, the density of the unexposed but processed emulsion, and the *base density*, which is determined by the transmission of the glass or film. The chemical fog level is in fact the equivalent of the dark current in electronic detectors. Next comes the non-linear part of the curve, which is called the *toe*. The mid-section of the characteristic curve is the linear part, where density is directly proportional to a power γ of the number of photons reaching the emulsion. The length of the linear part depends on the type of emulsion. The slope of the straight line has been termed γ, or contrast index, and indicates for a film the rate of change of density with exposure.

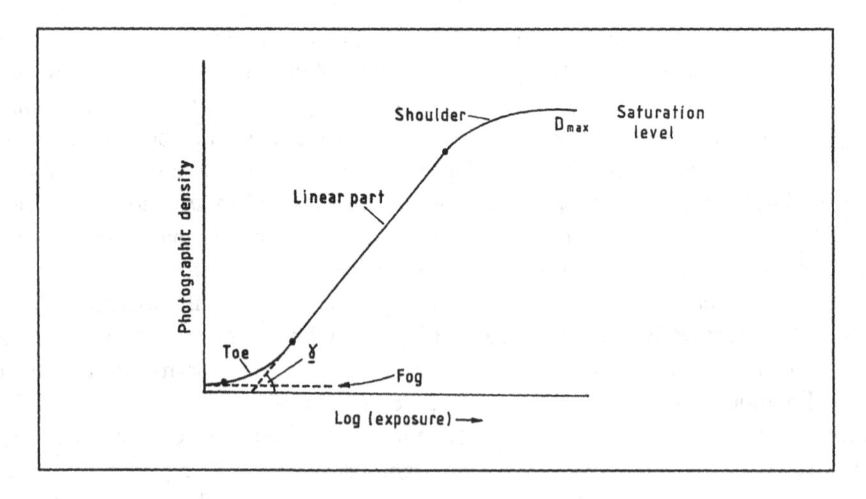

Fig. 14.1 Characteristic curve.

The upper portion is the *shoulder*, where the gradient decreases until a horizontal line, the *saturation level*, or maximum density is reached. There, exposure differences will no longer record as density differences, and further exposure will not provoke further darkening. The non-linear form of the H and D curve illustrates the level of difficulty encountered when measuring stellar images, where sharp gradients and diffraction patterns are being formed on different parts of the curve.

A widely used analytical approximation to the characteristic curve is the Honeycutt-Chaldu function (Tsubaki & Engvold 1975)

$$\log(E/E_0) = A_4 + A_1 D + A_2 \ln(\exp(A_5 D^{A_6}) - 1) + A_3 \exp(A_5 D^{A_7}) \tag{14.3}$$

where E/E_0 is the known relative irradiance of calibration spots, D is the measured calibration density, and A_i are the parameters to be determined by a least-squares fit to the calibration data.

This function represents well all parts of the characteristic curve; A_2 determines the shape of the toe, A_3 determines the shape of the shoulder, and A_4 shifts the curve along the axis of $\log E$. Table 14.1 gives values of A_6 and A_7 according to the form of the curve:

Table 14.1 Coefficients A_6 and A_7 of the Honeycutt-Chaldu function describing the characteristic curve of photographic emulsions

	A_6	A_7
no saturation	1	0
saturation of emulsion	1	1
saturation of densitometer	1	3

An iterative procedure for determining A_5 gives rapid convergence (Tucholke 1989). Kormendy (1973) proposes to construct an average characteristic curve from several plates for which the same emulsion, same filter and identical treatment (hypersensitization and dark room processing) were used. This can be done when the differences between the characteristic curves of one group are not larger than the effects of processing errors, and when the differences in the overall irradiance level of each set of spots can be eliminated by shifting in $\log E$ the whole sequence of spots. For an application of this principle, see Liu et al. (1992).

The S/N ratio in photography bears no simple relation to the number of photons that generated the recorded density. It depends instead on the slope γ of the characteristic curve ($\gamma = dD/d(\log E)$) and on the density noise characteristic σ_D of the plate. Hoag et al. (1978) showed that

$$(S/N)_{\text{out}} = 0.4343\gamma/\sigma_D \tag{14.4}$$

where γ and σ_D are taken for the D-values in question. The value of σ_D starts at a relatively low value at the density value of fog and increases slowly, in a first coarse approximation it is proportional to $D^{1/2}$. Furenlid et al. (1977) have determined σ_D as a function of D for a variety of emulsions of astronomical interest, so in order to obtain the S/N ratio one must only determine γ for the emulsion under study.

14.4 Post-observational processing of plates and films

Photographic densities are measured with a micro-densitometer. Though some analog devices are still in use, the bulk of photographic photometry is done with digital scanners. There are many differences in design between the various scanning machines. They use pixel size of the order of 5 to 20 μm in order to provide a resolution at least equal to that of the plate. Some use single-element detection (photomultiplier), others use multi-element detectors (1- and 2-D arrays). A review of the status and performances of existing densitometers and scanners in construction is given by MacGillivray (1990). They offer, in addition to highly precise density measurements, a positional accuracy of a few microns. These performances are combined with qualities like high speed and high repeatibility. They do, however, have a limited dynamic range (as does any copying process), and due to this they can only reach densities up to about 2.5.

14.5 Calibration of digitally recorded photographic densities

Photographic photometry is a two-stage process. First one must measure a sensitometric calibration plate (or strip on one of the edges of the science plate), and construct the characteristic curve. This problem of calibration is the most basic problem of photographic photometry. The second step consists of measuring the density of each pixel using *the same microdensitometer configuration*, and transforming these measured densities into a scale proportional to the incident radiant energy.

The characteristic curve is sensitive to differences in sensitization of the emulsion, temperature and humidity of storage before and during exposure, uniformity of the plate, elapsed time between exposure and development, and of course to the whole developing process. Calibration exposures are unfortunately most often put on the edge of the plate (in order to reserve a maximum field for the science exposure), and outer edges of plates (at least "real" edges as provided by the plate manufacturer) are sensitometrically useless (edges are often more fogged because they get more turbulent agitation in development, they are exposed to contaminants during storage, and they also dry faster). Some older sky surveys do not even have calibrations on each plate. In such a case one can use photometry of stars as a secondary base for calibration.

Kormendy (1973) described a calibration technique that uses the brightness profiles of field stars: their shapes calibrate relative intensities, and the zero point is given by their magnitude determined photoelectrically. This technique is independent of sensitometer calibrations, and can thus be used in the cases where such calibrations are not available.

14.6 Photometric analysis of photographic intensities

Once photographic intensities are available in digital form, further treatment and analysis is basically reduced to the one described in Sections 13.5 and further, with this difference that flat-fielding cannot be performed. On the other hand, spurious signals, like dust or emulsion defects, must be removed.

Special attention must be drawn to the consequences of the large fields which, as a rule, are being registered on photographic plates: not only the enormous yield in number of pixels (at least a linear factor of 100) causes trouble during the analysis, but one must be especially careful with the extinction corrections. Differential extinction (see also Section 15.3.2) may cause irradiance variations amounting to 10% or more within one plate, and this should be corrected by mapping the extinction over each plate (this could soon be a problem with big CCDs).

Chapter 15 The observations

15.1 Introduction

Photometric reductions can be repeated several times until one is sure that all elements have been taken into account and that no further improvements are possible. But conducting the photometric observations themselves is more critical, because one night of measurements can never be repeated. Observing does not allow an idle pace. Correct decisions have to be taken at once and back-pedalling is impossible.

The constraints met by the astronomer are the following:

- Telescope time is expensive and should not be wasted.
- Every lost measurement is effectively wasted and nothing can restore it. Particularly, irregularly variable objects will not repeat their performance when the observer is back at the photometer.
- Telescope time is usually very limited and one may have to wait months to years before getting another chance to make new observations.
- The best reduction procedures are unable to extract useful information from incorrect data.

In this chapter we shall examine the various steps one may take in order to make sure that only a minimum amount of time is lost, and that a maximum amount of information is extracted from an observing run.

15.2 Preparing the observing run

15.2.1 Feasibility of the programme

The proposed photometric programme should in the first place be feasible. This sounds obvious but unrealistic proposals are frequently submitted. Telescope-time allocation committees generally look into that matter so that the authors of such proposals will not always make it to the telescope. It is nevertheless wise to check the feasibility issues before proposing observations.

The observing programme has to be suited to the equipment. The objects should not be so bright that they saturate the detectors, nor so faint as to give unacceptably large errors or prohibitively long exposure times. Past experience with the same or with similar equipment is an asset in deciding about this issue. The operating manuals published by most large observatories are generally good sources of information (ESO, CTIO, Kitt Peak, CFHT...) but they can rapidly outdate, so it is wise to consult the technical or astronomical department of the observatory beforehand.

The observing programme should be adapted to the observing site. This is not only a matter of geographical location (northern objects are usually better observed from northern sites!) but also a matter of climate and of atmospheric conditions. Some observations require a number of consecutive perfect nights, a situation that cannot be met in temperate climatic regions. Other observations, such as those of long-period variables with large amplitude, may be perfectly possible in a less privileged environment. The acceptance of "second class" sites is often unavoidable, for example when one needs full longitude-coverage of observations of comets or short-period variable stars. Observations of dense stellar clusters may require the optimal seeing of very specific sites.

The observing programme should be adapted to the epoch of the year. The objects should be well placed in the sky during the night. One should manage to avoid gaps of a few hours where no objects are observable under good conditions because they are too low on the horizon. For equatorial objects one may have to choose between a northern or a southern observatory on the basis of the season at the time of best visibility.

The list of objects should be large enough so that no time is wasted.

15.3 Programme scheduling

An efficient and flexible observing programme should be devised before going to the dome. Priorities must be set. Lists of stars to be observed at given air mass (sidereal time) or at given phases (universal time) must be at hand. Short clearings within bad weather periods are to be dealt with efficiently. Specifically, the observer should schedule the observations in such a way as to avoid measurements at too high air mass. Unless one needs to cover specific phases of a variable object, programme-star observations in practice should not be carried out at air masses larger than ≈ 1.5.

Photometric observations are divided into two categories. The goal of *absolute* (or *all-sky photometry*) is to obtain accurate magnitudes and/or colors of all objects in a consistent photometric system. On the contrary, *differential photometry* is an expeditive and accurate way to obtain relative—or differential—measurements between programme stars and stable comparison stars. The latter method must be preferred when one only wishes to study variations.

15.3.1 Absolute photometry

In absolute photometry, one must determine all extinction and color-transformation parameters with great accuracy. This means observing lots of standard stars and/or lots of constant stars at various air masses, and this should be done during the whole night in order to detect possible drifts either of the atmospheric parameters or of the equipment.

Such an observing procedure allows reducing observations of programme stars made all over the sky—hence the term "all-sky photometry". As shown in Chapter 10, non-standard (but constant) stars are perfectly suited to calculate all parameters, except for the color transformation coefficients. Consequently, the astronomer ought to rely on some of his programme stars—which have the advantage of being observed throughout the nights—for determining the extinction and the various drifts. If many such constant stars are observed, genuine standard stars may not be needed continuously, but perhaps only during some convenient "gaps" in the scheduled programme. The very beginning and end of nights are often chosen to perform standard star measurements because standards are usually bright, and the varying sky background of twilight is less prejudicial to measurement. Care should be taken, however, to keep the background within control, and to avoid systematic effects that could be due, for instance, to temperature variations at the beginning of the night.

Standard stars should be generous in number, and they should very well cover the whole spectral region covered by the photometric system. In addition, photometric indices of standard stars should, whenever possible, resemble those of the programme stars. More details about the choice of standard stars are given in Chapters 10 and 11.

15.3.2 Differential photometry

The technique of differential photometry was simply adopted from visual photometry of variable stars: a variable star is compared to one or several nearby constant stars. Most CCD and photographic photometry is differential by nature, i.e., as soon as several stars are measured on a single frame.

The magnitude differences between nearby stars can be estimated with a good accuracy, even when the extinction coefficients are only approximately known, because the air masses are nearly identical. However, this is only true for the monochromatic extinction correction. The color effects of the atmosphere, which are expressed by the secondary extinction coefficients (see Section 6.5.1), depend on the color of the stars observed, and they can be quite large for nearby stars of different spectral types. For instance, CCD or photographic photometry of clusters involves very large second-order extinction terms, and these should be accounted for when establishing color diagrams. In variable-star work, using wide-band photometry, one should minimize the importance of color effects by selecting comparison stars with spectra similar to those of the variable stars. As already mentioned, this is not always possible, and a good knowledge of the secondary coefficients is necessary, as well as a good knowledge of the color indices (this eventually requires at least some multi-filter photometry).

The differential procedure also eliminates the influence of instrumental drifts, since they affect every star equally. On the other hand, one should not forget that all random errors, such as photon noise, are added when differential magnitudes and color indices are calculated. Very simple reduction procedures are then possible which only take into account raw-magnitude differences between a few successive stars and average extinction coefficients. The resulting light curves of the variable stars contain the essential information about the variations, but they may not be easily tied to other measurements made at other places, even with the same comparison stars. To do this the color transformation parameters must be determined through absolute photometry.

In fact both methods, absolute and differential, are not mutually exclusive. Differential measurements benefit from the better knowledge of the atmosphere and of the equipment resulting from absolute observations. On the other hand, trying to obtain millimagnitude accuracy from absolute measurements of variable stars is somewhat illusory, and it is always useful to have stable reference stars nearby.

There are many ways to proceed with differential measurements in classical aperture photometry. It is interesting to repeat measurements of programme and comparison stars in order to pick out anomalous values. For instance, instead of observing comparison C and variable V in a sequence CV (including sky measurements), one would better use a sequence CVCVCV (for sake of symmetry one may prefer CVC or CVCVCVC, etc.). Long sequences have the additional advantage that errors due to star centering, to rapid atmospheric fluctuations, or to bad determination of the sky background tend to average out. One should also use two or more comparison stars, because most stars have not been checked for variability and many comparisons finally turn out to be variable. Observing sequences may become $C_1VC_2VC_1VC_2$ or, if one prefers symmetry, $C_1VC_2VC_2VC_1$. When several variable stars have to be monitored in a same sky area, one could cleverly devise sequences such as $C_1V_1V_2C_2V_1V_2C_2V_1V_2C_1$. Such repetitive work is best handled by automatic telescopes with fast setting and centering time. Speed is critical in long sequences lest large variations of air mass, and of other parameters, occur between the first and the last measurements. However, speed means short integration times, and thus higher photon noise.

15.3.3 Spurious frequencies

A common problem in the analysis of variable star observations is the occurrence of aliasing, viz the introduction of spurious frequencies, a situation that arises when the observations are sampled at equally spaced time intervals. Common modulating (sampling) frequencies are the day (sidereal day when stars are measured close to the meridian), the lunar month (e.g., for observing runs systematically around New Moon), the year (objects always observed at same periods of the year), etc. To avoid this, one tries to carry out as many observations as possible outside the "best" periods. For example, measurements should be made at different times of the night. This means that the observations are performed at various air masses. If comparison stars are used, this leads to the interesting consequence of automatically providing constant-star observations which are extremely valuable for the reduction procedures.

15.4 Evaluation of the quality of the signal for non-imaging detectors

Photon counting photometers allow continuous monitoring of the quality of the signal through statistical tests. These tests are unfortunately impossible with integrating devices such as CCDs. Several statistical formulae can been applied that show the variations of the scintillation and the presence of trends or pulses in the data (Bartholdi et al. 1984). Bad data can be flagged or eliminated.

Suppose the signal is sampled in intervals ΔT_i ($i = 1, n$). During each interval N_i photons are recorded. The tests are provided by the analysis of the distribution of the N_i with time, and by the departure of this distribution from a random population of mean $N = \sum N_i/n$ and variance $\sigma^2 = \sum(N_i - N)^2/n$.

Scintillation. The photon flux is subject to both a Poisson noise and to additional random fluctuations, mainly caused by atmospheric scintillation. The latter can be described by a stationary random process $s(t)$ of mean 0. Over an interval ΔT_i, the observed N_i will be the sum of two components:

$$\begin{aligned} N_i &= N_{p,i}\,(1 + \int_{\Delta T_i} s(t)dt) \\ &= N_{p,i}\,(1 + s_i) \end{aligned} \tag{15.1}$$

where $N_{p,i}$ is the Poisson component, of mean N (which is also the variance), containing only photon noise.

Bartholdi et al. (1984) define a coefficient Q as the square of the ratio of the observed noise to the theoretical Poisson (photon) noise:

$$Q = \sigma^2/N$$

Q can be evaluated as

$$Q = 1 + s^2 N \tag{15.2}$$

where s^2 is the average of the s_i^2,

$$s^2 = < s_i^2 >$$

For a given integration time ΔT, s^2 is a measure of the scintillation and depends on the air mass and on the telescope aperture. Equation (15.2) illustrates that measurements of bright stars are scintillation-noise dominated ($s^2 \gg 1$), while those of faint stars are dominated by photon noise.

Slow changes. The parameter used to characterize slow changes in the data is defined as

$$R = \sum(N_i - N_{i-1})^2/(2n\sigma^2) \tag{15.3}$$

This coefficient is also known as "Allan variance". Depending on the trends present in the data, we have

$$\begin{cases} 0 < R < 1 & \text{data show a slow trend;} \\ R = 1 & \text{there is no trend;} \\ 1 < R < 2 & \text{there are slow oscillations.} \end{cases}$$

Values outside the interval (0,2) cannot occur. R is insensitive to the presence of scintillation in the data. The Q test, on the contrary, is influenced by trends, which show up as additional scintillation.

Spikes. Another disturbing problem confronting the photometrist is the presence of spikes. These can be caused by electrical interferences or cosmic rays. Spikes deform the distribution of the N_i in an asymmetrical way. The statistical test used by Bartholdy et al. is the "skewness" coefficient, or normalized third-order moment:

$$G = \sum (N_i - N)^3/(n\sigma^3) - 1/\sqrt{N} \tag{15.4}$$

The $1/\sqrt{N}$ term is included to correct for the Poisson distribution which is, by nature, asymmetric. We have the following possibilities:

$$\begin{cases} G > 0 & \text{presence of spikes;} \\ G = 0 & \text{neither spikes, nor drops;} \\ G < 0 & \text{drops.} \end{cases}$$

Trends in the signal do not affect the G test.

Those three tests can be calculated on-line. Ideas on how to program the formula efficiently and on how to set up confidence intervals can be found in Bartholdi et al. (1984).

In order to get significant statistical data in these tests a large n, and hence a short ΔT, is necessary. It is also sensible to have ΔT in a range where scintillation has considerable power, i.e., at high frequencies. Many pulse-counting data-acquisition systems, however, do not allow data collection with such short integration times: very often the shortest possible integration time is only a few seconds, and in such situations spikes cannot be recognized, and the observer can only rely on the standard deviation of the successive integrations which are combined into one measurement. In this respect it is much more advantageous to avoid single integrations of long duration, and to use multiple integrations of short duration. Bartholdi et al. use $\Delta T = 0.015$ sec.

15.5 Noise and error on sky background

The observer has no control of the signal-to-noise ratio of the light emitted by the celestial source, but, in the case of aperture photometry, one can optimize the resulting S/N in the output signal. A common procedure is to count photons from the star plus the sky until the associated shot noise $(n_{\text{star+sky}} t_{\text{star+sky}})^{1/2}$ is well below the desired level of accuracy. This gives an error on the count rate equal to

$$\begin{aligned} \epsilon_{\text{star+sky}} &= (n_{\text{star+sky}} t_{\text{star+sky}})^{1/2} / t_{\text{star+sky}} \\ &= (n_{\text{star+sky}} / t_{\text{star+sky}})^{1/2} \end{aligned} \tag{15.5}$$

Thereafter, the sky background is measured until the error on the count rate $\epsilon_{\text{sky}} = (n_{\text{sky}}/t_{\text{sky}})^{1/2}$ becomes smaller than the error of the star measurement.

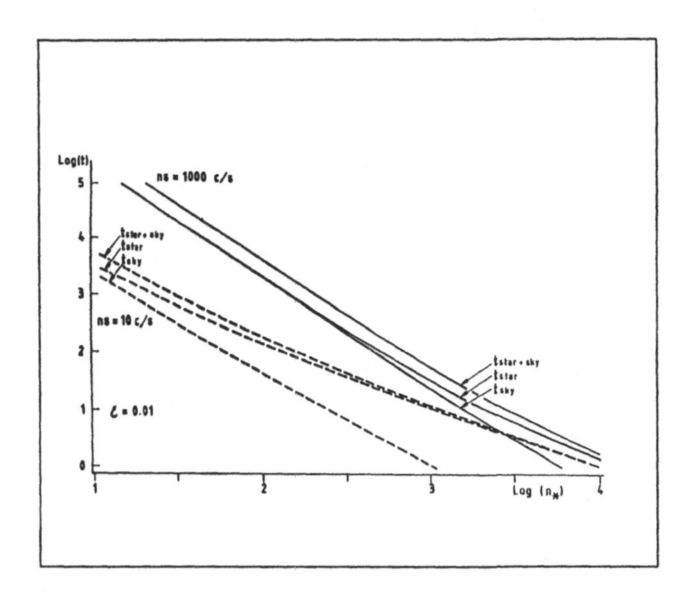

Fig. 15.1 *Integration time and count rate. The integration times* $t_{star+sky}$, t_{star} *and* t_{sky} *are plotted as a function of* $\log n_{star}$ *for sky levels* $n_{sky} = 1000$ *and 10 counts/sec and for an error level of 1%.*

A better estimate of the optimum integration times may be obtained. The count-rate of the star alone is

$$n_{\text{star}} = n_{\text{star+sky}} - n_{\text{sky}} \tag{15.6}$$

and the corresponding error is determined by

$$\begin{aligned} \epsilon^2_{\text{star}} &= \epsilon^2_{\text{star+sky}} + \epsilon^2_{\text{sky}} \\ &= n_{\text{star+sky}}/t_{\text{star+sky}} + n_{\text{sky}}/t_{\text{sky}} \end{aligned} \tag{15.7}$$

If we call t the total integration time ($t = t_{\text{star+sky}} + t_{\text{sky}}$) and p the fraction of time spent on the star ($p = t_{\text{star+sky}}/t$), then Eq. (15.7) becomes

$$\epsilon^2_{\text{star}} = n_{\text{star+sky}}/p\,t + n_{\text{sky}}/(1-p)\,t \tag{15.8}$$

This error is minimal when the derivative with respect to p vanishes. One finds the condition

$$(1-p)/p = (n_{\text{sky}}/n_{\text{star+sky}})^{1/2} \tag{15.9}$$

which allows to make the best use of a given total integration time t. For faint stars, $n_{\text{star+sky}} \simeq n_{\text{sky}}$, star and sky integrations should be of equal duration. See Fig. 15.1 for a graphical representation.

15.6 Some practical hints

15.6.1 Instrument checking

Before the observing run, and eventually more often during the observing mission, all instrumental functions should be checked. Optical alignments and focusing of various eyepieces should be performed. Checks of filter and diaphragm positioning, high voltage, and level of dark current are necessary in order to ensure that the observations will be carried out in good conditions. In CCD observations the proper calibration frames (flat fields, bias) should be taken. Simple tests—such as checking that an adequate supply of listing paper, diskettes, magnetic tapes, etc., is available—are necessary to avoid time losses during the observations. As soon as observing starts it is advisable to check the data. This could reveal, for instance, the presence of incorrect filters.

15.6.2 Control of seeing

Good seeing is critical for many photometric applications. In order to ensure good seeing conditions, proper ventilation of the dome must be made before starting the observations, so that temperature equilibrium can be reached. This depends on the particular environment. When the day-night temperature differences are large, or/and when the dome is well insulated, it is not advisable to open the dome much in advance. A necessary, but not sufficient condition for good seeing is that the mirror should not be warmer than the ambient air.

15.6.3 Sky background measurements

With aperture photometers, measurement of the sky background should be done regularly according to the magnitude of the stars being observed. Faint stars demand higher accuracy in the sky measurements. Bright stars may not need repeated background measurements (see Section 15.5). In crowded fields, the sky should be measured far enough from bright objects, so they do not add an unwanted contribution. For binaries, the sky should be taken at symmetrical positions so as to include the right amount of scattered light from the companion.

15.6.4 Focal-plane diaphragm

The choice of the diaphragm size is guided by several factors: seeing, density of the stellar field, level of background, size of telescope, accuracy of tracking. For what concerns the latter, one should check that stars remain correctly centered for a sufficiently long interval of time (this generally depends on the direction to which the telescope is pointing). Small diaphragms are the most critical.

It is recommended to use a single diaphragm throughout the night, unless one very accurately calibrates the effect of changing diaphragms (this is perfectly possible, see

Section 10.3.5). The reason is that the time-averaged image profile is the same for all stars, independent of brightness. The cut-off by a diaphragm is equal for all stars, irrespective of magnitude, but it is a function of the diaphragm size. This is an issue where novice observers commonly make mistakes. An observer looking first at a faint star, and subsequently at a bright star (at the same zenith angle and under identical atmospheric conditions) might be tempted to choose a smaller diaphragm to observe the faint star, simply because a "smaller" image is seen. What really happens is that the light surrounding the bright core is bright enough to be seen, whereas in the case of the faint star this is not so. In reality the wings of the stellar image extend farther than the eye would suspect (see Fig. 3.3). For small diaphragm sizes, seeing effects make accurate calibration imposible, but for very faint stars, the error so introduced must be compounded by the fact that reducing the aperture correspondingly diminishes the error on the background. It is up to the observer to appreciate where his interest lies.

If, for some specific reason, one is forced to measure with diaphragms of different sizes (for example when measuring dense clusters or double stars), one should include observations of standard or constant stars in all diaphragms so that accurate calibration is possible.

15.6.5 Front diaphragm

The lack of good-quality neutral-density filters (see Section 5.7) in a photometer is often compensated by the use of a diaphragm in front of the telescope to stop down the aperture (entrance pupil). Such stops, however, do not act as neutral attenuators when the pupil is imaged on a spectrally non-uniform detector (case of a non-imaging detector only). The spatial variations in spectral response of the detector produce a non-neutral effect. Front attenuators in the form of a rotating sector are sometimes used in order to use the same detector area as in full light, thus eliminating the color effect. Front diaphragms yield other problems:

- When the entrance pupil is imaged onto the detector (non-imaging detectors in combination with Fabry optics, see Section 3.1), the local irradiance on some areas of the detector is not reduced by the use of front diaphragms, and local saturation cannot be eliminated that way. Of course, saturation only due to subsequent components of the data-acquisition chain (digital counters) can be overcome since the overall number of photoevents is decreased.

- The image quality is degraded (increased diffraction effects because of the reduced aperture; increased scintillation) and the image profile is broadened, so the proportion of light entering the focal plane diaphragm is modified.

- There can be additional noise due to the poor mechanical stability of such devices.

- The entrance diaphragm cannot be repeatedly removed or replaced in a short time interval (e.g., for measuring bright and faint stars).

- The diaphragm may change the F-ratio of the telescope, with the associated problems of filter passband definition.

- Also in the case of non-imaging detectors, the use of front diaphrams may modify the overall response curve of the telescope-detector system, because of spatial non-uniformities of the latter. To overcome this effect, a rotating sector-shaped diaphragm can be used, which averages these effects. The detector surface is then used in a modulated (chopped) regime, and one should make sure this does not introduce other unwanted effects.

For all these reasons, front diaphragms should be avoided at all cost, unless there are absolutely no other ways to carry out observations of bright stars.

15.6.6 Dead time

The values quoted for the photometric dead times τ (see Section 4.5.3) in observatories manuals and documentation are often inaccurate, and must be determined by the observer. One method is to carry out the reduction with the dead times considered as free parameters which must then be determined together with all the other parameters, but this complicates enormously the reduction algorithms (see Chapter 10). Alternatively, the reduction can be done for a few discrete values of τ and the value of τ which yields minimal residuals can be adopted. If based on insufficient data, this method may not lead to very satisfactory results (see Tobin & Wadsworth 1991).

A reasonable determination of τ can be done by measuring standards of different magnitudes (including bright ones giving clear non-linear effects). Ideally these standards should have accurate values, they should be of the same spectral type, and be measured at about the same air mass, and at about the same time, so effects other than those due to dead time are identical. Unfortunately, bright standards may have calibration errors related to non-linear effects in the original system. Hence this method is better suited for medium-sized telescopes with which fainter and better calibrated stars produce non-linear effects. On the other hand, too large telescopes suffer from a dearth of well-calibrated primary standards, so the dead-time determination by this method is again unreliable.

A set of accurately calibrated neutral densities can also be used during the observation of a bright star. One then relies upon the filter calibrations instead of that of standard stars.

Chapter 16 Photometric systems

16.1 Introduction

Measurements in a given photometric system should yield parameters describing stellar radiation that are not subject to changes when new observing techniques are introduced. Modern systems use parameters such as

- a "monochromatic" magnitude at a visual wavelength (e.g., V);
- a "gradient" (frequently around V);
- one or several parameters to describe the departure of the actual stellar radiation from the assumed blackbody distribution (e.g., around the Balmer Jump);
- a parameter indicative of the degree of interstellar reddening.

Bandwidth effects should be small in order to avoid second-order terms. At the same time the limiting magnitude should be of the highest degree. The optimum multicolor system should allow the most accurate determination of spectral type, luminosity class, population type and interstellar reddening, and this can only be achieved when the functional relations between parameters and color indices are established. It is clear that any useful photometric system must be a compromise. In addition, an ideal system, if it would exist, would certainly not be equally optimized for such different tasks as cluster work, spectral class determination, etc. Formally, a photometric system would be precisely described if the function $s(\lambda)$ (Eq. 1.16) is known. The system would then be defined in terms of a specific receiver attached to a unique telescope working at a definite site with specific atmospheric extinction at a particular air mass. Besides the impracticable requirements, small changes in $s(\lambda)$ would never yield exactly the same results for different observers, or even for the same observer at different times. Therefore, any photometric system is being defined in terms of a system of magnitudes of selected "standard" stars.

Each photometric instrument has its own natural (or instrumental) system. The choice of passbands is guided by astrophysical reasons (inclusion or exclusion of certain spectral features) as well as by practical reasons (cost of the filters, high transmission, spectral response of the available detector, avoidance of bands with specific atmospheric features of natural or artificial origin...). The size of the photometric passbands of the filters places existing photometric systems into one of three classes: wide-band systems (e.g., the *UBV* system), covering at least 30 nm in each filter, intermediate band-widths (e.g., *uvby* system), with bands of about 10 to 30 nm, and narrow-band systems, with transmission curves with an extent of no more than a few nm, which transmit only a very small part of the spectral energy distribution of a star, and even isolate specific spectral lines. Each of these classes of systems entails particular reduction schemes, each of them being a special case of the general reduction schemes discussed in previous chapters.

A few systems have emerged as particularly interesting and have been duplicated in many observatories. This standardization has, in principle, the advantage of allowing direct comparisons among observations made at various places. Unfortunately, in many cases the response functions of the original systems (mainly filter passbands and spectral response of detector) have not been published with sufficient accuracy so that an exact reconstitution is not possible.

In the remainder of this chapter we discuss only the most common photometric systems, i.e., the Johnson-Morgan *UBV(RI)*, the infrared *JHKLM* system, the Geneva $UBVB_1B_2V_1G$ system, the Walraven *VBLUW* system, the Strömgren *uvby* system, some Hα systems, and finally the Hubble Space Telescope WF/PC and the IHW cometary system. For information about other valuable systems (e.g., DDO, Vilnius, 13-color system...) we refer the reader to Straižys (1977) and Lamla (1982).

Before discussing the photoelectric systems we shall describe the systems using the first photometric detectors, the eye and the photographic plate.

16.2 The visual system

Visual photometry is the oldest branch of observational astrophysics. The visual system was the first of all photometric systems, based on the eye's sensitivity. The very wide passband introduces severe color effects, and this is one of the reasons for the difficulties encountered in comparing measurements obtained by different observers. Other factors affecting the measurements are the location of the site (altitude, latitude, climatic conditions) and, last but not least, the personal equation (which itself depends on health, age and other factors, see Section 1.10 and also Sterken & Manfroid (1991) for a more complete discussion).

The greatest asset of existing visual data is that they have been collected over a very long time (in principle over thousands of years). But one must be careful: old visual data are not directly comparable to more recent estimates, even though the detector (the human eye) has not changed since stellar magnitudes were first recorded. There are two basic reasons why modern visual estimates are not comparable to ancient ones. First of all the ancient astronomers (like Ptolemy, Al-Sufi, Brahe) used discrete classes, and the associated errors are of the order of 0.5 to 1 magnitude. Second, the definition of the magnitude itself has changed (see Section 1.6): until quite recently, all famous catalogues used magnitude classes based only on the psychophysical perception of equal brightness steps. Herschel has shown (see Young 1990) that this more or less corresponds to magnitudes being inversely proportional to the square root of stellar brightness. Nowadays, since the work of Steinheil and Pogson, magnitude is defined as being proportional to the logarithm of brightness (see Section 1.6).

The most comprehensive catalogue of visual magnitudes is the great Bonner Durchmusterung. But it is generally known that, particularly for faint stars, there are considerable deviations from the logarithmic scale. Personal errors, as well as systematic errors introduced by the use of a multitude of instruments with own optical properties, can be smoothed out by averaging, but one could, by applying proper homogenization techniques, eliminate the systematic errors and obtain average values with reduced mean errors. In such a way one can build catalogues of homogeneous character.

A nice illustration of the systematic differences between observers is the magnitude limit of unaided vision. Older catalogues disagree on these limits. It is generally accepted that the faintest stars visible are of 6th magnitude. Curtis (1901) listed the faintest magnitude in different catalogues in terms of the magnitude scale of the Harvard Photometric Durchmusterung. The faintest Almagest star has magnitude 5.38, but Al-Sufi notes 5.64. Later catalogues score better. Argelander reached 5.74, whereas Heis recorded 1400 stars more than Argelander, and noted 5.84 for the faintest. Houzeau could see stars down to magnitude 6.4, whereas Gould (at Cordoba) seems to have seen stars of 7th magnitude. Besides the personal condition of the observer's eye, factors such as the altitude of observation, the clarity of the air and the intensity of the diffuse sky background light are important. It is also easier to see a faint star when you know its position, and this may explain the reason why Houzeau and Gould could reach fainter magnitudes than the earlier observers. The absolute registered record probably comes from Curtis (1901) who, in an experimental setting, could detect a star of 8th magnitude. Besides Comet Halley at magnitude 19.5, O'Meara saw 20.1 mag stars, with a 24 inch telescope (at an elevation of 13800 feet on Mauna Kea), and from the rest facility at 9000 feet he consistently saw 8.4 mag stars with the naked eye (Sky & Telescope 1985). These observations give an indication of the absolute detection threshold of the eye.

The accuracy of a visual estimate is quite poor: a trained observer can reach a tenth of a magnitude, when close comparison stars are available with colors and magnitudes similar to that of the program star. The optical equipment (telescope, binoculars, naked eye) plays a role also since it acts on many parameters (size of eye's pupil, sky background, scintillation effects, color aberrations, spectral transmission...).

The visual system is still used by thousands of amateur observers who provide a wealth of data on many variable stars. These data supplement photometry performed by professionals.

Many catalogues of visual magnitudes estimates have been compiled since Ptolemy's work. Among the best known and the most extensive are the *Bonner Durchmusterung*, the *Bonner Südliche Durchmusterung*, the *Uranometria Argentina* and the *Cordoba Durchmusterung*. In addition, there is the *Potsdamer Durchmusterung* (14,000 stars, Müller

& Kempf 1907), its extension (2,000 stars, Tass & Terkan 1916) and the *Potsdamer Polkatalog* relative to the polar cap (5,000 stars, Müller et al. 1927). The latter catalog had the advantage of making the same standard stars available throughout the year for observing sites at moderate or high northern latitudes. The polar sequence (North Polar Sequence, NPS, or International Polar Sequence, IPS) has also been used in other photometric systems, as we shall see in the following sections.

16.3 Photographic systems

The photographic plate has been the first detector other than the eye, and half a century ago photographic magnitudes were the best magnitudes available. Its particular blue chromatic response led to the photographic system of magnitudes $m_{\rm pg}$. This corresponds to emulsions without dyes such as in the well-known Kodak spectroscopic series. The spectral response is limited on the UV side by the atmospheric cutoff and by the transmission of the optics. One should realize that the spectral ranges for photographic photometry were chosen without regard to astrophysical requirements (very little was known about stellar radiation in these days).

With the apparition of emulsions treated to reach the yellow, red and even infrared domains, a variety of photographic systems have been set up. In particular a proper combination of orthochromatic emulsion and yellow filter has been used to reproduce the visual system. This led to the "photo-visual" system of magnitudes $m_{\rm pv}$. The zero points of the photographic and photo-visual magnitudes were defined through the International Polar Sequence (IPS, *Trans. IAU*, **6**, 1938, p. 215). These internationally defined magnitudes are sometimes called *IPg* and *IPv* and are the basis for the IS, the International System. From the $m_{\rm pg}$ and $m_{\rm pv}$ magnitudes, a single color index *CI* is derived. Photo-visual magnitudes can be compared directly to the visual ones (keeping in mind all difficulties linked to the large bandwidth, and remaining differences in spectral response). However, transformations to the International system were ambiguous because the Polar sequence did not contain enough different types of stars to permit highly accurate transformations to the system (absence of red and highly-reddened stars, supergiants, etc.), and also because the system provided no information in the ultraviolet spectral region (Johnson 1952). In fact, both visual and photographic magnitudes have much too wide passbands, so that the derived magnitudes vary too much with brightness and color of the stars.

The advantages of photography over visual observing are considerable. Photography offers a permanent record of the observations. It is less subjective than visual observing, many more objects can be recorded in a single observing session, and they can be measured at leisure in the laboratory.

However the analysis of the stellar images in order to derive magnitudes is a very difficult task and the results are most often inaccurate. Major problems are the non-linearity of the medium (with only a narrow useful range), the spatial inhomogenities, and the difficulty of adequate calibrations. A precision of 0.02 magnitude seems to be the very best that can be achieved in differential photometry with long-focus telescopes.

In non-differential work, accuracy of 0.1 mag can be obtained in optimal conditions. Still, a generation ago, photographic magnitudes were the best available.

Other photographic systems were devised. The oldest and most extensively studied is the *RGU* system (see Lamla 1982). The *RGU* system was introduced by Becker (1946) with a *U*-filter at 359 nm, a *G*-filter at 466 nm and an *R*-filter at 641 nm, with half-widths of about 40 to 50 nm (see also Becker 1965). A photoelectric equivalent system was established by Trefzger et al. (1983). However, the photographic equivalent of the photoelectrically defined *UBV* (which we discuss later) and a few other systems, have also been much used. A complete reference to the photographic photometric systems may be found in Lamla (1965, see Fig. 16.1).

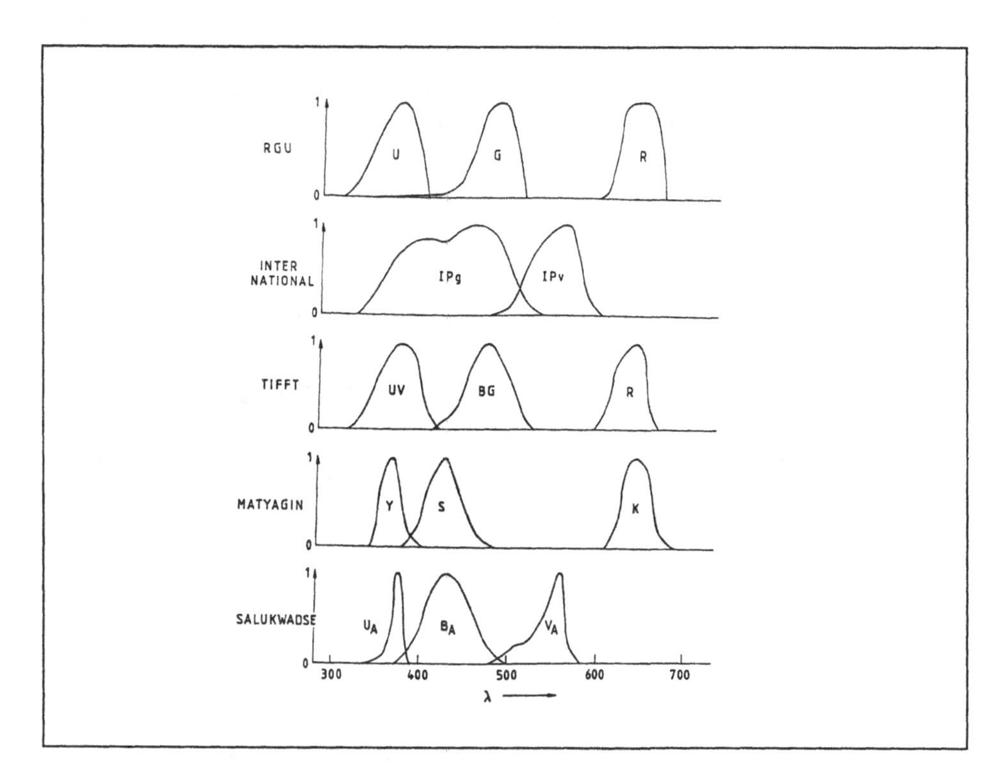

Fig. 16.1 Photographic systems. Passband of five major photographic photometric systems (from Lamla 1965).

A specially important, de-facto, photographic system is defined by the Palomar Sky Survey, which has recorded all of the northern hemisphere and the equatorial regions up to magnitudes close to 20, on two different emulsions: UV-blue sensitive Eastman Kodak 103a-O, and red-sensitive Eastman Kodak 103a-E. Color indices derived from this widely distributed atlas are invaluable for many astrophysical studies. Other atlases obtained with various filter and plate combinations at several big Schmidt telescopes (Palomar II, ESO and Anglo-Australian) are completed or in the making.

16.4 Photoelectric systems

The visual system, by its own character, has a wide passband. So do the photographic systems by the fact that narrow-band filters did not exist when photographic photometry was applied. Anyway, the low quantum efficiency of the photographic emulsions is a poor match for narrow-band filters. This situation changed with the introduction of the photocell and the photomultiplier. Photoelectric photometry yielded tens of different systems. Many of them have received only restricted interest (see Lamla 1965, 1982).

The leading principle in designing photometric systems is that the choice of the number of passbands, and the position of these bands, must be guided by the properties of stellar radiation. One cannot increase at will the number of passbands in wide-band systems: at most, three to four independent wide-band colors can be established in the visual region. One may get more physical information by at the same time reducing the width of the passbands and increasing the number of passbands. It is important to keep a correct sampling of the spectrum. This condition is necessary to allow extinction and color correction, and is thus an absolute prerequisite for precision photometry. The sampling requirements imply strong overlaps between adjacent bands, the more so when pass-bands are steep-sided (see Chapters 6 and 8, and Young 1974b, 1988).

How narrow may a wavelength band be? Too narrow bands (of the order of a wavelength unit or less) would lead to a situation similar to the one depicted above: the amount of starlight available would be so small that it would be impossible to secure enough measurements to build a standard photometric system, and the gain in the number of colors would be compensated by a loss in response. The sampling condition would be impossible to meet, and strictly monochromatic magnitudes would be strongly affected by the presence or absence of even weak absorption (or emission) lines in the specified wavelength range. For such isolated bands, extinction correction would be possible, and even easy, but only in wavelength ranges where the atmospheric absorption is smooth. Color corrections would often be impossible.

Photometrists must thus search for an acceptable compromise somewhere midway between two opposing situations: too narrow bandwidths leading to impractical photon statistics and undersampling effects; and too wide bands introducing insensitivity to important spectral features. Last, but not least, the foregoing remarks may help to create the best possible color system; however, such a color system only becomes a standard system when it is accompanied by a number of reliable measurements of standard stars.

16.4.1 The Johnson-Morgan *UBV(RI)*

The *UBV* system was established by Johnson & Morgan (1951, 1953) as a substitution for and an improvement and a further development of the ambiguous International Photographic and Photovisual systems, and it is the system which is the most similar to the photographic *RGU* system. The basic idea was to have a system that was closely tied to the Morgan-Keenan spectral classification scheme. It became increasingly clear that two-color photometry was not able to define a magnitude system adequately precise for the needs of modern astronomical photometry (Stoy 1954). Individual data on one two-color system simply cannot be converted to another system, whether one uses linear or non-linear transformations, since at least two parameters (involving surface temperature and luminosity) are required. Johnson and Morgan exclude the ultraviolet light from the photographic blue magnitude, and observed it independently in the *U*-band, which covers the spectral region which includes the Balmer Discontinuity with the prominent drop in the spectral energy distribution caused by absorption by ionized hydrogen (see also Fig. 16.6).

Spectral classification by means of photometry has a complementary aspect, since M-K spectral types alone do not give any information on the magnitude nor on the interstellar absorption: spectra give a lot of information on the intrinsic color of a star, but not on the observed colors.

Two independent color indices, standing for luminosity or absolute magnitude and for effective temperature, are required. Hence, three distinct filters were needed. The definition of the system entirely rests on three broad-band colored-glass combination-filters that give a visual magnitude *V*, a blue magnitude *B*, and an "ultraviolet" magnitude *U*, and on the spectral response curve of the historical RCA 1P21 photomultiplier with its S-4 photocathode (see Section 4.3.2 and Fig. 4.3).

Note that the *UBV* system can only be used with aluminized reflectors, but refractors and reflectors with silvered mirrors will also reproduce the *B*- and *V*-bands. Unfortunately, Johnson never published the numerical data of the response functions; most *UBV* work now refers to "reconstructed" response functions, such as those published by Matthews & Sandage (1963), or even better, Lamla (1982).

The *V*-filter has a peak transmission around 550 nm and yields a *V*-magnitude which is nearly identical to the photovisual magnitude of the International System. Unfortunately, the red cutoff is determined by the response of the photomultiplier and not by the filter (see Fig. 5.1). The *B*-filter centered on 430 nm was introduced to match the blue photographic magnitudes (that the *UBV* system became so popular must partly be ascribed to the fact that *UBV* photometry could be carried out photographically as well as photoelectrically). Some *B*-filters, however, have red leaks (see Section 5.3). The *U* filter peaks short of 350 nm and has (besides a red leak) a blue cutoff completely defined by the ultraviolet transmission of the 1P21 tube envelope (see e.g., Bessel 1986). Note also that the canonical altitude for the *UBV* system is 2150m, because the atmospheric extinction is included in U–B. The color index B–V is not inconsistent with the International System (except for a zero-point difference), and U–B is a new color index.

It is unfortunate that the *UBV* system is not a purely filter-defined system. As can be seen in Fig. 5.1, the passbands do intersect, but not enough to come close to satisfying the sampling condition. In addition, a multitude of "clone" systems unfortunately exist today (for a comparison of such passbands, see Fig. 1 of Buser 1978, and Fig. 3 of Bessell 1986). These are major drawbacks of this system.

The zero points of the *UBV* system were defined by putting to zero the color indices of a normal and unreddened main-sequence star of spectral type A0. Originally the system had only 10 standard stars, and their indices were the (extinction-corrected) readings in the basic instrumental system after a zero-point shift to the A0 standard (thus, this list defines, in terms of stars, the physical characteristics of the photometer-telescope combination). Since then elaborate lists of secondary standard measurements have been published.

B–V correlates well with effective temperature (see Jaschek & Jaschek 1987). *U–B* measures the intensity ratio on both sides of the Balmer discontinuity, and hence is a measure of the hydrogen line strength. Determination of interstellar reddening and luminosity class from *U–B* alone is not possible, but when the luminosity class of a star is known, the Johnson *Q*-method—first used by Becker (1938), and developed by Johnson & Morgan (1953), and Johnson (1958) (see also Serkowski 1963)—allows dereddening of the color indices of early-type stars. For stars later than A0, one cannot extract intrinsic colors from the (*U–B*,*B–V*) diagram. If only *B–V* is available (many stars in *UBV* catalogues have only *V* and *B–V* entries), one can only extract information concerning the spectral class if—from another source—we know that the star is unreddened. Figure 16.2 illustrates the value of *Q*,* when used together with $(B\text{–}V)_0$.

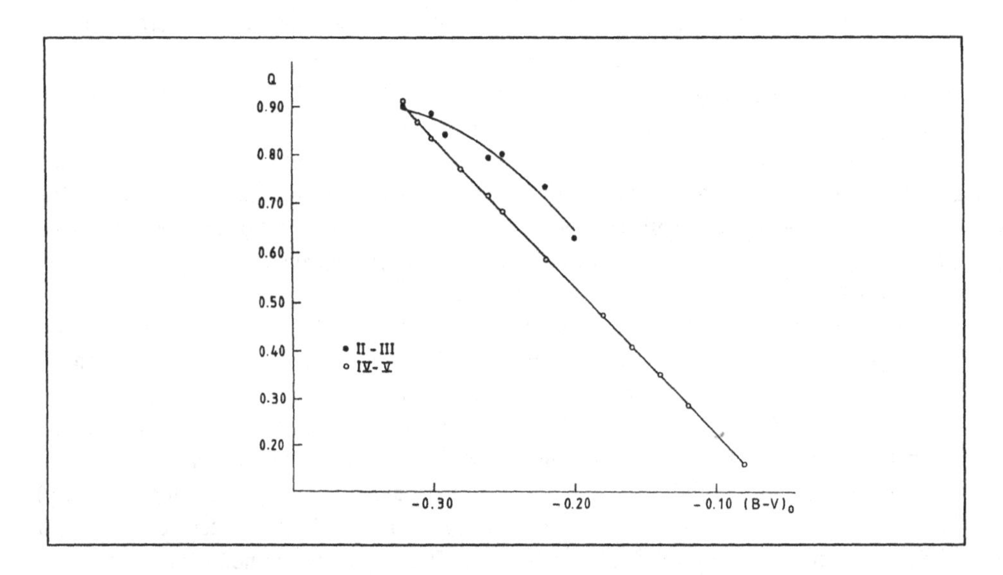

Fig. 16.2 Johnson's reddening-free parameter Q. The parameter Q is shown as a function of $(B\text{–}V)_0$ *(based on Serkowski 1963).*

R- and *I*-bands, at 700 and 900 nm respectively (to be used with S-1 or extended S-20 photocathodes) were added later (Johnson 1965), and this firmly extended the usefulness of *B–V* as temperature indicator for late-type stars. *R* and *I* are, furthermore, less affected by interstellar extinction. It is clear that, if one wishes to measure *U*, *B*, *V* and *R*, *I*, with the same detector (e.g., extended S-20), one must cut off the long-wavelength tail and red leaks of the *U*, *B* and *V* filters by adding appropriate color-blocking filters.

* The Johnson reddening-free parameter *Q* should not be confused with the magnitude measured in the infrared *Q*-band at 20 μm.

Table 16.1 Characteristics of the Johnson *UBVRI* system

Symbol	λ_0 (nm)	$\Delta\lambda$ (nm)
U	365	70
B	440	100
V	550	90
R	720	220
I	900	240

Table 16.1 gives some approximate characteristics of the *UBVRI* filters

Another system with similar (but distinct in terms of the discussion in Chapter 11) properties to the Johnson-Morgan system is the Cousins *VRI* system (Cousins 1976, 1978). The *V*-band (complemented with *U* and *B*) is alike the classical *UBV* passbands, but the *R* and *I* bands have mean wavelengths, respectively, at 670 nm and 810 nm, thus shortwards of the *R* and *I* bands of the Johnson-Morgan system. The Cousins *R* and *I* are commonly indicated by $(RI)_C$ (C stands for "Cape").

16.4.2 The infrared *JHKLM* system

A system of broad photometric bands in the infrared, analoguous to the Johnson-Morgan *UBV* system, has also been defined by Johnson and his collaborators. The bands match the regions of optimal atmospheric transmission and are labeled *J*, *H*, *K*, *L* and *M* (designations are in increasing wavelength); the associated magnitudes yield corresponding color indexes *J–H*, *H–K*, *K–L*, etc. Table 16.2 gives the particulars of these bands; Fig. 16.3 gives the transmission curves. Since the atmospheric transmission in the different bands is a strong function of altitude and climatic conditions of the site, those transmission curves should not be taken at face value since they all contain wavelengths where there is atmospheric absorption. Therefore, any computation of effective wavelength of such bands has to take atmospheric transmission into account; this means that λ_0 is at most valid for a single site (even for a single season or observing run). For a detailed discussion, see Low & Rieke (1974).

Subsequent band designations (such as *N*, *O* and *P*) have also been introduced, but they are less standardized. Photometry up to the *M* band is carried out with an InSb detector, whereas at longer wavelengths a different kind of detector (usually a bolometer) is needed (*M*-band photometry is difficult due to the poor atmospheric window). There are currently few good-quality bolometer systems in operation. Since the early definitions of the system, several other photometric systems related to the original system have been conceived. Therefore, one refers to the "Arizona" system, "ESO" system, "SAAO" system, etc. (see Bersanelli et al. 1991, Bouchet et al. 1991).

Also in the infrared a zero point to the system of magnitudes must be adopted. It was fixed according to Johnson's definition: a normal and unreddened star of spectral

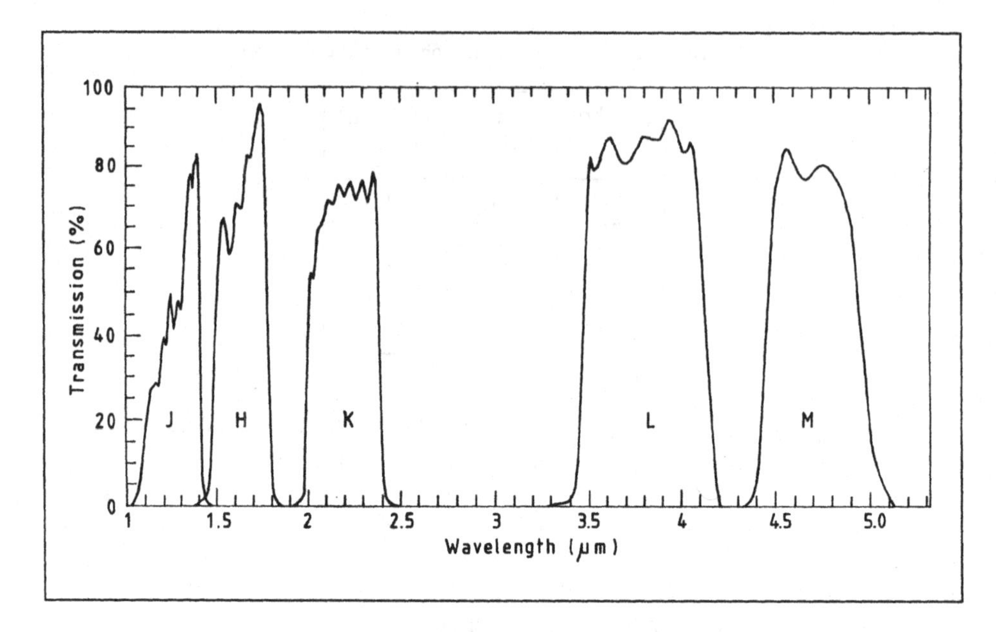

Fig. 16.3 Passbands of the JHKLM *system.*

Table 16.2 Characteristics of the *JHKLM* system

Band	λ_0 (μm)	$\Delta\lambda$ (nm)	waveband (μm)
J	1.25	380	1.1–1.4
H	1.65	480	1.5–1.8
K	2.2	700	2.0–2.4
L	3.5	1200	3.0–4.0
M	4.8	5700	4.6–5.0

type A0 has the same numerical magnitude at all wavelengths. In the infrared Vega (α Lyrae) is used, but at 20 μm Vega is too faint, so other standards had to be taken.

16.4.3 The Geneva $UBVB_1B_2V_1G$ system

The Geneva 7-color system has been introduced by Golay (1963), and is intermediate between a wide-band system and an intermediate-band system. Its *(U)*, *(B)* and *(V)* filters are close to the Johnson-Morgan *U*, *B* and *V* filters; in addition there are four other filters (B_1, B_2, V_1, G) that cover the regions occupied by *(B)* and *(V)*. The passbands are delimited by glass filters and by the spectral response of a S-11 photocathode, and the original response curve was published in detail by Rufener & Maeder (1971). The newest version of a Geneva 7-filterset is described by Rufener & Nicolet (1988). The Geneva photometric system is about the only existing photometric system for which the defining passbands are accurately known throughout the history of the system. Moreover, the fact that observation and data-reduction procedures have not changed since its coming into opearation, has added considerably to the stability of the system.

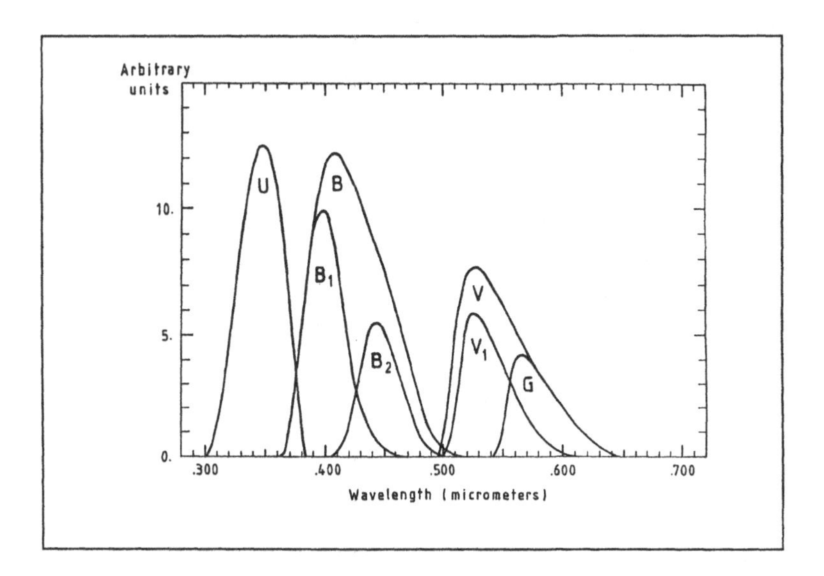

Fig. 16.4 *Geneva system. Spectral responses (filter transmission combined with photocathode response curves) taken from Rufener (1988). Note that the red wing of the* V- *and* G-*bands are completely determined by the red response beyond 530 nm of the S-11 photocathode (cf. Johnson-Morgan* V *filter in Fig. 5.1), which is not a big problem, as EMI S-11 is probably more reproducible than RCA S-4.*

The published transmission curves are of fundamental importance for calculations of synthetic color indices. The fact that glass filters in the Geneva photometers are thermally controlled also contributes to the stability of the system by exclusion of seasonal effects due to thermally induced passband changes (see Fig. 5.2).

Users of Geneva catalogues (Rufener 1980, 1988) often do not realize that *by definition* the B-magnitude has been set to zero (B is the reference color because it is the least sensitive to atmospheric extinction effects; any measurement or reduction error in B is thus transferred to all colors), so that each color is reduced to one index and only the 6 independent *color indices* U, V, B_1, B_2, V_1 and G are available (this procedure was adopted because the condition of stability of the measuring equipment was not guaranteed by the technology available in the sixties). Besides the gradients $B_1 - B_2$, $B_2 - V_1$

Table 16.3 Characteristics of the Geneva $UBVB_1B_2V_1G$ system (Rufener & Maeder 1971, Rufener & Nicolet 1988)

	1971		1988	
Symbol	λ_0 (nm)	$\Delta\lambda$ (nm)	λ_0 (nm)	$\Delta\lambda$ (nm)
U	345.8	17.0	346.4	15.9
B_1	402.2	17.1	401.5	18.8
B	424.8	28.3	422.7	28.2
B_2	448.0	16.4	447.6	16.3
V_1	540.8	20.2	539.5	20.2
V	550.8	29.8	548.8	29.6
G	581.4	20.6	580.7	20.0

and $V_1 - G$, the catalogue also gives the reddening-free parameters

$$\begin{aligned} d &= [U - B_1] - 1.430[B_1 - B_2] \\ \delta &= [U - B_2] - 0.832[B_2 - G] \\ g &= [B_1 - B_2] - 1.357[V_1 - G] \\ m_2 &= [B_1 - B_2] - 0.457[B_2 - V_1] \end{aligned} \tag{16.1}$$

(see also Golay 1972, 1980). d and δ measure the Balmer discontinuity, but they do respond differently to line blocking and to various extinction laws; g measures blocking (Golay 1974).

Cramer & Maeder (1979) defined a set of reddening-free orthogonal coordinates X, Y, Z which are linear combinations of Geneva photometric system color indices, and which are given by:

$$\begin{pmatrix} X \\ Y \\ Z \end{pmatrix} = \begin{pmatrix} .3788 \\ -.8288 \\ -.4572 \end{pmatrix} + \begin{pmatrix} 1.3764 & -1.2162 & -.8498 & -.1554 & .8450 \\ .3235 & -2.3228 & 2.3363 & .7495 & -1.0865 \\ .0255 & -.1740 & .4696 & -1.1205 & .7994 \end{pmatrix} \begin{pmatrix} U \\ B_1 \\ B_2 \\ V_1 \\ G \end{pmatrix} \tag{16.2}$$

The X-parameter is an indicator of effective temperature and has been calibrated for luminosity classes III, IV and V stars. The XY-plane was calibrated in regions of equal luminosity. The Z-coordinate is an indicator of peculiarity and is related to the strength of the surface magnetic field. For O, B and early A stars, X and Y can be transformed to a quasi-Hβ index (the β-index is defined in Section 16.4.5), $\beta(X, Y)$, defined by (Cramer 1984)

$$\beta(X,Y) = 2.5909 + .0667X - .6801Y - .2559XY + .1748X^2 - 2.4676Y^2 + .1448XY^2 + .2582X^2Y - .0612X^3 + .4418Y^3 \tag{16.3}$$

$\beta(X,Y) - \beta$ then seems to be a sensitive index to departures from the normal Hβ line-width versus spectral energy distribution relationship. Also the intrinsic Johnson-Morgan colors $(U\text{–}B)_0$ and $(B\text{–}V)_0$ can be expressed in terms of X and Y.

16.4.4 The Walraven *VBLUW* system

There exists only one "Walraven photometer"; it was first attached to the Dutch 90 cm "light collector" in South Africa (Walraven & Walraven 1960), and later at the same telescope relocated at ESO La Silla, Chile. It was designed and built by Walraven with the purpose of studying early-type stars. Passbands of the system were published by Lub & Pel (1977). The passbands are quite stable, except for *V* and *W* which exhibit minor changes since the early days.

Table 16.4 Characteristics of the Walraven *VBLUW* system

Band	λ_0 (nm)	$\Delta\lambda$ (nm)
V	547	72
B	432	45
L	384	23
U	363	24
W	325	14

A very interesting aspect of the Walraven photometer is that it measures all five colors simultaneously (the spectrum is formed by quartz lenses and prisms), which not only means economy in observing time, but also enhanced accuracy of the photometric indices (see also the Danish *uvby* photometer, Chapter 3). Three of the regions have effective wavelengths which are approximately the same as those of the standard *UBV* photometry; *W* and *L* are situated, respectively, in the more remote ultraviolet, at 322 nm, and at 390 nm (between *B* and *U* where the higher members of the Balmer series crowd together). The five bands define four color indices *V–B, B–L, B–U* and *U–W*, which are physically relevant. Only *V–B* is transformable to a similar index in another system. All Walraven indices are not given as magnitude differences, but as plain decimal logarithms.

16.4.5 The Strömgren *uvby*β system

The Strömgren system is an intermediate-bandwidth system characterized by an astrophysically sound selection of passband widths and locations (Strömgren 1966). It is almost completely filter-defined. The system was designed in such a way that it would be independent of the spectral response of the detector, but practice shows that this is not completely so (see Manfroid & Sterken 1987 and Manfroid et al. 1991). Second-order color terms play a minor role, and the system overcomes many difficulties of the *UBV* system. Because at around 550 nm (i) few absorption lines occur, (ii) the atmospheric transmission is smooth and almost constant and (iii) the detectors used have smooth and almost constant response curves, the *y* (for "yellow", but actually green!) filter magnitudes transform very well to Johnson-Morgan *V*-magnitudes. The *b*-filter sits 30 nm redward of Johnson-Morgan *B*, and encompasses Hβ, but no other strong spectral absorption lines, and is (especially in the case of late-type stars), less sensitive to line-blanketing effects. *v* is centered on Hδ, a region of strong blanketing (but still far from the Balmer limit), and *u* is blueward of the Balmer discontinuity. *b* and *y* are equally influenced by blanketing (if present at all), and hence *b–y* is a sensitive effective-temperature indicator. Two other color indices, *v–b* and *u–b*, are very useful. The index

$$m_1 = (v - b) - (b - y) = v - 2b + y \tag{16.4}$$

is a good blanketing-indicator and is called the *metallicity-index*. *u* is affected by blanketing and by the Balmer jump, and *v* contains only blanketing (which is approximately 50% weaker than in *u*). The quantity

$$c_1 = (u - v) - (v - b) = u - 2v + b \tag{16.5}$$

eliminates all blanketing from *u*, so that c_1 measures the Balmer discontinuity.* Though *uvby* standard star catalogues (e.g., Grønbech et al. 1976, Grønbech & Olsen 1977) only list *b–y*, m_1 and c_1 values, *uvby* photometry yields an independent measure of visual magnitude.

"New" *uvby*-filters of the square-passband type (see Fig. 3.6) with cut-off wings have slightly reduced bandwith and higher transmission. This improves the color-independency of the bandwidth; for instance, the *u*-band is better isolated from the Balmer discontinuity, but dramatically degrades the sampling. These filters are by far not compatible with the original Strömgren *uvby* system. Those incompatibilities have been demonstrated by calculations and by critical observations (see e.g., Manfroid 1985a,b, Manfroid & Sterken 1987, Young 1988, and also Section 11.2). Hence indices obtained with those instruments cannot be transformed to the standard system and, even worse, available astrophysical calibrations have no value if applied to such data.

Strömgren photometry is often supplemented with 2 Hβ filters: the H$\beta_{\mathrm{w(ide)}}$ and H$\beta_{\mathrm{n(arrow)}}$ bands are centered on the wavelength of the Hβ line at 486 nm, see Fig. 3.7). A measurement through both filters (preferably simultaneously in order to cancel out atmospheric variations) yields the Crawford & Mander (1966) β-index. Calling m_{w} and m_{n} the magnitudes in the wide and narrow filters, the β-index is defined as

$$\beta = m_{\mathrm{w}} - m_{\mathrm{n}} \tag{16.6}$$

* Whereas *u–v*, *v–b* and *b–y* measure the slope, indices like m_1 and c_1 estimate the curvature of a star's spectrum.

β is free from interstellar and—in the case of simultaneous measurements—from atmospheric extinction effects.* It constitutes a luminosity index for O to A-stars, and an effective-temperature index for A to G stars, and of course it is independent from the *uvby* color indices. The whole range in magnitude that is spanned by β only amounts to a couple of tenths of a magnitude. A nominal precision of a couple of thousandths of a magnitude on β still represents only several per cent of the physically available range in β.

For transforming *uvby*β indices to astrophysical parameters, we refer to the papers by Crawford (1975, 1978, 1979). Standard stars for β-photometry can be found in Grønbech & Olsen (1977).

Table 16.5 Characteristics of the Strömgren *uvby* system

Band	λ_0 (nm)	$\Delta\lambda$ (nm)
u	350	34
v	410	20
b	470	16
y	550	24
$H\beta_w$	486	15
$H\beta_n$	486	3

16.4.6 Hα systems

In contrast with Hβ, where the β-index defines a standard system, there exist several Hα systems; some of them even require spectrophotometers instead of filters. The basic principle is the same: a wide filter (in which position and width is not very critical) together with a narrow filter (with very critical properties, hence the narrow filter *defines* the system) at the same effective wavelength (656 nm). Since differential extinction (atmospheric and interstellar) is less critical at longer wavelengths, the exact matching of the effective wavelengths of the narrow and wide filter is less crucial than it is in the case of Hβ. Though photometric measurements of Hα and Hβ are very useful in the study of certain types of stars (e.g., supergiants), few of the filter-defined photometric Hα systems have been used for the measurement of more than about 100 stars. The best-known systems are those described by Dachs & Schmidt-Kaler (1975), Cester et al. (1977), Price (1966) and Strauss & Ducati (1981). The best known spectrophotometric Hα system is the one used by Andrews (1968).

* This statement rests on the asumption that both filters are symmetric on exactly the same central wavelength. In practice, though, this condition is seldom met (see Fig. 3.7).

16.4.7 The WF/PC filter system of the Hubble Space Telescope

A *uvby*-like photometric system for use at the Wide Field/Planetary Camera (CCD) of the Hubble Space Telescope exists (Griffiths 1985). Kiselman et al. (1990) discuss the Strömgren-like indices $(B-Y)_1$, C_{12} and M_1.

16.4.8 The IHW filter system for cometary research

Photometry of comets is mainly the photometry of (moving) extended objects with an emission-line spectrum and a continuum spectrum. The continuum spectrum is caused by sunlight that is scattered by dust, and the emission lines are caused by de-excitation of atoms and molecules.

The use of broad-band filters to isolate a portion of a cometary spectrum is very limited because broad passbands will smear out contributions from the continuum and from the strong emission lines. Broad-band techniques are only useful in the case of distant comets which are not very active. Therefore a working group established by IAU Commission 15 at the 1979 General Assembly recommended nine specific narrow-band filters which isolate particular spectral regions. Figure 16.5 shows the spectrum of Comet Kohoutek at 1 AU (pre-perihelion) with the bandwidths of the International Halley Watch (IHW) filters superimposed.

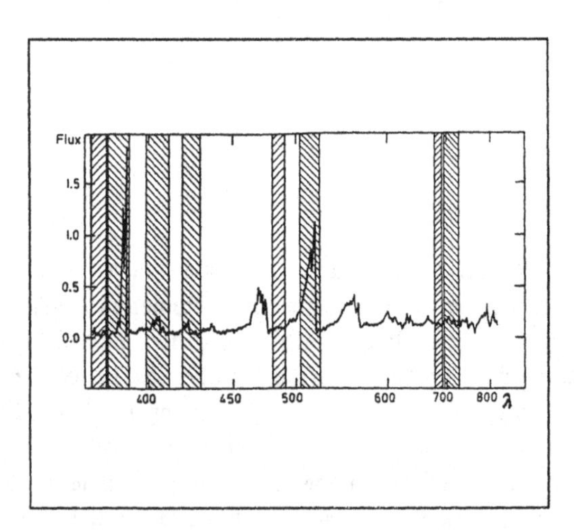

Fig. 16.5 IHW cometary system. The position of the passbands of the IHW cometary filters, superimposed on a spectrum of comet Kohoutek at 1 AU pre-perihelion (adapted from A'Hearn 1975).

Cometary photometry is different from stellar photometry in the sense that the primary limitation of accuracy is not the detector, but the choice of sky-background reading, standard stars (which have to be changed along the path of the moving comet), extinction,

Table 16.6 Characteristics of the IHW cometary system

Band	λ_0 (nm)	$\Delta\lambda$ (nm)
H_2O^+	700	17.5
continuum	684	9
C_2	514	9
continuum	485	6.5
CO^+	426	6.5
C_3	406	7
CN	387	5
continuum	465	8

the exact knowledge of the sizes of the diaphragms, etc. The measurements are frequently collected at excessively large air masses and often happen in twilight, when sky brightness changes rapidly, and when few standard stars can be observed. In addition, two totally different classes of standard stars are required: G-type standard stars similar to the Sun, for accurately determining the continuum due to the reflectance of the cometary dust, and standard stars for absolute flux calibration of the emission features, usually chosen as B-stars because of their few spectral lines. For more details, we refer to A'Hearn (1983); a list of standard stars suitable for photometry of comets was published by Osborn et al. (1990).

16.5 Absolute calibration

Stellar irradiances are expressed in terms of magnitudes, which only are ratios of irradiances of one object relative to another, or (in the case of a color index) in one spectral band relative to another. For physical interpretation of photometric measurements, one must calibrate the photometric magnitudes in terms of true spectral irradiance units. This, for example, allows comparison to a black-body energy distribution.

In order to compare photometric observations with fluxes predicted by a theoretical model, one must know precisely not only the passbands of the system, but also the combined chromatic response of the whole instrumental setup (telescope and photometer mirrors, filters and photomultiplier). Direct laboratory calibration, as is done in the case of experimental radiometry, is very difficult, since

- one cannot be sure that the laboratory calibrations are made under exactly the same conditions as those that intervene while observing (especially for what concerns the optical path),
- direct laboratory calibration involves costly blackbody radiators,

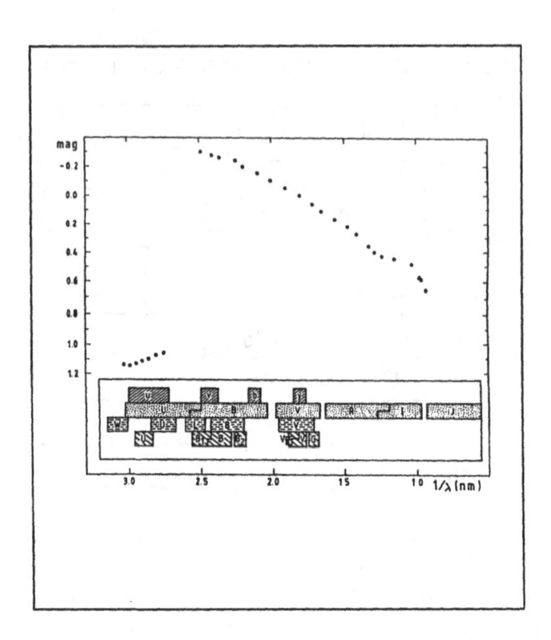

Fig. 16.6 Spectral energy distribution of α Lyrae, based on data from Hayes & Latham (1975). Vertical scale is in magnitudes, horizontal scale is 1 / λ where λ is in nm. The extent of the passbands of some photometric systems is indicated.

- calibrations must be done-over repeatedly since one is not sure that the instrumental configuration is stable over a long period of time. Proper calibration yields the irradiance of (at least one) standard star in watts per square meter per wavelength unit, and this scale defines the zero point of the magnitude scale. In other words: absolute standards become a kind of "standard candle", by which magnitude measurements of any other star can be converted to absolute energy units.

Absolute calibration is a fundamental problem of observational astronomy. The original calibration by Johnson (1965) was based on the observation of a number of solar-type stars under the assumption that the colors of the Sun are the same as the average colors of the calibration stars. The zero points of the magnitude scale were determined in such a way that for an average A0V star all numerical magnitudes are the same at all wavelengths, so all color indices are zero (a direct consequence of this is that the additive constant C in Eq. (1.11) must be determined only in one band, and that the additive constants for all other bands follow immediately from Johnson's constraint that color indices for an A0V star are zero). Johnson's calibrations were only preliminary, since they were not observationally defined at wavelengths beyond 5 μm. More reliable calibrations have been defined since, and can be found in several publications with standard star lists (see Table 16.3). The most direct method is to compare observed

Table 16.7 Absolute calibrations: monochromatic irradiance E_λ (in 10^{-11} W nm^{-1} m^{-2}) for stars of spectral type A0V and O V with magnitude $V = 0$.

Band	A0V	OV
U	4.22	16.22
B	6.40	8.59
V	3.75	3.75
R	1.70	
I	0.83	
J	0.307	
H	0.12	
K	0.041	
L	0.0064	
M	0.0019	
u	3.25	18.00
v	7.18	10.17
b	5.81	6.60
y	3.70	3.66

irradiances with a terrestrial source, whereby the amount of radiation emitted by the blackbody source was properly attenuated in order to match the much lower spectral irradiances measured for the stellar source. Absolute calibration measurements are beset with many problems, such as, for example, time variations in the extinction (vertical as well as horizontal extinction). α Lyrae (Vega) has spectral type A0V, and is the brightest early-type star in the northern sky, and is often used as standard candle.

So for each waveband a calibration exists giving the flux for a zero-magnitude star. These calibrations are the result of intensive thought and observational labour; however, they remain unreliable because we simply cannot know with great precision the flux from a zero-magnitude star: we do not know the atmospheric extinction, the interstellar extinction, and even not the transmission properties of telescope and photometer. These facts put an uncertainty level of about 10% on existing calibrations. However, several calibrations of the energy distribution of α Lyr published since 1960 have shown a range amounting to 0.1 mag (Hayes 1970, Hayes et al. 1975, and references therein). Table 16.7 gives absolute calibrations for all photometric systems discussed above (taken from different sources). Figure 16.6 gives the spectral energy distribution of α Lyrae. Neckel & Labs (1984) published absolute radiation data (from 330 nm to 1.25 μm) for the Sun.

Appendix A. References

A'Hearn, A.T.: 1975, *Astron. J.*, **80**, 861

A'Hearn, A.T.: 1983, Photometry of Comets, in *Solar System Photometry Handbook*, (R.M. Genet, Ed.), Willmann-Bell, Richmond, Chap. 3

Andrews, P.J.: 1968, *Mem. R. Astron. Soc.*, **72**, 35

Bartholdy, P., Burnet, M., Rufener, F.: 1984, *Astr. Astrophys.*, **134**, 290

Barwig, H., Schoembs, R.: 1987, *The Messenger*, **48**, 29

Barwig, H., Schoembs, R., Buckenmayer, C.: 1987, *Astr. Astrophys.*, **175**, 327

Becker, W.: 1938, *Zeitschr. f. Astrophys.*, **15**, 225

Becker, W.: 1946, *Veröff. Univ. Sternwarte Göttingen*, **79**

Becker, W.: 1965, *Zeitschr. f. Astrophys.*, **62**, 54

Bemporad, A.: 1904, *Zur Theorie der Extinktion des Lichtes in der Erdatmosphäre*, Grossh. Sternwarte Heidelberg, **4**, 1

Bersanelli, M., Bouchet, P., Falomo, R.: 1991, *Astr. Astrophys.*, **252**, 854

Bessell, M.S.: 1986, *Pub. Astron. Soc. Pac.*, **98**, 354

Biberman, L.M.: 1971, in *Photoelectronic Imaging Devices*, (L.M. Biberman, S. Nudelman, Eds.), Plenum Press, New York

Blanco, V.M.: 1953, *Astrophys. J.*, **123**, 64

Blanco, V.M.: 1955, *Astrophys. J.*, **125**, 209

Born, M., Wolf, E.: 1964, *Principles of Optics*, 2nd Edition, Pergamon Press, Oxford

Bouchet, P., Manfroid J., Schmider F.X.: 1991, *Astr. Astrophys. Suppl. Ser.*., **91**, 409

Bouguer, P.: 1729, *Essai d'optique sur la gradation de la lumière*, C. Jombert, Paris

Bouguer, P.: 1760, *Traité d'optique sur la gradation de la lumière*, H.L. Guerin & L.F. Delatour, Paris

Branham, R.L.: 1982, *Astron. J.*, **87**, 928

Brou, P., Sciascia, T.R., Linden, L., Lettvin, J.Y.: 1986, *Scientific American*, **255**, No. 3, 80

Buonanno, R., Buscema, G., Corsi, C.E., Ferraro, I., Iannicola, G.: 1983, *Astr. Astrophys.*, **126**, 278

Buser, R.: 1978, *Astr. Astrophys.*, **62**, 411

Cester, B., Giuricin, G., Mardirossian, F., Pucillo, M., Castelli, F., Flora, U.: 1977, *Astr. Astrophys. Suppl. Ser.*, **30**, 1

Clark, R.N.: 1991, *Visual Astronomy of the Deep Sky*, Cambridge University Press, Cambridge

Compton, A.H.: 1933, *Physical Review*, **43**, 387

Coulman, C.: 1985, *Ann. Rev. Astron. Astrophys.*, **23**, 19, Annual Reviews Inc., Palo Alto

Cousins, A.W.J.: 1976, *Mem. R. Astron. Soc.*, **81**, 25

Cousins, A.W.J.: 1978, *Mon. Not. Astron. Soc. S. Africa*, **37**, 8

Cramer, N.: 1984, *Astr. Astrophys.*, **132**, 283

Cramer, N.: 1991, private communication

Cramer, N., Maeder, A.: 1979, *Astr. Astrophys. Suppl. Ser.*, **78**, 305

Crawford, D.L.: 1975, *Astron. J.*, **80**, 955

Crawford, D.L.: 1978, *Astron. J.*, **83**, 48

Crawford, D.L.: 1979, *Astron. J.*, **84**, 1858

Crawford, D.L., Barnes, J.V.: 1974, *Astron. J.*, **79**, 687

Crawford, D.L., Mander, J.: 1966, *Astron. J.*, **71**, 114

Crawford, D.L., Mandwewala, N.: 1976, *Pub. Astron. Soc. Pac.*, **88**, 917

Crawford, D.L.: 1988, *Pub. Astron. Soc. Pac.*, **100**, 887

Curtis, H.D.: 1901, Lick Obs. Bull., **38**, 67

Dachs, J., Schmidt-Kaler, Th.: 1975, *Astr. Astrophys. Suppl. Ser.*, **21**, 81

De Groot, W.: 1934, *Nature*, **134**, 494

de Vaucouleurs, G.: 1968, *Applied Optics*, **7**, 1513

Ditchburn, R.W.: 1963, *Light*, Wiley (Interscience), New York

Djorgowski, S., Dickinson, M.: 1989, *Highlights of Astronomy*, **8**, (D. McNally, Ed.), Kluwer Academic Publishers, Dordrecht, p645

Dobrowolski, J.A.: 1978, in *Handbook of Optics*, Optical Soc. of America, (W.G. Driscoll, W. Vaughan, Eds.), McGraw-Hill

Eccles, M.J., Sim, M.E., Tritton, K.P.: 1988, *Low light level detectors in astronomy*, Cambridge University Press, Cambridge
Eddington, A.S.: 1926, *The internal constitution of the stars*, Cambridge University Press, Cambridge
Edgeworth, F.Y.: 1887, *Philos. Mag.*, **24**, 222
Engstrom, R.W.: 1980, *Photomultiplier Handbook*, RCA Corporation, p.54
Eppeldauer, G., Schaefer, A.R.: 1984. Proc. Second Workshop Improvements to Photometry, NASA Conf. Publ., **10015**, 111
The ESO Users Manual 1989, (H.E. Schwartz, J. Melnick, Eds.)
Evans, D.S.: 1968, *Observation in modern astronomy*, The English University Press, London
Evans, J.: 1987a, *Journal Hist. Astron.*, **18**, 155
Evans, J.: 1987b, *Journal Hist. Astron.*, **18**, 233
Fabry, Ch.: 1910, *Astrophys. J.*, **31**, 394
Fechner, G.T.: 1859, *Abh. K. Sächs. Ges. d. Wiss.*, **4**, 455
Fechner, G.T.: 1860, *Elemente der Psychophysik*, Breitkopf und Härtel, Leipzig
Fernie, J.D.: 1976, *Pub. Astron. Soc. Pac.*, **88**, 969
Fernie, J.D.: 1982, *J. Roy. Astron. Soc. Can.*, **76**, No. 4, 224
Florentin Nielsen, R.: 1983, *Inst. Theor. Astrophys. Oslo, Rep.*, **59**, 141
Florentin Nielsen, R., Nørregaard, P., Olsen, E.H.: 1987, *The Messenger*, **50**, 45
Forbes, J.D.: 1842, *Phil. Trans.*, **132**, 225
Fort, B.: 1988, in *Very Large Telescopes and their Instrumentation*, ESO Conference, (M.H. Ulrich, Ed.), 929
Furenlid, I. Schoening, W.E., Carder, B.E.: 1977, *AAS Photo-Bulletin*, **16**, 14
Garrison, R.F.: 1985, in *Calibration of Fundamental Stellar Quantities*, IAU Coll. **111**, (D.S. Hayes et al., Eds.), Reidel Publ. Co., Dordrecht, p17
Genet, R.M., Hall, D.S.: 1989, *Photoelectric Photometry of Variable Stars*, Willmann-Bell, Richmond
Genet, R.M., Hayes, D.S.: 1989, *Robotic Observatories*, Autoscope Corporation, Mesa, Arizona
Geyer, E.H., Hoffmann, M.: 1975, *Astr. Astrophys.*, **38**, 359
Golay, M.: 1963, *Publ. Obs. Genève, Série A*, **64**, 199
Golay, M.: 1972, *Vistas in Astronomy*, **14**, 13
Golay, M.: 1974, *Introduction to Astronomical Photometry.*, Reidel Publ. Co., Dordrecht
Golay, M.: 1980, *Vistas in Astronomy*, **24**, 141
Görlich, P.: 1962, *Photo-effekte, I*, Akad. Verlagges. Geest & Portig, Leipzig
Grasshoff, G.: 1990, *The History of Ptolemy's Star Catalogue*, Springer-Verlag, Berlin
Griffin, D.R., Hubbard, R., Wald, G.: 1947, *Journal Opt. Soc. Am.*, **37**, 546
Griffiths, R.: 1985, *Wide Field and Planetary Camera Instrument Handbook*, Space Telescope Science Institute
Grønbech, B., Olsen, E.H.: 1977, *Astr. Astrophys. Suppl. Ser.*, **27**, 433
Grønbech, B., Olsen, E.H., Strömgren, B.: 1976, *Astr. Astrophys. Suppl. Ser.*, **26**, 155
Gunn J.E., Stryker L.L.: 1983, *Astrophys. J. Suppl. Ser.*, **52**, 121
Guthnick, P.: 1913, *Astron. Nach.*, **196**, 357
Hall, J.S., Jerzykiewicz, M., Riley, L.: 1975, *Journal of the Air Pollution Control Association*, **25**, 1045
Halm, J.: 1917, *Mon. Not. R. astr. Soc.*, **77**, 243
Hardie, R.H.: 1962, in *Astronomical Techniques*, Stars and Stellar Systems, (W.A. Hiltner, Ed.), University of Chicago Press, Chicago, 178
Harris, W.E., Fitzgerald M.P., Reed. B.C.: 1981, *Pub. Astron. Soc. Pac.*, **93**, 507
Hayes, D.S.: 1970, *Astrophys. J.*, **159**, 165
Hayes, D.S., Latham, D.W.: 1975, *Astrophys. J.*, **197**, 593
Hayes, D.S., Latham, D.W., Hayes, S.H.: 1975, *Astrophys. J.*, **197**, 587
Hearnshaw, J.B.: 1991, *Southern Stars*, **34**, 33
Henden, A.A., Kaitchuk, R.H.: 1982, *Astronomical Photometry*, Van Nostrand Reinhold, New York
Herschel, W.: 1800, *Phil. Trans. Roy. Soc. London*, **90**, 255
Hertz, H.: 1887. *Ann. Phys. und Chemie*, **31**, 983
Hoag, A.A., Furenlid, I., Schoening, W.E.: 1978, *AAS Photo-Bulletin*, **19**, 3
Hoffleit, D.: 1982, *The Bright Star Catalogue*, 4th rev. edit., Yale Univ. Obs., New Haven

Huber, P.J.: 1981, *Robust Statistics*, John Wiley & Sons, New York
Hurter, F., Driffield, V.C.: 1890, *J. Soc. Chem. Ind. London*, **9**, 455
Jaschek, C., Jaschek, M.: 1987, *The Classification of Stars*, Cambridge University Press, Cambridge
Jerzykiewicz, M., Sterken, C.: 1987, *Proc. 27th Liège International Astrophysical Colloquium*, 49
Johnson, H.L.: 1952, *Astrophys. J.*, **116**, 272
Johnson, H.L.: 1958, Lick Observatory Bulletin, **4**, 37
Johnson, H.L.: 1962, *Astronomical Techniques*, Stars and Stellar Systems, (W.A. Hiltner, Ed.), Univ. Chicago Press, Chicago, 158
Johnson, H.L.: 1965, *Comm. Lunar and Plan. Lab.*, **3**, 73
Johnson, H.L., Coleman, I., Mitchell, R.I., Steinmetz, D.L.: 1968, *Comm. Lunar Plan. Lab.*, **113**, 83
Johnson, H.L., Morgan, W.W.: 1951, *Astrophys. J.*, **114**, 522
Johnson, H.L., Morgan, W.W.: 1953, *Astrophys. J.*, **117**, 313
Kasten F., Young, A.T.:1989, *Applied Optics*, **28**, 4735
KenKnight, C.E.: 1984, *Proc. Workshop Improvements to Photometry*, NASA Conf. Publ., **2350**, 222
King, I.R.: 1952, *Astron. J.*, **57**, 253
King, I.R.: 1971, *Pub. Astron. Soc. Pac.*, **83**, 199
Kiselman, D., Oja, T., Gustaffson, B.: 1990, *Astr. Astrophys.*, **238**, 269
Kormendy, J.: 1973, *Astron. J.*, **78**, 255
Kristian, J., Blouke, M.: 1982, *Scientific American*, **247**, No. 4, 48
Kurilienė, G., Straižys, V.: 1987, *Bull. Vilnius Obs.*, **77**, 54
Lamb, T.D.: 1990, in *Night Vision*, (R.F. Hess, L.T. Sharpe, K. Nordby, Eds.), Cambridge University Press, Cambridge
Lambert, J.H.: 1760, *Photometria, sive de mesura et gradibus luminis colorum et umbrae*, Augsburg
Lamla, E.: 1965, in *Landolt-Börnstein, NS*, Vol VI/1, Springer-Verlag, Berlin (H.H. Voigt, Ed.)
Lamla, E.: 1982, in *Landolt-Börnstein, Zahlenwerte und Funktionen aus Naturwissenschaften und Technik*, Vol. VI/2b, Astronomie und Astrophysik, (K. Schaifers, H.H. Voigt, Eds.), Springer-Verlag, Berlin
Laulainen, N.S., Taylor, B.J., Hodge, P.W.: 1977, *Atmospheric Environment*, **11**, 21
Léna, P.: 1988, *Observational Astrophysics*, Springer-Verlag, Berlin
Lennie, P.: 1991, *Optics and Photonics News*, August 1991, 10
Levander, F.W.: 1889, *Mon. Not. R. astr. Soc.*, **50**, 33
Liu Zongli, Sterken, C., Hensberge, H., De Cuyper, J.-P.: 1992, in *Digitised Optical Sky Surveys II*, (H.T. MacGillivray, E.B. Thomson, Eds.). Preprint.
Lontie-Bailliez, M., Meessen, A.: 1959, *Ann. Soc. Sci. Bruxelles*, **73**, 390
Low, F.J., Rieke, G.H.: 1974, in *Methods of Experimental Physics*, Academic Press, New York (N. Carleton, Ed.), **12**, 454
Lub J., Pel, J.W.: 1977, *Astr. Astrophys.*, **54**, 137
MacGillivray, H.T.: 1990, *Digitised Optical Sky Surveys Newsletter*, **2**, 11
Mackay, C.D.: 1986, *Ann. Rev. Astron. Astrophys.*, **24**, 255
McLean, I.S.: 1989, *Electronic and Computer-aided Astronomy: from eye to electronic sensor.* The Ellis Horwood Library of Space Science and Space Technology Series in Astronomy., John Wiley & Sons, Chichester. Chap. 9
Mandwewala, N.J.: 1976, *Arch. Sc. Genève*, **29**, 119
Manduca, A., Bell, R.A.: 1979, *Pub. Astron. Soc. Pac.*, **91**, 848
Manfroid, J.: 1984, *Astr. Astrophys.*, **141**, 101
Manfroid, J.: 1985a, *Traitement numérique des données photométriques*, Liège
Manfroid, J.: 1985b, in *Calibration of Fundamental Stellar Quantities*, IAU Coll. **111**, (D.S. Hayes et al., Eds.), Reidel Publ. Co., Dordrecht, 505
Manfroid, J., Heck, A.: 1983, *Astr. Astrophys.*, **120**, 302
Manfroid, J., Heck, A.: 1984, *Astr. Astrophys.*, **132**, 10
Manfroid, J., Sterken, C.: 1987, *Astr. Astrophys. Suppl. Ser.*, **71**, 539
Manfroid, J., Sterken, C.: 1992, *Astr. Astrophys.*. In press
Manfroid, J., Sterken, C., Bruch, A., Burger, M., de Groot, M., Duerbeck, H.W., Duemmler, R., Figer, A., Hageman, T., Hensberge, H., Jorissen, A., Madejsky, R., Mandel, H., Ott, H.-A., Reitermann, A., Schulte-Ladbeck, R.E., Stahl, O., Steenman, H., vander Linden, D., Zickgraf, F.-J.: 1991, *Astr. Astrophys. Suppl. Ser.*, **87**, 481

Matthews, T.A., Sandage A.R.: 1963, *Astrophys. J.*, **138**, 30
Melbourne, W.G.: 1960, *Astrophys. J.*, **132**, 101
Merkle., F., Kern, P., Léna, P., Rigaut, F., Fontanella, J.C., Rousset, G., Boyer, C., Gaffard, J.P., Jagourel, P.: 1989, *The Messenger*, **58**, 1
Merriam G. & C.: 1976, *Webster's Third New International Dictionary*, Encyclopædia Britannica, Inc.
Müller, G., Kempf, P.: 1907, *Publ. Astron. Obs. Potsdam*, **52.**
Müller, G., Kron, E., Kohlschütter A., Hassenstein W.: 1927, *Publ. Potsdam*, **26**, 85
Neckel, H., Labs, D.: 1986, *Solar Phys.*, **90**, 205
Neckel, Th.: 1966, *Z. für Astroph.*, **63**, 221
Neckel, Th., Klare, G.: 1980, *Astr. Astrophys. Suppl. Ser.*, **42**, 251
Newcomb, S.: 1906, *A Compendium of Spherical Astronomy*, Macmillan, New York
Nicolet, B.: 1991, personal communication
Olsen, E.H.: 1977, *Astr. Astrophys.*, **58**, 217
Olsen, E.H.: 1983, *Astr. Astrophys. Suppl. Ser.*, **54**, 55
Osborn W.H., A'Hearn, M.F., Carsenty, U., Millis, R.L., Schleicher, D.G., Birch, P.V., Moreno, H., Gutierrez-Moreno, A.: 1990, *Icarus*, **88**, 228
Osthoff, H.: 1900, *Astronomische Nachrichten*, **153**, 3657
Osthoff, H.: 1908, *Astronomische Nachrichten*, **178**, 4252
Pecker, J.-C., Schatzman, E.: 1959, *Astrophysique Générale*, Masson, Paris
Pickering, E.C.: 1888, *Memoirs of the American Academy of the Arts and Sciences*, **11**, 202
Pogson, N.: 1856, *Mon. Not. R. astr. Soc.*, **17**, 12
Popper, D.M.: 1982, *Pub. Astron. Soc. Pac.*, **94**, 204
Posternak, J.: 1948, *Helv. Physiol. Acta*, **6**, 516
Price, M.J.: 1966, *Mon. Not. R. astr. Soc.*, **134**, 171
Reddish, V.C.: 1966, *Sky & Telescope*, **32**, 124
Reiger, S.H.: 1963, *Astron. J.*, **68**, 395
Roddier, F.: 1981, *Progress in Optics*, **19**, Chap. 5
Rodman, J.P., Smith, H.J.: 1963, *Applied Optics*, **2**, 181
Roig, J.: 1967, *Optique Physique*, Masson et Cie., Paris
Roosen, R.G., Angione, R.J., Klemcke, C.H.: 1973, *Bull. Am. Met. Soc.*, **54**, 307
Rufener, F.: 1964, *Publ. Obs. Genève* A, **66**, 413
Rufener, F.: 1967, *Archives des Sciences, Genève*, **20, 3**, 425
Rufener, F.: 1980, *Third Catalogue of stars measured in the Geneva Observatory photometric system.* Observatoire de Genève.
Rufener, F.: 1985, in *Calibration of Fundamental Stellar Quantities*, IAU Coll. **111**, (D.S. Hayes et al., Eds.), Reidel Publ. Co., Dordrecht, 253
Rufener, F.: 1986, *Astr. Astrophys.*, **165**, 275
Rufener, F.: 1988, *Catalogue of stars measured in the Geneva Observatory photometric system* (4th edition), Observatoire de Genève.
Rufener, F., Maeder, A.: 1971, *Astr. Astrophys. Suppl. Ser.*, **4**, 43
Rufener, F., Nicolet, B.: 1988, *Astr. Astrophys.*, **206**, 357
Schaefer, A.R.: 1984, *Proc. Workshop Improvements to Photometry*, NASA Conf. Publ., **2350**, 193
Scheffler, H.: 1982, in *Landolt-Börnstein, Zahlenwerte und Funktionen aus Naturwissenschaften und Technik*, Vol. VI/2c, Astronomie und Astrophysik, (K. Schaifers, H.H. Voigt, Eds.), Springer-Verlag, 45
Schild, R.E.: 1977, *Astron. J.*, **82**, 337
Schmidt-Kaler, Th.: 1982, in *Landolt-Börnstein, Zahlenwerte und Funktionen aus Naturwissenschaften und Technik*, Vol. VI/2b, Astronomie und Astrophysik, (K. Schaifers, H.H. Voigt, Eds.), Springer-Verlag, 451
Schnapf, J.L., Baylor, D.A.: 1987, *Scientific American*, **256**, No. 4, 32
Serkowski, K.: 1963, *Astron. J.*, **138**, 1035.
Shannon, C.E.: 1949, *Proc. Inst. Radio Eng.*, **37**, 10
Sharpe, J.: 1970, *EMI Doc. Ref.* R/P021 Y70
Sky & Telescope: 1985, **69**, 376

Smith, C.J.: 1960, *Optics*, Arnold Publishing, London
Sterken, C.: 1983, *The Messenger*, **33**, 10
Sterken, C.: 1986, in *Instrumentation and research programmes for small telescopes*, (J.B. Hearnshaw, P.L. Cottrell, Eds.), IAU Symp. **118**, 255. Reidel Publ. Co., Dordrecht
Sterken, C.: 1992, *Vistas in Astronomy*, In press
Sterken, C., Manfroid, J.: 1987, *Proc. 27th Liège International Astrophysical Colloquium*, 55
Sterken, C., Manfroid, J.: 1991, in *Variable Star Research: An international perspective* (J.R. Percy, J.A. Mattei, C. Sterken, Eds.). Cambridge University Press, Cambridge, 75
Sterken, C., Manfroid, J.: 1992, Proc. Joint Commission 9 and 25 meeting on automated telescopes (S.J. Adelman, R.J. Dukes, Eds.). In press
Sterken, C., Snowden, M., Africano, J., Antonelli, P., Catalano, F.A., Chahbenderian, M., Chavarria, C., Crinklaw, G., Cohen, H.L., Costa, V., de Lara, E., Delgado, A.J., Ducatel, D., Fried, R., Fu, H.-H., Garrido, R., Gilles, K., Gonzalez, S., Goodrich, B., Haag, C., Hensberge, H., Jung, J.H., Lee, S.-W., Le Contel, J.-M., Manfroid, J., Margrave, T., Naftilan, S., Peniche, R., Peña, J.H., Ratajczyk, S., Rolland, A., Sandmann, W., Sareyan, J.-P., Szuskiewicz, E., Tunca, Z., Valtier, J.-C., vander Linden, D.: 1986, *Astr. Astrophys. Suppl. Ser.*, **66**, 11
Stetson, P.B.: 1987, *Pub. Astron. Soc. Pac.*, **99**, 191
Stetson, P.B.: 1989, *Highlights of Astronomy*, **8**, 635, (D. McNally, Ed.), Kluwer Academic Publishers, Dordrecht
Stigler, S.M.: 1977, *Ann. Stat.*, **5**, 1055
Stock, J.: 1968, *Vistas Astron.*, **11**, 127
Stock, J., Keller, G.: 1960, *Telescopes*, Stars and Stellar Systems, (G.P. Kuiper, B.M. Middlehurst, Eds.), Univ. Chicago Press, Chicago, 138
Stoy, R.H.: 1954, *Trans. IAU*, **8**, 381
Straižys, V.: 1977, *Multicolor Stellar Photometry* (in Russian), Mokslas Publishing House, Vilnius
Strauss, F.M., Ducati, J.R.: 1981, *Astr. Astrophys. Suppl. Ser.*, **44**, 337
Strömgren, B.: 1937 in *Handbuch der Experimentalphysik*, (W. Wien, F. Harmes, Eds.), Akad. Verlagsgesellschaft, Leipzig, **26**, 321
Strömgren, B.: 1966, *Ann. Rev. Astron. Astrophys.*, **4**, 433. Annual Reviews Inc., Palo Alto
Tass, A., Terkan, L.: 1916, *Publ. Ogyalla*, **1**
Taylor, B.J., Joner, M.D., Johnson, S.B.: 1989, *Astron. J.*, **97**, 1798
Taylor, B.J., Lucke, P.B., Laulainen, N.S.: 1977, *Atmospheric Environment*, **11**, 1
Tobin, W., Wadsworth, A.: 1991, in *Third New Zealand Conference on Photoelectric Photometry*, (E. Budding, J. Richards, Eds.), Southern Stars, **34**, 3, 49
Toomer, G.J.: 1984, *Ptolemy's Almagest*, Duckworth, London
Trefzger, C.F., Cameron, L.M., Spaenhauer, A., Steinlin, U.W.: 1983, *Astr. Astrophys.*, **117**, 347
Trumpler, R.J.: 1930, *Lick Obs. Bull.*, **14**, 154
Tsubaki, T., Engvold, O.: 1975, *AAS Photo-Bulletin*, **9**, 17
Tucholke, H.-J.: 1989, personal communication
Van Isacker, J.: 1953, *Publ. Inst. R. Météor. Belg.*, Ser. B, **8**
Vasyutin, V.V., Tishchenko, A.A.: 1989, *Scientific American*, **261**, No. 1, 66
Wade, R., McLean, I.S.: 1989, in *New Technologies for Astronomy*, SPIE **1130**, 166
Walker, E.N.: 1988, *Proc. Second Workshop on Improvements to Photometry*, NASA Conference Publication **10015**, 57
Walker, G.: 1987. *Astronomical Observations*, Cambridge Univ. Press, Cambridge, 225
Walraven, J.H., Walraven, T.: 1960, *Bull. Astron. Inst. Neth.*, **15**, 67
Walsh, J.W.T.: 1958, *Photometry*, Third Edition, Dover Publications, New York
Warner, B.: 1962, *J. Brit. astron. Ass.*, **72**, 177
Warner, B.: 1988, *High Speed Astronomical Photometry*, Cambridge University Press, Cambridge
Weaver, H.F.: 1946, *Popular Astronomy*, **54**, 211
West, R.M.: 1991a, ESO Preprint Ser., **757**
West, R.M.: 1991b, *The Messenger*, **65**, 45
Whitford, A.E., Kron, G.E.: 1937, *Rev. Sci. Inst.*, **8**, 78
Widorn, Th.: 1955, *Die Sterne*, **31**, 217

Wood, F.B.: 1963, *Photoelectric Astronomy for Amateurs*, Macmillan, New York
Young, A.T.: 1963, *Applied Optics*, **2**, 51
Young, A.T.: 1966, *Rev. Sci. Instr.*, **37**, 1472
Young, A.T.: 1967a, *Mon. Not. R. astr. Soc.*, **135**, 175
Young, A.T.: 1967b, *Astron. J.*, **72**, 747
Young, A.T.: 1968, *Observatory*, **88**, 151
Young, A.T.: 1969a, *Icarus*, **11**, 1
Young, A.T.: 1969b, *Applied Optics*, **8**, 869
Young, A.T.: 1970, *Applied Optics*, **9**, 1874
Young, A.T.: 1971, *Sky & Telescope*, **189**, 139
Young, A.T.: 1974a, *Astrophys. J.*, **189**, 587
Young, A.T.: 1974b, in *Methods of Experimental Physics*, (N. Carleton, Ed.), **12A**, Academic Press, New York, Chap. 1, 2 and 3
Young, A.T.: 1984, *Proc. Workshop Improvements to Photometry*, NASA Conf. Publ., **2350**, 8
Young, A.T.: 1988, *Proc. Second Workshop on Improvements to Photometry*, NASA Conf. Publ., **10015**, 215
Young, A.T.: 1990, *Sky & Telescope*, **79**, 311
Young, A.T.: 1992, *Astr. Astrophys.*. In press
Young, A.T., Genet, R.M., Boyd, L.J., Borucki, W.J., Lockwood G.W., Henry, G.W., Hall, D.S., Smith, D.P., Baliunas, S.L., Donahue, R., Epand, D.H.: 1991, *Pub. Astron. Soc. Pac.*, **103**, 221
Young, A.T., Irvine, W.M.: 1967, *Astron. J.*, **72**, 945
Zinner, E.: 1926, *Veröff. der Remeis Sternwarte Bamberg*, **2**, 12
Zinner, E.: 1939, *Veröff. der Remeis Sternwarte Bamberg*, **4**, 88

Appendix B. Glossary

Absolute magnitude. Magnitude an object would have if it were at a distance of 10 pc.

Absolute photometry. All-sky photometry.

Absorptance. Ratio of the radiant flux lost by absorption to the incident flux.

ADU. Analog-to-digital unit, see *data number.*

Aerosol. Suspension of solid or liquid particles in a gas.

Air mass. A measure of the path length traversed by starlight in the atmosphere before it reaches the detector, and taken relative to the vertical path length.

All-sky photometry. Measurement of magnitudes on some standard scale, as opposed to differential photometry.

Analog-to-digital unit. ADU, see *data number.*

Anode. An electrode through which a stream of electrons leaves.

Anode pulse rise time. Time difference between the 10% and 90% amplitude level of a pulse for a test stimulus of specific duration. It is the time to rise from 10% to 90% of the peak amplitude.

Aperture photometry. Photometry using a diaphragm to isolate a small sky area, either directly with a focal-plane diaphragm, or with an image processing system.

Apparent magnitude. Magnitude observed from the Earth (opposed to *Absolute magnitude).*

Atmospheric extinction. Dimming of light due to the absorption and scattering by atmospheric constituents.

Averted vision. Looking a few degrees off a faint object in order to increase its visibility.

Bad column. Column of a CCD array that does not read out charge.

Band-pass filter. Filter isolating (or rejecting) a limited spectral region.

Bandwidth. The width of passband at specified points such as 50% of the peak transmittance. For optical filters, bandwidth is sometimes expressed as a percentage of the central wavelength.

Bias. In CCDs, electronic offset which prevents negative signal.

Bias frame. CCD frame with exposure time set to zero and giving the bias level.

Binning. Combining a few adjacent pixels during readout; this improves signal-to-noise ratio at the expense of spatial resolution.

Blocking. Elimination of transmission outside the filter's bandpass.

Blooming. *Charge bleeding.*

Blue leak. Fluorescence phenomenon in a filter, causing an unwanted response to blue or green light.

Bolometer. Thermal detector used to measure irradiance, usually at infrared wavelengths, and characterized by a temperature-dependent electrical resistance.

Bolometric magnitude. Magnitude defined over the whole electromagnetic spectrum.

Brightness. Designation avoided by radiometrists because it is so commonly used in everyday life for all different aspect of light, and it is quite confusing; an obsolete meaning of *photometric brightness* was equivalent to luminance.

Candela. Basic SI unit of *luminous intensity*; the intensity, in a perpendicular direction, of a surface of 1/600000 square meter of a black body at the temperature of freezing platinum under a pressure of 101325 pascal.

Candle-power. Obsolete term for *luminous intensity.*

Cavity. Spacer layer between two reflective stacks in an interference filter (solid Fabry-Perot etalon).

CCD. Charge-coupled device; 2-D solid-state detector that stores the electrons produced by incident photons as discrete packets of charge in potential wells maintained by an electric field.

Center wavelength. Center of passband measured at 50% of peak transmittance

Channel stop. Strips in a CCD that prevents sideways movement of charges.

Charge bleeding. Electrons from a heavily saturated pixel in a CCD, spilling over up and down the column (also known as *blooming*).

Charge transfer efficiency (CTE). Fraction of the original charge which is succesfully tranferred from one pixel to the next in one CCD cycle; it is usually expressed as a fraction per transfer.

Charge trap. Impurities or defects in a CCD structure which prevent charges in certain columns from being transfered to the next pixel.

Circular Variable Filter (CVF). Circular band-pass interference filter whose thickness, and central wavelength, vary along the perimeter. They are used in low-resolution monochromators or spectrophotometers, mainly in the IR.

Clocking. Successive raisings and lowerings of voltage on the electrodes of a CCD in order to move the electrons from one pixel to the next.

Cold-box. Common name for the cryostat used to cool a photomuliplier tube.

Color excess. Modification brought to a color index by the interstellar absorption.

Color index. Linear combination of magnitudes at different wavelengths.

Color temperature. The temperature of that black body which has the same spectral energy distribution in a limited spectral region, as the object under study has.

Color transformation. Empirical mathematical transformation applied to the observed magnitudes in order to convert them into a standard system ("normal" transformation), or into a different system ("special" transformation).

Comparison star. A constant star used to monitor the variations of another nearby star.

Conduction band. Energy range allowed to the freely moving electrons in a solid; metals have many electrons in this range, insulators have none; in semiconductors the conduction band contains few electrons provided by impurity atoms or ejected from the valence bands by thermal energy or photon absorption; in the latter cases, corresponding holes with positive charges appear in the valence band and participate to the electrical conduction.

Cryostat. Device for maintaining an enclosed area at a stable low temperature; both passive and active cryostats are used. Active cryostats may be refrigerators or Peltier effect units. Passive cryostats contain a cryogenic fluid (helium, nitrogen or both) or solid (dry ice, nitrogen) according to the required temperature.

CTE. *Charge transfer efficiency.*

Cut-off filter. Filter rejecting all light with wavelengths on one side of the *cut-off wavelength.*

Cut-off wavelength. Wavelength at which the transmittance of a filter, or the detectivity of a detector, has fallen to one-half its peak value.

Dark adaptation. Automatic gain control of the eye, due to a chemical process in the retina.

Dark current. Current generated in a detector by thermal effects, even in the absence of input signal.

Dark exposure. CCD frame obtained with closed shutters—and preferably in the absence of all light sources—in order to estimate the dark current of the detector.

Dark integration. Integration with a photomultiplier, obtained with closed shutters—and preferably in the absence of all light sources—in order to estimate the dark current of the detector.

Data number. (DN) unit of the analogue-to-digital conversion system of a CCD apparatus. For example a 16 bit system may use a maximum of 65536 DN. The acronym ADU, for *analog-to-digital unit,* is also used.

Dead time. Minimum time interval between two pulses larger than the input threshold for which two output pulses will be counted.

Deferred charge. Phenomenon caused by charge traps or potential pockets in a CCD which prevent electrons from being released to the adjacent pixel; eventually the electrons may be released in a subsequent cycle.

Detective quantum efficiency (*DQE*)**.** The square of the ratio of the output S/N to the input S/N ($DQE = (S/N)^2_{\text{out}}/(S/N)^2_{\text{in}}$).

Detectivity. Either spectral, or blackbody detectivity; reciprocal of the corresponding noise-equivalent power ($D = 1/NEP$).

Dewar. Insulated bottle containing a cryogenic fluid. The infrared detectors operated at very low temperature are mounted inside a dewar.

Dichroic filter. Dielectric interference filter that transmits certain wavelength regions and reflects the others. Dielectric filters are practically absorption free.

Differential photometry. Measurement of magnitude differences between programme stars and nearby reference stars.

DN. *Data number.*

Drift. A slight change of a quantity with time, for example the atmospheric extinction, or the sensitivity of a photomultiplier continuously operated at high output current.

DQE. Detective quantum efficiency.

Dynode. An electrode that performs electron multiplication by means of secondary emission.

Electron-volt (ev). Energy acquired by an electron being accelerated through a drop in potential of 1 Volt (1 ev= 1.6×10^{-19} J).

Entrance pupil. In an optical configuration, the image of the aperture stop formed by the elements preceding it.

Equivalent wavelength. Wavelength at which a heterochromatic magnitude is best approximated by a monochromatic magnitude.

Etendue. *Throughput.*

Exit pupil. In an optical equipment, the image of the aperture stop formed by the elements following it.

Extinction. Dimming of light by an intervening medium (the atmosphere or the interstellar medium); extinction is usually due to both scattering and absorption.

Extinction coefficient. Gradient of apparent magnitude with air mass.

Extinction stars. Star specifically observed at selected air masses in view of determining the atmospheric extinction coefficients.

Flat-field. Exposure of a diffuse and uniform source in order to calibrate the non-uniformity of an imaging detector such as a CCD.

Flux or flux density. Illumination (in astronomy).

Forbes effect. Increased reddening and monochromaticity of light as the path length in the air (air mass) increases.

Free atmosphere. Part of the atmosphere where the effects of the ground on the turbulence conditions are negligible.

Fringes. Wavy patterns due to the layered structure of CCDs; these interference effects are prominent when emission lines such as the night-sky emissions are present.

FWHM. Full Width at Half Maximum

F-ratio. Ratio of the focal length to the diameter of the entrance pupil of an optical system.

Gray filter. *Neutral filter.*

Haze. Condition wherein the scattering properties of the atmosphere are greater than that attributable to gas molecules, but less than fog.

Heterochromatic magnitude. Magnitude measured over a wide range of wavelength.

Hysteresis of a detector. Change in absolute output signal when input light signal is moved up and then brought back to the original level.

Illuminance. Luminous flux incident on unit area of target surface. It is measured in lux.

Image intensifier. Device that produces an observable image that is brighter at output than the image at input. It is often mounted on a photometer diaphragm eyepiece for centering faint objects.

Imaging detector. Detector with 2-D capability, such as a CCD.

Instrumental system. Photometric system corresponding to the equipment used.

Intensity. Vague concept used for almost any measurable photometric or radiometric quantity, and it should be avoided; the terms *luminous intensity* and *radiant intensity* are well-defined photometric and radiometric concepts; in astrophysics *specific intensity* is widely used to designate spectral radiance.

Interference filter. Filter based on the phenomenon of optical interferences between plane-parallel semi-transparent reflectors; those filters may have very narrow passbands.

Interstellar extinction. Dimming of light due to the scattering by the interstellar dust particles; the extinction rapidly increases with frequency so that one speaks of *interstellar reddening*.

Interstellar reddening curve. Relation between interstellar absorption (in magnitudes) and wavelength

Inversion layer. A colder layer of air under a warmer layer; these conditions are found when the air cooled at night flows down the hills into the lower areas.

Irradiance. Radiant flux incident on unit area of surface. It is measured in Watts per square meter

Limiting magnitude. The faintest magnitude reachable by an instrument (or by the dark-adapted eye).

Line index. A linear combination of magnitudes obtained at the wavelength of a spectral line, but with different passbands.

Lumen. SI unit of *luminous flux*.

Luminance. The *luminous intensity* in a given direction of a small element of surface area divided by the orthogonal projection of this area onto a plane at right angle to the direction. It is measured in candelas per square meter. Luminance is often called *surface brightness* of the object.

Luminosity. *Radiant flux* (in astronomy).

Luminous flux. A measure of the rate of flow of luminous energy, evaluated according to its ability to produce a visual sensation. It is measured in lumens.

Luminous intensity. A measure of the amount of light that a point source radiates in a given direction. It is expressed by the *luminous flux* per unit leaving the source in that direction per unit of solid angle.

Luminous sensitivity. Ratio of photocathode current (in A) to the incident *luminous flux* in lm. This sensitivity is meaningless ouside of the visible domain.

Lux. SI unit of illumination equal to a *luminous flux* of 1 lumen per square meter.

Magnitude. A measure of the *illuminance* or *irradiance* of astronomical objects on a logarithmic scale; faint stars have higher magnitudes; a ratio of 100 in illuminance (irradiance) is equivalent to a difference of 5 magnitudes.

Monochromatic magnitude. Magnitude measured in a narrow spectral band.

Natural system. *Instrumental system.*

NEP. Noise-equivalent power.

Neutral filter. Also *neutral-density filter* or *gray filter* : filter having a flat response over the range of wavelengths of interest.

Nit. A unit of luminance equal to 1 candela per square meter.

Noise. Any unwanted disturbance in a signal (e.g., background noise, readout noise, photon noise...). Photon noise is fluctuation in photon arrival rates. Strongest signals are photon noise limited. Receiver noise is caused by events generated in the detection system which are not caused by the incident photons.

Noise-equivalent power. (*NEP*) is defined either at a given wavelength or for a blackbody at a given temperature. The *spectral NEP* at wavelength λ is the rms radiant power at λ of a sinusoidally modulated signal falling on a detector which gives rise to an rms output voltage equal to the rms noise voltage of the detector in a bandwidth of 1 Hz. This technical specification depends on the chopping frequency, the detector area, and the electrical bandwidth. The *blackbody NEP* is the same as the spectral *NEP*, but for a blackbody at a given temperature.

Non-imaging detectors. Detectors which do not discriminate photons arriving at different locations.

Offset. In CCD systems, another term for the *bias*. On the celestial sphere, displacement between two nearby objects.

Offset guiding. Guiding an astronomical exposure on a star, when the object of interest is nearby, but invisible.

Opacitance. Complement to one of transmittance.

Optical density. Common logarithm of the reciprocal of transmittance.

Output S/N. Ratio of mean output to its fluctuations (usually measured as rms fluctuations).

Passband. Wavelength range of significant transmission.

Photocathode. Electrode capable of releasing electrons when illuminated.

Photodiode. *p-n* junction detectors (photovoltaic detectors); the term has also been applied to various photoemissive detectors such as gas or vacuum diode phototubes and photocells.

Photoelectric effect. The release of electrons from certain materials due to impinging photons.

Photometric brightness. *Luminance.*

Photometric intensity. *Luminous intensity.*

Photometric system. System of magnitudes, each of them characterized by a response curve. The system is defined by the values given for the standard stars.

Photometry (astronomical). Measuring and studying the radiant flux emitted by celestial bodies.

Photomultiplier tube. Electronic tube which converts photons into electrons, multiplies the electrons via a series of electrodes, and produces a measurable current from a very small input signal.

Photovisual magnitude. Magnitude defined for the combination of a photographic plate and a yellow filter, approximating the spectral sensitivity of the eye.

Photovoltaic detector. A detector usually constituted by a *p-n* junction. Upon irradiation, the electron-hole pairs which are created, are immediately separated by the strong electric field across the junction, and a current is generated, which is proportional to the number of incident photons per second.

PIN photodiode. Abbreviation for "positive-intrinsic-negative" photodiode (the semiconductor has intrinsic conductivity between the *p* and *n* regions).

Pixel. Abbreviation for "picture element", i.e., the smallest useful element of image information.

Point-spread function. (PSF) this function represents the distribution of the irradiance due to a point source as a function of the position relative to the centroid of its image.

Preflash. Carefully chosen uniform exposure of a detector; for CCDs this can be used to overcome the deferred charge phenomenon; in photography this helps bring dim images to a comfortably high density.

Profile fitting. Modelling stellar images by a least-squares algorithm.

PSF. *Point-spread function.*

Quantum efficiency. (*QE*) ratio of the number of photoelectrons to the number of incident photons at wavelength λ. *QE* refers to the interaction of the incident radiation and the primary sensor. The *QE* of the photomultiplier is indicated by its gain: number of photoelectrons emitted/number of incident photons $\times$ 100 (ratio of emitted photoelectrons for each photon received at photocathode, expressed in %).

Radiance. A measure of the amount of radiation leaving or arriving at a point on a surface. It is the radiant intensity in a given direction of a small element of surface area divided by the orthogonal projection of this area onto a plane at right angles to the direction.

Radiant flux. Rate of flow of energy as radiation. It is measured in Watts.

Radiant intensity. A measure of the amount of radiation emitted from a point expressed as the radiant flux per unit solid angle leaving this source.

Radiant sensitivity. Ratio of photocurrent (A) flowing in the photocathode to the intensity of the incident light (in W).

Radiometer. Instrument for measuring spectral irradiance.

Radiometry. Measurement of electromagnetic radiation without regard to its seeability.

Ratio of total-to-selective absorption. ($\mathcal{R}$) usually the ratio of the Johnson-Morgan V absorption to the B–V excess, due to the interstellar absorption; this ratio depends on the nature of the absorbing grains.

Read-out noise. (RON) noise added in the process of reading a detector such as a CCD.

Real time. Refers to a data processing system in which a computer processes data as it is generated.

Reddening-free index. A linear combination of magnitudes which is not affected by "normal" interstellar extinction.

Reddening vector. Vector indicating the direction in which interstellar reddening moves the position of a star in a multi-dimensional space of color indices.

Red leak. Unwanted secondary window in a filter band pass, on the red side of the main window.

Reflectance. Ratio of reflected radiant flux to the incident flux.

Rejection ratio. Ability of an interference filter to block light at unwanted wavelength; it is defined as the ratio of the maximum transmittance to the minimum transmittance.

Rejection region. Wavelength domain over which the transmittance of an interference filter is low.

Resolution. In the space domain, a measurement of the ability of an instrument to distinguish between separate parts of an image. In spectroscopy it is the ratio of the smallest wavelength difference between discernible features, to the wavelength. The concept exists also in times series, where it corresponds to the time interval between successive discernible signals.

Responsive quantum efficiency. (*RQE*) *quantum efficiency.*

Responsivity/sensitivity. Ratio of output signal to the input signal.

RON. *Read-out noise.*

Saturation. A detector or a pixel is saturated when it is submitted to a signal so strong that it cannot handle it properly; the result is a non-linear useless response.

Scintillation. Twinkling of stars, caused by rapidly moving changes in the density of the earth's atmosphere, producing uneven refraction of starlight.

Secondary emission coefficient. Average number of secondary electrons (N_s) emitted for N_p primary electrons: $\delta = N_s/N_p$.

Seeing. Image degradation by air turbulence.

Selective absorption. Apparent modification of a color index (usually the Johnson-Morgan *B–V*) of a star due to interstellar absorption (opposed to the *total absorption*).

Semiconductor. A material which has conducting properties intermediate between metals and insulators.

Shadow pattern. Irregular patchwork of illumination caused by scintillation.

Shot noise. Random noise produced by the statistical fluctuations of the electrons being produced by a detector.

Signal-to-noise ratio (S/N). Concept used to quantify the effects of noise. It is the ratio of a signal to the standard deviation of the signal.

Sky brightness. Atmospheric (airglow, auroral emission, artificial light) or extraterrestrial (scattered sunlight from moon, scattered starlight, interplanetary dust) foreground light that interferes with observations.

Specific intensity. Monochromatic radiance (in astrophysics).

Spectral response. Domain of the electromagnetic spectrum over which a detector is sensitive.

Standard stars. Stars for which accurate color indices and/or magnitudes exist, defining a *standard system*.

Standard system. Photometric system used as a reference.

Standard values. Photometric values of selected stars in a standard system.

Thermal detector. Detector sensing the change of temperature due to the absorption of photons.

Thermionic emission. Electrons gaining enough thermal energy to escape spontaneously from the cathode or dynodes and mimic photoelectrons.

Throughput. Geometrical quantity characterizing a light beam (the product of the effective surface area and solid angle).

Time constant. Speed of response of a detector, usually measured as $1/(2\pi\nu)$, where ν is the chopping frequency at which the responsivity falls to $1/\sqrt{2}$ of its maximum value.

Total absorption. Apparent increase of the magnitude (usually in the Johnson-Morgan *V* band) of a star, due to interstellar absorption (opposed to the *selective absorption*).

Trailing. In a CCD, charges which had been trapped in some pixels, may be released later and show up as signal in subsequent pixels.

Transit time. Time interval between the release of a electron at the photocathode and the arrival of an electron at the anode. Transit time is not a single-valued quantity, but has a bell-shaped distribution.

Transmittance. Ratio of transmitted light to total incident light.

Visual magnitude. Magnitude defined with the spectral sensitivity of the eye.

Wave number. Reciprocal of wavelength, expressed in cm^{-1}.

Zero point. Independent term in the expression of a magnitude or color index.

Appendix C. Symbols and notations

A_1 interstellar absorption in V
BC bolometric correction
c velocity of light 2.997925×10^8 m s^{-1}
cd candela (lm/sr) unit of luminous intensity
CI color index
CTE charge transfer efficiency
D optical density of a filter
DQE detective quantum efficiency
e electron charge 1.602192×10^{-19} C
E irradiance, illuminance; also energy
E_m measured irradiance
ev electron-volt 1.602192×10^{-19} J
F radiant flux, luminous flux
h Planck's constant 6.6262×10^{-34} J s
H linear operator describing the transformation between the magnitudes and the color indices
$I(x, y)$ point-spread function
I radiant intensity, luminous intensity
Jy Jansky unit of monochromatic irradiance = 10^{-26} W Hz^{-1} m^{-2}
k Boltzmann's constant 1.38062×10^{-23} J K^{-1}
k extinction coefficient
$\hat{k}$ second-order extinction coefficient
K zero point
L radiance, luminance
$\mathcal{L}$ radiant flux of a star ("luminosity" in astronomer jargon)
$\mathcal{L}_\odot = 3.9 \times 10^{26}$ W radiant flux of the Sun
LLD lower level discriminator
lm lumen, unit of luminous flux
lx lux (or meter-candle), visual unit of illuminance (lm m^{-2})
m magnitude
$m_{\rm bol}$ bolometric magnitude
$m_{\rm out}$ extra-atmospheric magnitude
$m_{\rm pg}$ photographic magnitude
$m_{\rm pv}$ photovisual magnitude
$m_{\rm v}$ visual magnitude
M absolute magnitude
$M_{\rm bol\odot} = 4.72$, absolute bolometric magnitude of the Sun
M linear operator approximating the color transformation
$P(i)$ probability distribution
pc (parsec) 3.086×10^{16} m
PSF point-spread function
Q(K) general operator describing the color transformation
QE quantum efficiency
$\mathcal{R}$ total-to-selective absorption ratio
RQE quantum efficiency
$s(\lambda) = s_t(\lambda)s_s(\lambda)s_r(\lambda)$ response curve of the observing system
s_i, s_e, s_t, s_s spectral transmissions due to the interstellar medium, the Earth's atmosphere, the telesco[
the photometric system
s_r sensitivity curve of a detector
$S = Es_e$ non-instrumental part of the signal arriving to a photometric detector
S/N signal to noise ratio
T transmittance
T temperature

ULD upper level discriminator
X air mass
y_s standard magnitude of star s
z zenit angle
$\vec{Z}$ reddening vector
γ slope of the characteristic curve of a photographic emulsion
$\zeta(\lambda)$ interstellar extinction law
$\kappa(\lambda)$ absorpion coefficient per unit mass
λ wavelength
λ_i isophotal wavelength
λ_0 mean, effective or equivalent wavelength
μ_2^2 normalized second moment of instrumental function
ν frequency
σ Stefan-Boltzman constant 5.67×10^{-8} W m^{-2} deg^{-4}
τ dead-time
ϕ merit function
Φ work function in photoelectric effect
Ω solid angle

Appendix D. Index

9 780792 316534